Lecture Notes in Computer Science 998

Edited by G. Goos, J. Hartmanis and J. van Leeuwen

Advisory Board: W. Brauer   D. Gries   J. Stoer

Springer
Berlin
Heidelberg
New York
Barcelona
Budapest
Hong Kong
London
Milan
Paris
Santa Clara
Singapore
Tokyo

Anne Clarke Mario Campolargo
Nikos Karatzas (Eds.)

Bringing Telecommunication Services to the People – IS&N '95

Third International Conference
on Intelligence in Broadband Services and Networks
Heraklion, Crete, Greece, October 16-19, 1995
Proceedings

 Springer

Series Editors

Gerhard Goos, Karlsruhe University, Germany

Juris Hartmanis, Cornell University, NY, USA

Jan van Leeuwen, Utrecht University, The Netherlands

Volume Editors

Anne Clarke
HUSAT Research Institute,The Elms, Elms Grove
Loughborough LE11 1RG, United Kingdom

Mario Campolargo
European Commission
Av. de Beaulieu 9 4/86, 1160 Brussels, Belgium

Nikos Karatzas
Alpha Systems Analysis Integration Ltd.,Alpha Systems S.A.
Xanthou 3, Tavros 177 78, Greece

Cataloging-in-Publication data applied for

Die Deutsche Bibliothek - CIP-Einheitsaufnahme

Bringing telecommunication services to the people :
proceedings / IS&N '95, Third International Conference on
Intelligence in Broadband Services and Networks, Heraclion,
Crete, Greece, October 16 - 20, 1995. Anne Clarke ... (ed.). -
Berlin ; Heidelberg ; New York ; Barcelona ; Budapest ; Hong
Kong ; London ; Milan ; Paris ; Tokyo : Springer, 1995
  (Lecture notes in computer science ; Vol. 998)
  ISBN 3-540-60479-0
NE: Clarke, Anne [Hrsg.]; IS&N <3, 1995, Hērakleion>; GT

CR Subject Classification (1991): C.2, B.4.1, D.2, H.4.3, H.5, K.4, K.6

ISBN 3-540-60479-0 Springer-Verlag Berlin Heidelberg New York

Typesetting: Camera-ready by author
SPIN 10485951   06/3142 – 5 4 3 2 1 0   Printed on acid-free paper

Preface

This book contains the Proceedings of an International Conference (IS&N95) addressing the technologies required for the effective engineering of telecommunications service software - a discipline which is called *Service Engineering*.

A key enabler for applications addressing the future needs of the 'information society' is the deployment of an integrated telecommunications service infrastructure which can support multi-media, multi-party, interactive and distributive communications. This telecommunications service infrastructure needs to operate across heterogeneous fixed and mobile networks and hide the complexity of the underlying network technologies from the users of this infrastructure, be they telematic applications or human users. Major constituents of the telecommunications service infrastructure are the distributed processing software systems which implement the communications services and provide the control and management of the underlying networks.

In the increasingly competitive environment developing for telecommunications services, new services will need to be conceptualised, and created rapidly and cost-effectively in response to user requirements. To meet this challenge, service engineering has now emerged as a new engineering discipline. As such it is strongly related to software engineering, adapting the appropriate best practices from that discipline. In particular, Object-Orientation and Open Distributed Processing, are being adopted to meet the needs of the new telecommunications services market place.

Many of the papers in the book reference the results of a group of RACE* projects which have been undertaking research and development in the area of *Intelligence in Services and Networks (IS&N)*. However, it is not solely the purpose of this book to record the results of the RACE IS&N projects. Many of the selected papers describe related work in other programmes (in particular, TINA), together with a contribution from Australia. The scope of the work is extensive and complex, and all interested parties need to share their experiences and converge to a common understanding, if the goal of *"openness"* is to be achieved.

This scope of the IS&N area in RACE includes:

Service Architecture - defining and validating a framework architecture which can be used to constrain and support the design of re-usable, interoperable components of user and management communication services. An open distributed processing object-oriented architecture, embedding concepts of IN and TMN, is the goal of this work.

Usability - addressing user requirements for advanced telecommunications services, usability aspects of existing and new services, and design guidelines to ensure usability of any and every new service.

---

* Further information about the RACE and ACTS programme can be obtained from the European Commission DG XIII-B/BU9 Rue de la Loi, 200 B-1049 Brussels Belgium. e-mail: aco@postman.dg13.cec.be   tel: +32 2 296 3415   fax: +32 2 295 0654

Communications Management - the evolution of a standardised TMN reference configuration with particular emphasis on service management (including the management of IN services) and its relationship with network management, inter-domain management, and customer access to service and network management facilities.

Advanced Communication Service Concepts - in particular, the definition of Personal Service Communication Space (PSCS) which encompasses facilities for personal mobility, advanced UPT, and personalised call management.

Security of telecommunications services - the development of validated specifications, guidelines and technology for practical and effective information security for the realisation and operation of advanced telecommunications-based applications and services.

Service Creation - definition and validation of processes, methods, and tools for the development of telecommunications service software.

RACE is a collaborative, pre-competitive research and development programme into advanced telecommunications technologies in Europe, partially funded by the European Union. The RACE programme, which started in 1987, is coming to an end in 1995, to be followed by the ACTS programme. The theme for ACTS is *"making it work"* with the focus of the IS&N related activities on validating new service engineering concepts in the context of user-driven experiments and trials of advanced telecommunications services. Whilst RACE has been a predominantly European programme, ACTS recognises the global nature of telecommunications services, with participation from non-European countries and active encouragement of linkage between ACTS related trials and similar trials in other countries.

The presentation of these papers at IS&N95 led to the crystallisation of concepts and methodologies on issues and technologies for future development of the global telecommunications service environment. Some of these issues will be tested in the forthcoming ACTs programme. It is expected that the next conference in this series will report on further advances in this exciting field.

The book is a result of a call for papers for the conference, which resulted in a total of 88 contributions received for evaluation. A careful selection process has resulted in 44 full papers and 8 poster papers presented at the conference and printed in this book. Introductions to the topics are provided by the session chairmen, who were active in editing the technical material for their sessions.

The editors would like to thank all contributors and session chairmen for their effort and co-operation, which has made this book possible. A special thank you goes to Allison White who was vital in bringing the papers together into a common format.

Mario Campolargo
Anne Clarke
Nikos Karatzas

August 1995

List of Referees

Serious and detailed reviews are an essential foundation for the quality of the papers selected. The reviewers for the papers presented in this book have fulfilled their role diligently. It is therefore a great pleasure to thank them as well as the authors for their efforts.

Marco de Alberdi
Gordon Allison
Anne Clarke
Chris Condon
Keith Dickerson
Phil Fisher
Panos Georgatsos
Carmelita Goerg
David Griffin
Carsten Gym
Nick Hine
Kersten Keil
Malcolm Key
Ulf Larsson
Joanna McCaffrey
Patrick McLaughlin
Al Mullery
Rune Nilsson
Rick Reed
James Reilly
Robin Smith
Munir Tag
Staale Wolland

Table of Contents

Challenges for Service Engineering

Session 1 Introduction

Chairman - Anne Clarke

The HUSAT Research Institute
The Elms, Elms Grove, Loughborough, LEICS LE11 1RG, UK
Tel: +44 (0)1509 611088
Fax: +44 (0)1509 234651
e-mail: a.m.clarke@lut.ac.uk

This book, the Proceedings of the IS&N95 Conference, covers the state of the art in Service Engineering, the core discipline in the development and deployment of new, global telecommunications services. The origins of Service Engineering can be traced back to the early days of the EU RACE Programme. The trend in the telecommunications services sector for new developments to focus more heavily on the software systems provided the main impetus, and it was from software engineering that the initial concepts and paradigms were drawn.

Over the past few years, the efforts of many researchers have shown that the telecommunications services sector has at the same time very specific requirements, as well as needing to draw on the skills from a very wide range of disciplines - human factors, communications management, security, service requirements analysis, software specialisations, etc. Much of this work was undertaken as part of the Project Line addressing Intelligence in Services and Networks of the RACE II Programme. As Chairman of the Project Line for the past year, I have been privileged to listen to many project presentations and demonstrations, and to participate in the detailed debate which has provided the foundations for the discipline of Service Engineering.

The IS&N Conferences have also provided an international context for these development activities, and allowed participants from outside the current RACE community to understand the role of Service Engineering in the future development of the telecommunications services environment. Thus we now see Service Engineering emerging globally as the core discipline in the development of new multimedia, multiparty, multidomain teleservices based on advanced technologies. Service Engineering is recognised as the framework for service development, deployment, execution and management in the telecommunications services sector.

As a new discipline which has evolved over the past few years, it is useful to begin with a session which covers the current state of the art in Service Engineering, and to provide an appraisal of the challenges to be faced over the next few years. The pervasive nature of the Information SuperHighway has drawn widespread attention to the need for an engineering approach to new service development and deployment. Much of what is currently available over the present instantiation of the

SuperHighway (the Internet) is acknowledged as falling below the standards of quality and performance recognised in the telecommunications sector.

The first paper in this conference session provides an overview of the experiences of the past few years in the deployment of new technologies. Covering much of the development of the telecommunications services environment in recent years, the paper analyses the lessons to be learned from the work on IN, TMN, and in TINA. The conclusions drawn highlight the need for an adaptive approach in order to meet the particular needs of the telecommunications services sector.

The second paper develops a theme from last years conference, and which has been discussed extensively over the past year. The need for integration of TMN and IN is generally accepted; both ETSI and ITU-T groups and others have been looking at the issues for some time. The paper reviews the options which derive from this work, and points to several approaches and their resulting technical problems.

The IS&N95 conference was attended by many participants engaged in current R&D projects, including those funded under the EU ACTS programme. Meeting the challenges described in these papers will no doubt occupy the attention of many researchers over the next few years. We wish them well, as they move forward the theoretical concepts and practical experience of Service Engineering.

Requirements for Rapid Technological Deployment and Exploitation[1]

Dr. J. P. Chester,
Regional Technical College,
Carlow, IRELAND
+353 503 31324

Mr. B. V. Dentskevich, ETIC
165 Bld du Souverain, 1160 Brussels, BELGIUM
+32 2 6748534
bvd@ric.iihe.ac.be

Abstract. This paper proposes essential characteristics for successful advanced technology deployment and exploitation. The conclusions are based on a review of some of the most recent results in the TMN and IN areas, and early trials with TINA-like systems. The specific requirements of the telecommunications services environment for integration of de facto and de jure standards and technologies, and for adaptation of distributed computing technology to legacy systems can be identified from this work. In addition, the crucial requirement to meet functional and non-functional (ie quality) requirements of the telecommunications services environment suggests a future strategy based on integration, adaptation, and careful assignment of services to the most suitable technology platform.

1. Introduction

Over the next few years, the pace of technology development and implementation in the telecommunications services environment is expected to accelerate. The deployment of IN and TMN based systems is already well underway. Initial experience with these systems have highlighted the specific requirements of the operators, complicated by the present massive investment in the switched infrastructure (legacy systems). In addition to specific performance requirements, there is a need for integration of components, preferably within an overall framework.

The move to deployment of distributed computing components within the telecommunications services environment is expected to proceed rapidly over the next few years. The process of integrating such components within the present infrastructure of switch based services will give rise to the same concerns.

This paper will highlight the requirements imposed by the present infrastructure which have become apparent during the implementation of TMN and IN environments. It will also show the need for integration of components within an overall framework. The results of this analysis will be applied to the deployment of TINA like components.

---

[1] Completed as part of EU RACE Project R2083 Consensus Management

2. Implementation of IN

The basic functions of the Intelligent Network Architecture have remained unchanged for many years[Ref1]. Implementations based on this model have increased rapidly through the early 90s. The most extensive experience of IN implementation can be found in the US. Two aspects of this experience are noteworthy.

From the technical perspective, operators have found that new services can be delivered much faster using IN, provided the required SSP functionality exists. However, there has been relatively slow progress in the type of services that can be developed using IN[Ref2]. The main reason for this is switch limitations - new services continue to require new IN switch software. In addition, technical issues, such as feature interaction, have limited IN capabilities. As a result, a full feature IN switch may not be available for some time. In addition, although considerable success has been achieved in the independence of switches and SCPs due to the interface standards (AIN and ETSI Core INAP), service logic is not yet portable between SCPs. Service Creation Environments remain closely linked to the SCPs for which they were designed due to the lack of any form of API in the SCPs.

The other main result from the experiences of early implementors of IN is the nature of the investment decision. There is a clear distinction between implementations based on the making of sound business cases, and those where executive management were committed to the technology. In this context it has been noted that the RBOCs who had achieved most success in IN had done so because of management commitment rather than to narrowly defined business cases[Ref3].

This section has outlined the experience of implementing IN components by operators. This has shown that the main success criterion is not necessarily business case based, and that a key success factor in implementation of new technology within the current technical infrastructure requires a strong commitment within the organisation to a technology push approach rather than a market pull. It should also be noted that attempts to develop standards from the current practice in this area have been somewhat difficult. The problems with Q1200 series of ITU recommendations on IN have been highlighted by work within RACE[Ref4].

3. Development of TMN based systems

Major telecommunications network operators, service providers and suppliers, through ETSI and ITU, have developed the concept of TMN (Telecommunications Management Network) for the management of the current and next generation services [Ref5]. TMN is a powerful tool for achieving a highly co-ordinated implementation of services for telecommunications management. It provides a highly structured framework for development and deployment of telecommunications management services.

One of the key characteristics of TMN deployment is the strong need for conformance to standards. The implementation of the M3000 series standards to produce TMN conformant management platforms has proceeded in both RACE [Ref6] and in EURESCOM[Ref7]. The efforts of current R&D projects are towards implementation of a standards compliant, functional adaptive platform architecture. RACE projects in this area have made valuable contributions to this experience[Ref8].

Suppliers are currently marketing the first generation of TMN conformant management platforms, e.g. HP Open View, Ericsson TMOS, IBM Netview, and DEC TMIP. These platforms, and others, which can be characterised as being strongly based on de jure standards, with additional capability being derived from de facto technologies, will be used by operators and service suppliers to develop specific TMN management applications. The use of TMN ensures compatibility with equipment from different sources, and creates a uniform development environment for service management applications.

The results from these areas show the difficulties of defining management services *ab initio* through standards. A major requirement of operators and service providers is the capability to develop a wide range of management services. Current RACE work even suggests that some of these management services will need to be made available to subscribers and end-users[Ref9].

A key result from this work is that there is a concurrent need to rapidly incorporate de facto standards and emerging IT technologies, as well as the de jure standards from ITU and ETSI into the emerging management platforms. The incorporation of the de facto technologies allows development to proceed at an accelerated pace. In addition, the time to market is reduced, while at the same time, because of the use of the de jure standards base, products remain compatible into the longer term.

4. Integration of TMN and IN

The current ITU recommendation for TMN management of IN[Ref10] raises many issues. The integration of TMN and IN may be approached from a number of perspectives. In the ITU-T TMN JCG, the main emphasis has been on the physical layer. The mapping of IN devices as TMN Managed Objects has also received attention in the joint NA4/NA6 Working Group in ETSI[Ref11]. As yet unresolved, due in large part to lack of appropriate definition, is the relationship between the IN SMS and TMN OS.

In this area, a major problem has been the impact of current approaches to service creation, and in particular, to the SCEF of the IN Q1200 series recommendations from ITU. The current view of the telecommunications services environment (TSE) contains three main functional blocks - SCEF, SMF (TMN conformant), and SCF/SSF/SDF (known as the Execution Platform EP, or Advanced Service Platform). Attempts over the past few years to realise the SCEF have been almost exclusively based on the reusable component approach, implemented as SIBs. Trials, and detailed formal analysis of this approach, including the work of BOOST [Ref12], and SCORE [Ref13], and others [Ref14] have shown clearly the weaknesses of the SIB based approach.

The requirement of the operators for rapid service creation, as described in Q1200 series on IN, still remains a major strategic objective. This includes the need for rapid downloading, instantiation, provisioning, execution and management. To achieve this objective in a multivendor environment, the current technical strategy generally accepted, and discussed most recently at ETSI [REF15] and ITU [REF10], and indeed in TINA, can be described as -

1   development of service logic (SL)

2   deployment of the service logic via the management functional entity

3   instantiation of the service logic on the execution platform

The ability to design, deploy, instantiate and execute service logic on a single vendors equipment range is currently widely available. To realise this in a multivendor SCF/SSF environment, two features are needed. Firstly, the EP must have agreed distribution transparency in-built. This is necessary to guarantee transparent operation of distributed services. Secondly, the EP must implement a common agreed API for the SL. In this way, SL from any one vendors SCE can be downloaded to any EP. There are of course management and policy issues (e.g. charging, security, etc.) to be resolved for this to be successful across administrative domains.

It follows from this approach, that the internal SCEF technologies, processes and procedures are subject to competition, as long as the output (i.e. the SL) conforms to the agreed API. It is for SCE providers to adapt technologies to realise the SCE by whatever means. This would allow realisation of many different SCEs, in line with the different organisational requirements (as detailed in BOOST D26, for example).

The impact of this on both IN and TMN as currently defined is only beginning to be discussed in standards bodies and elsewhere. Whereas, there is some understanding of, for example, TMN management of IN services at the physical layer, and of the need for Q3 information models of IN functional entities such as SCF and SSF, the technical resolution of how the IN SCEF interacts with and through the TMN needs considerable study. From the viewpoint of the operators of the current service platforms (ie the present switch based infrastructure), it is clear that the downloading of new service logic to, say, the switch will have to be under management control. The relationship of these issues to the deployment of distributed computing technologies will also need careful study.

The problems for integration of TMN and IN which derive from the service creation issues, highlight the need for an overall integration framework. One attempt to address this issue resulted in the SMP Integration Framework developed jointly by a team drawn from suppliers and operators[Ref20]. The definition of roles among the network components which results provides key pointers for IN and TMN applications implementors. The separation of roles, and consequent definition of interaction boundaries, enables the work of different groups to be better co-ordinated. In addition, and assuming adequate attention to boundary functionality, the subsequent integration of components into an overall systems framework will proceed more smoothly. This approach also addresses the legacy systems issue more easily.

The emerging need for integration of TMN and IN derives mainly from the need to develop a consistent evolution plan for the infrastructure. An overall framework is needed within which technology deployment and exploitation can proceed. This framework will also need to address the speed of technological development, the integration of both de facto and de jure standards, and the constraints (or opportunities) imposed by legacy systems. The overall trend highlighted by these efforts shows that to achieve rapid technology deployment an evolutionary model is needed.

5. Distributed computing and TINA

There is a general understanding that the next major technology evolution within the public switched network will be the incorporation of distributed computing technologies. In recognition of this trend, operators and suppliers have established the TINA-C initiative. The early results emerging from this effort support the general consensus on the use of advanced IT technologies to enhance the services of the public network, both in the user and management planes.

Although the TINA-C project is not due to finish until 1997, the main components of TINA architecture are well known[Ref16]. The main characteristics are the focus on ODP conformance, and in particular the distribution transparency essential for a substantial increase in the range of services, multimedia, multiparty, and multidomain, supportable by the telecommunications services infrastructure. It is confidently expected, that the required features can be implemented in the DPE through CORBA and its associated computing technologies (IDL, etc.).

What remains unclear is the performance implications of such systems in the telecommunications infrastructure. While the performance of distributed computing systems in specific applications is well understood, the performance network operators and service providers can expect from large scale deployment of this technology within the telecommunications services environment is not as yet well defined. Nor is possible at this time to relate the know performance in particular cases, to the more general case of multiparty, multimedia, multidomain services.

Based on the early architectural studies, there have been a number of early trials. It is essential to be able to assess the performance of an architecture or design at an early stage of development of a TINA-like system. One study [Ref17] applied a measurement technique called layered queuing analysis to management of a small scale telepresence conference engineered as a TINA-like engineering model. The conclusions, while validating the measurement technique, also noted that it was too early to draw conclusions about the scalability of TINA-like systems. This case featured a platform engineered to deliver a single service.

A second example selected the freephone service, implemented in a TINA consistent network[Ref18]. The conclusions clearly show that there are problems in providing overload control for DPE Traders which are as performant as traditional methods in current generation switches. In addition, the expected rate of DPE Trader access to support freephone (or even POTS in a TINA-like network), is so high compared with the throughput of current database technologies that use of a single Trader to support a potential calling population the size of a small town would appear impractical. In another case [Ref19] of a multiservice platform, other performance problems with current trader implementations were identified.

It is clear that these early experiences do more than record the current technology limitations when such technology is applied to the specific problems of the telecommunications services environment. The experience of implementing TINA-like distributed computing environments for use as telecommunications services platforms highlights the need to address the functional and non-functional aspects specific to the telecommunications services environment. This once again points to the need for an integrated approach, based on an agreed integration framework, to implement the specific required functionality in the most appropriate technology.

It further highlights the need, in advanced technology development, to adapt the best practice from previous programmes. In this case, the experiences with TMN and IN provide valuable lessons to the distributed computing community in their attempts to expand the capability of the present telecommunications services infrastructure.

6. Requirements for rapid deployment and exploitation

From the previous sections, conclusions can be draw about the characteristics and requirements for rapid technological deployment and exploitation which will be of particular relevance to ACTS projects in this area. Implementation of TMN, IN and TINA have delivered valuable lessons on the likely bottlenecks to rapid deployment and exploitation of new technical developments. This paper highlights issues which relate to the technical domain, and the technical problems of adopting new technologies, adapting to specific environments and integrating them with existing installed base. It is for others to comment on the impact of other factors (the regulatory framework, etc.).

The issue identified in the preceding analysis are summarised in the following sections.

6.1 The role of technology push and early adoption

The IN experience highlights the role of technology push in the early deployment of new technologies. It was found that those companies who had made the decision to implement IN were the ones who most benefited from the implementation. The business case approach of only adapting existing systems to market demands, was perceived as less successful in the longer term.

This also argues for an approach based on interface definition, early agreement on basic functionality, and the role of vendor independence in such an environment. This is the only way in which new technology can be moved into the field rapidly. The alternative, based on the detailed specification of components, message sets, protocols and standards delays indefinitely roll out of technology in sufficient quantity to justify production.

In the specific case of IN, it is the early adopters who now have the experience to recognise the limitations of, for example, the standardised SIB approach, and who can most benefit from the clustering of components to enable development of a common API. In the case of distributed computing it is the organisations with the early experiences, even if these are negative, who will most likely in future benefit from new developments based on their results. These organisations are best placed to see the possibilities, and to be in a position to exploit the new technologies more effectively.

6.2 Integration of *de facto* and *de jure* components and standards

In the case of TMN, the developments of TMN platforms accelerated rapidly towards the end of RACE2 through incorporation of de facto standards, overlaid on a base of de jure standards from ETSI and ITU. Currently, suppliers are marketing TMN conformant platforms which also have major industry standard components installed.

This points to the role of the standards organisations as providing the overall framework, and elucidating the requirements for interoperability across boundaries, both technical and administrative. What has not been developed in the standards meetings are specifications for management applications and services. This is left to individual operators and service providers. The guarantee of interoperability derives from the use of the de jure standards base. The overlay of the de facto standards enhances the capability of the platforms, and the ease of development of the necessary applications.

TINA recognises the need for its management architecture to conform to TMN, as well as exploiting the other technologies form the de facto area. Clearly there are technical problems with this. However, the work on integration of IN and TMN shows the usefulness of having an agreed integration framework in mind throughout this work. For full adoption of distributed computing technologies into the telecommunications services environment, these two aspects are essential.

6.3 Requirement for an overall technology integration framework

The need for integration of TMN and IN has been recognised for some time. The problems involved are described in more detail above. The results of these effort show clearly the need for early agreement on exactly what is to be integrated, the reasons for so doing, the goals, especially performance goals, to be reached, and the requirements on the overall integrated result. Without agreement on these, the opportunity for misunderstanding, and for divergence is very real.

It is in this context, that an overall framework for technical integration activities is most useful. While no one framework will satisfy all, it is clear that progress can be made if effort is given to detailing interactions between functional entities at a sufficiently high level. Most importantly, there can be confidence that whatever technical development takes place, the results will fit within the overall agreed framework.

The problems with TMN and IN are now well known. With TINA coming on stream soon, we face the possibility of two additional integration tasks - TINA with IN and TINA with TMN. The issues in TINA/IN integration are already under discussion. There is already work in hand to develop TINA to IN gateways, to allow interworking of both IN controlled and TINA controlled services.

Legacy systems will not be decommissioned for many years to come. The benefits of early agreement on an overall framework within which such integration activities can take place are clear. This in turn will lead to a more evolutionary approach to implementation.

6.4 Functional and non functional requirements specific to the telecommunications services environment must be met

The telecommunications services environment imposes specific requirements on its technology. These requirements derive from years of practise and well established user expectations, backed up in some cases by tools to guarantee quality of service based on measurable results. There are business imperatives here also. Subscribers will be reluctant to pay for new services if there is a major derogation from the current service quality expectations. It is only necessary to attempt to call to (or worse from) those places where quality of service goals are lower (or higher) than one is used to, in order to see the effect of this.

Early experience with new technology always identifies performance bottlenecks. This was the case with IN and with TMN; it is also the case with the early TINA trials. What is clear is that adoption of new distributed computing technologies to enhance the telecommunications service infrastructure must recognise two real requirements.

The first is the heavy investment in current equipment. This is sometimes seen as a constraint. However, the present technical infrastructure is capable of delivering a range of services to very stringent performance requirements (e.g. post dialling delay less than a few hundred milliseconds). The second is the need for new services to approach the kind of performance that subscribers associate with current services.

What is needed is a way to adapt distributed computing technology so that the best features of both the new platforms and the existing systems can be utilised. In this way, the benefits become very clear - enhanced service capability, low risk to current service operations, and most significantly, a lower barrier to early adoption of new technology in the telecommunications services sector.

7. References

[1] ITU-T Rec Q1200 on IN

[2] Lauer, Dr. G. Intelligent Networks Summit, Chicago, 1993

[3] Ludlam, D. Proceedings of ICIN94, Bordeaux, 1994

[4] EU project SCORE, contribution to ITU-T "Problems with Q1210 based on an SDL model"

[5] ITU-T Recommendations - M3000 Series on Telecommunication Management Network

[6] RACE94, RACE95; Research and technology development in advanced communications technologies in Europe, RACE Central Office

[7] EURESCOM Project P208 in P208 Newsletter "TMN Operations Systems Platform"

[8] RACE ICM in Proceedings of TMN Platform Workshop, Watford, 1994

[9] RACE MOBILISE Deliverable "PSCS Concepts", December 1994

[10] Minutes of ITU-T JCG on TMN, 1994

[11] Minutes, NA4/NA6 Meeting, Tuusaala, Finland, 1994

[12] RACE BOOST Deliverable D14, 1995, and others

[13] RACE SCORE Deliverable D4.13 and others, 1994, 1995

[14] T. Chang, "The weaknesses of the standard SIBs"; Proceedings of INCM94, Bordeaux, p67

[15] Minutes of NA4/NA6, Cambridge, 1994

[16] Proceedings, TINA Conference 95, Melbourne, 1995

[17] Jogalekar, P. P, et al, "TINA architectures and performance, a telepresence case study", Proceedings, TINA Conference 95, Melbourne, 1995, p415

[18] Parhar, A, Rumsewiez, M, "A preliminary investigation of performance issues associated with Freephone service in a TINA consistent network", Proceedings, TINA Conference 95, Melbourne, 1995, p431

[19] Speirs-Bridge et al, "MSP: practical experiences in the application of TINA", Proceedings, TINA Conference 95, Melbourne, 1995, p685

[20] Chester J, Dickerson, K, "Standards for Integrated Services and Networks", Proceedings of IS&N94, Aachen, 1994; publ as lecture Notes in Computer Science 851, Springer-Verlag

Issues in the Integration of IN and TMN

George Pavlou - University College London, UK
David Griffin - FORTH-ICS, Crete, Greece

Abstract. Over recent years, the need to introduce rapidly new telecommunications services has led to the development of the Intelligent Network (IN). These services and the increasingly complex supporting network infrastructure need to be managed. The Telecommunications Management Network (TMN) provides the framework for their management. Now, it is becoming clear that future sophisticated services, diverging from the simple telephony call model, will need to be deployed, operated and managed in an integrated fashion. Target, long term architectures such as TINA are being developed to support these services. This paper considers the issues behind the co-existence of IN and TMN, contrasts their philosophies and architectures and explains the nature of operation in the control and management planes. It considers the use of the TMN to manage or even replace the IN and discusses issues for their integration in a unifying target framework such as TINA. The role of the supporting technologies is also examined.

Keywords: Control, Management, Service, IN, TMN, TINA, OSI, ODP

1. Introduction

Over the last few years, the increasing complexity and sophistication of telecommunication network infrastructures has led to the Telecommunication Management Network (TMN) [M3010] as the framework for their management. At the very same time, the need for sophisticated services based on the telephony call-model, such as Universal Personal Telecommunications (UPT), free-phone, Virtual Private Networks (VPN), etc. has led to the Intelligent Network (IN) framework [Q1200] in order to achieve their rapid introduction and operation. In the future, more sophisticated services breaking away from the simple call model, e.g. multi-media, multi-party conferencing etc., will need to be rapidly and efficiently introduced, deployed, operated and managed. Long term service architectures such as the Telecommunications Information Networking Architecture (TINA) [TINA] try to provide frameworks to make this possible.

Such service architectures address only the long term integration of the service creation, execution and management infrastructure. In the mean time, the traditional IN evolution is continuing while TMN systems have started being deployed. They both constitute substantial investment which cannot be neglected. In fact, new integrated architectures should provide for a smooth migration. What has certainly become clear is that the IN, sometimes also referred to as Advanced IN (AIN), adopts a functional, centralised approach while distributed object-oriented approaches would be more suitable. The modernisation of IN and its extension to support future sophisticated services is an important issue and this has led to the exploration of target integrated architectures. On the other hand, TMN principles should be used to manage the IN and, as TMN has a distributed object-oriented nature, there is the possibility of using TMN principles to realise IN services.

This paper considers the issues behind IN and TMN co-existence and integration and the evolution to target service architectures in the long term. As IN operates in the control while TMN in the management plane, this distinction is sometimes confused and the terms *control* and *management* are used in the wrong context. Here, similarities and differences between operation in the control and management planes are examined in detail while the current IN and TMN architectures are contrasted. The use of the TMN to manage the IN infrastructure in the medium term is considered while the possibility of using TMN object-oriented distributed principles to replace the IN by operating also in the control plane are discussed. The latter points to an eventual integration in a unifying framework using common underlying mechanisms e.g.a supporting Distributed Processing Environment (DPE) and bridging the gap between the computing and telecommunications worlds. The role of supporting technologies in this integration i.e. OSI Management / Directory [X701] [X500] and ODP / Object Management Group (OMG) Common Object Request Broker Architecture (CORBA) [X900] [CORBA] are considered.

2. Scope of IN and TMN in the Operation and Management of Communications Networks

The ITU-T have distinguished between the management and control planes in the operation of communications networks [I.320] [I.321] and introduced the TMN [M.3010] as a means of provisioning management systems.

IN addresses the separation of service control logic from core-network routing logic and has two aspects: the off-line service creation process and the service operation aspects in which new logic is used to intercept in the call establishment process and interpret/redirect it accordingly. The latter procedure takes place in the control (signalling) plane. Service creation is concerned with the initial generation of the logic involved while deployment procedures are used to plant it in the IN infrastructure, the process referred to in IN as "service management". An interpreted scripting architecture is used so that existing compiled logic does not need to be updated.

On the other hand, the TMN is a conceptual separate data network overlaid on the telecommunications infrastructure being managed. This monitors network/service resources through object-oriented abstractions and may perform intrusive actions to modify the way the network operates. The key difference to IN is that normal network operation (e.g. signalling procedures) are not affected at all as the whole operation takes place "outside" the managed network. The TMN should compliment and enhance the control plane functions by configuring operational parameters and, in general, it has less stringent requirements on real-time response

The two approaches have a lot of similarities and some important differences but are complementary in general and as such there is scope for their integration. Control affects the way the network operates and although the current IN architecture operates without the need to change the underlying signalling mechanisms, there are limits as to how far this approach may reach before such changes are necessary. On the other hand, the TMN operates outside the managed network and can be infinitely extended in functionality, as far as adequate provision for such functionality exists through abstractions of all the possible IN resources. The disadvantage of the management approach is potential lack high real-time reactions to conditions. However, this does not mean that the TMN approach is inferior in any way to the approach of IN, the management systems provided by the TMN are complimentary to the signalling plane and in general are not involved in real-time decision making processes to the same extent as the control plane features.

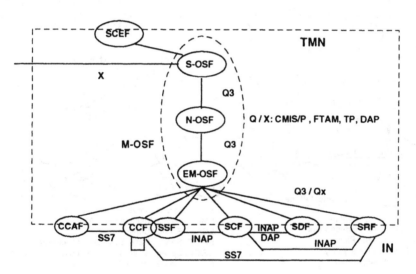

Fig. 1. Relationship between IN (control) and TMN (management)

The separation between control plane and management functionality is shown in Figure 1, which depicts the IN infrastructure operating in the control plane using signalling mechanisms and the TMN managing IN Functional Entities (FEs) operating as a logically different, overlaid management network. The Call Control Agent and Call Control Functions (CCAF / CCF) communicate over the standard signalling mechanisms while the Service Switching Function (SSF) intercepts in the call establishment process and communicates with other IN functions [Service Control Function (SCF), Specialised Resource Function (SRF)] using the IN Application Protocol (INAP) [INAP]. The TMN Operations System Functions (OSFs) manage the IN entities through q reference points (CMIS/P) [X710] in an object-oriented fashion. Note that currently the Service Creation Environment (SCE) does not yet communicate with IN FEs through the TMN. The various aspects of the IN and TMN architectures are presented in section 3.

3. Comparisons between the IN and TMN Architectures

Both the IN and TMN architectures follow logical hierarchical models and they both try to make physical (i.e. implementation specific) aspects independent of the logical/ functional aspects. A key concept in IN is its conceptual model which comprises four planes: the service plane, global functional plane, distributed functional plane and physical plane. The TMN comprises logical and physical architectures, while the management functionality is hierarchically decomposed into element, network, service and business management layers. The IN and TMN functional and logical layered architectures are shown in Figures 2 and 3 respectively.

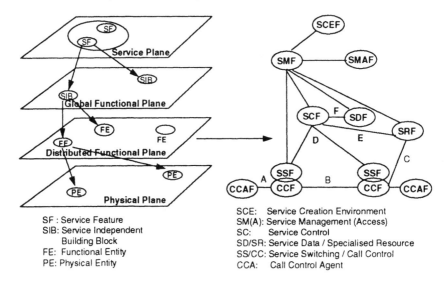

SCE: Service Creation Environment
SM(A): Service Management (Access)
SC: Service Control
SD/SR: Service Data / Specialised Resource
SS/CC: Service Switching / Call Control
CCA: Call Control Agent

SF : Service Feature
SIB: Service Independent
    Building Block
FE: Functional Entity
PE: Physical Entity

Fig. 2. The IN Layered and Functional Architecture

The TMN management services can be thought as analogous to IN services in the service plane as the latter represent an exclusively service oriented view. The global functional plane models the IN as a single entity and can also be considered to be analogous to the management services in the TMN as the management service definitions do not decompose the services into functional or physical components. The distributed functional plane and physical plane can be mapped onto the TMN logical and physical architectures.

An important distinction here is that the IN services are telecommunications services supplied to the end users or customers of the network operator. The management services provided by the TMN are primarily for the use of the operators and human managers of the telecommunications network. Despite that, it is possible to offer through the TMN services other than management e.g. international "leased" lines on demand (ATM-based VPN) etc. In principle though, the scope of the TMN and IN are different, however the comparison of their architectures is useful in the light of their long-term integration in a unifying framework.

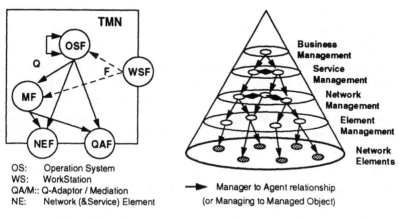

OS: Operation System
WS: WorkStation
QA/M:: Q-Adaptor / Mediation
NE: Network (&Service) Element

➤ Manager to Agent relationship
(or Managing to Managed Object)

Fig. 3. The TMN Functional and Layered Architecture

Considering this difference in scope, one aspect of the service management layer of the TMN is service deployment and service provisioning. The deployment and provisioning of services must consider the issues related to co-existence of newly deployed services with those already existing and therefore involves the network layer functions of performance management to ensure that the degradation of network performance is minimised and that the QoS targets for new and existing services is not disrupted. The TMN approach, via the hierarchy of network and element management layers, ensures that performance management capabilities of the TMN are involved in the deployment of services via the configuration management facilities of the TMN.

One of the major strengths of the TMN approach to network operation and management is its hierarchical nature. First of all, the functions related to the day-to-day control of the network such as call set-up, switching and signalling are considered to be separate from those related to network management. This fundamental hierarchy distinguishes the real-time nature of the control plane from the management operations associated with the management plane and the TMN. Although a distinction exists between the two aspects of network control and network management, they are related. The TMN influences the way the control plane behaves by configuring operational parameters, such as routing table entries, according to management decisions. The TMN monitors the network, makes decisions based on network conditions and other information, such as management policy and knowledge of future events, and feeds back management actions to the control plane of the network to influence its future behaviour. This architecture allows the network to operate as intelligently as possible without burdening the network elements with sophisticated features.

The second aspect of the hierarchical nature of management is within the TMN itself. Management functionality is distributed over a number of components, both horizontally and vertically. In the horizontal direction different management components exist for different network elements or sub-networks; and for different management services operating on the same element, sub-network, network or service. However the important distribution is the vertical one. In this manner element level managers are themselves managed by sub-network level managers or network level managers, and so on. There are different object models at each layer of the hierarchy according to the level of concern and abstraction at that layer. Information flows move through the hierarchy in both directions via well defined object interactions. For monitoring, statistical analysis, billing, etc., elementary information retrieved from the Network Elements (NE) is transformed, in higher level objects, via higher abstraction and summarisation. In the reverse direction, management actions (e.g., a service deployment request at the service layer) can be decomposed according to the intelligence of the management components into lower level management actions until finally configuration changes are made in the NEs themselves. Cascading is achieved through an ordered sequence of management activity in the form of operations on managed objects at various management layers. In short, the TMN projects a hierarchical object-oriented model.

4. Use of the TMN to Manage or Replace the IN Infrastructure

4.1 Managing the IN by the TMN

The IN infrastructure needs to be managed in order to support the smooth operation of IN-based services. In addition, support to the service creation environment can be offered through the provision of deployment mechanisms that ensure the rapid and efficient deployment of IN services. Considering the Service Creation, Management and Execution Platform model (SMP) [Dic94], it is certainly beneficial if the interaction of the service creation and execution environments is through the management environment in order to maintain information related to IN services.

The main advantage of using the TMN for IN management has to do with a common management philosophy to other networks (e.g. ISDN, SDH). This will result in reusability of management functions and associated logic (management service components) and the unification of management processes.

In addition to the management services provided by the TMN for all network technologies the TMN can manage specifically the following aspects of IN:

 • *service deployment:* the installation of service logic and data to the network and to the management systems associated with the management of that service

• *service provisioning:* the collection of service specific data and the installation of this data in subscriber and contact databases

• *service operation control:* software maintenance and information update

• *billing:* the collection and storage of usage records and tariffing

• *service monitoring:* the measurement, analysis and reporting of service usage and performance

To achieve management of the IN, its SSF, SRF, SDF and SCF functional entities have to be modelled as TMN NEFs, providing control of the associated resources through managed objects. The IN SMF functional entity will be modelled as a set of layered TMN OSFs, offering at the service level an interface to the Service Creation Environment Function (SCEF). Note that the various IN entities will still use the INA Service/Protocol across the IN reference points but they will also offer TMN q reference points. This architecture has been shown in Figure 1.

The important aspect of using the TMN to perform the above tasks is the existence of generic management functions that perform most of these tasks in their totality while they provide reusable generic capabilities to be specialised with respect to others. These generic management capabilities are offered through the OSI Systems Management Functions (SMFs[1]) and are used by TMN management service components and management functions [X701].

4.2 Replacing the IN in a TMN-based Framework

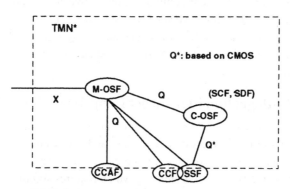

M-OSF: Management OSF
C-OSF: Control OSF (replaces SCF, SDF)
CMOS: CMIS/P over Signalling

Fig. 4. Realising the IN through the enhanced TMN

1. To avoid confusion with the IN SMFs, the acronym SMF will be prefixed by IN or OSI to refer to Service Management Functions and Systems Management Functions respectively.

One of the features of the current IN is that it adopts a functional approach, identifying the capabilities required in the IN and allocating them to functional and physical entities. On the other hand the TMN is object oriented by nature while distribution is also an important aspect. Since the TMN will be used to manage the IN, the (co-)existence of its richer framework leads to the consideration of replacing IN functional entities by equivalent TMN object-oriented functional blocks, communicating with each other over the signalling plane e.g. using CMIS/P over a signalling protocol instead of INAP [Mag95]. This approach in fact integrates the mechanisms for control and management and is in line with the spirit of projected future service architectures such as TINA.

In this approach, the SSF will have to be modelled as a lightweight TMN OSF or NEF. It will offer managed objects to be configured and managed while it will access other OSFs modelling the Service Control and Data Functions (SCF/SDF). This access may be peer OSF to OSF or as a subordinate NEF to OSF, as in Figure 4. In the latter case, an event to the control OSF will trigger the normal IN procedures with the call proceeding after a subsequent operation to the SSF, passing the necessary information.

In order to achieve this, efficient implementations of CMIS/P over lightweight transport mechanisms and efficient OSF Management Information Base (MIB) implementations are required. Research in investigating such mappings is under progress [Pav95b]. The overall architecture for such a TMN-based IN realisation is shown in Figure 4. Note that there are different OSFs for management and control. The former will have to configure the latter with the necessary customer profile information etc. over a standard Q interface while the communication between the latter and the SSF NEF will take place over Q^* using CMIS/P Over Signalling (CMOS). Note also that the TMN is now referred to as TMN^* as it is extended to perform both the standard management but also control functions using the same object-oriented (CMIS/P, GDMO) hierarchical architecture.

5. IN and TMN Integration in a Target Long Term Architecture

The traditional IN uses a centralised service control and service data model. This centralisation together with the relative simplicity of the SCF causes problems when considering services breaking away from the simple call model. An example of such a service is multi-media, multi-party conferencing which requires significant connection and session control and management.

One of the driving forces behind the TINA initiative was to modernise the IN and the traditional control plane functions. The TINA approach for the future IN resolves both the above issues, by adopting object oriented techniques, the ODP modelling approach and making use of a Distributed Processing Environment as a ubiquitous supporting infrastructure which encapsulates the transport network. The TINA approach has similarities with the approach presented in section 4.2 for replacing IN with a TMN-based framework but it does the opposite: it creates a new framework for the future IN and applies it also to the management of the services, network and supporting resource infrastructure. A simplified view of the TINA proposed layered architecture is shown in Figure 5.

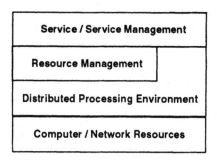

Fig. 5. The TINA Layered Architecture

An important consideration in the long term integration of IN and TMN is that the advantages of the hierarchical TMN approach are retained. Components related to real-time, on line decisions should be made as lightweight as possible and located near to the network elements, and at the other extreme, management components involved in sophisticated decision making activities over a longer time scale should be allowed to work in an off-line mode without burdening the network elements or the time-critical control and management functions. This principle holds whether a pure TMN approach based on OSI systems management, the TINA approach based on CORBA and DPE, or a hybrid approach is taken.

By integrating the methods for the control and management of INs into a single framework, the interactions between the control and management planes are simplified. Common components can be used, in particular the data used by the management components (OSFs, etc.) can be integrated with the data used by the IN components (SDFs). This will aid the manipulation of, for example, customer data records, by the service management layers. No longer will specific adaptors (QAFs) or mediators (MFs) be needed, as the control plane functions will be integrated into the same DPE as the management components.

The base technologies for TMN have been the OSI management and directory [X701] [X500], supported by file transfer and transaction processing. On the other hand, technologies for the future IN will be ODP-based [X901]. For example, OMG CORBA [CORBA] forms the basis of the TINA Distributed Processing Environment. As there are analogies and complementary aspects between IN and TMN, the same is true for the supporting technologies. OSI management is object-oriented technology and the impact of the emerging ODP-RM is currently under study, the result being the Open Distributed Management Architecture (ODMA). On the other hand, the management aspects of the TINA DPE use the OSI management operational model with managed objects specified in GDMO/GRM from an information viewpoint while they become CORBA IDL computational objects.

An open-minded view to the current TMN architectures and supporting technologies shows many similarities to the ODP-influenced approaches for the future IN. The TMN functional, information and physical architectures map directly onto the ODP computational, information and engineering viewpoints. OSI management projects an object-oriented model which could be supported by ODP-based technologies such as CORBA in the future. The necessary additions are object clustering; bulk data retrieval capabilities based on sophisticated queries that enable to traverse object relationships; and fine-grained notification capabilities based on sophisticated assertions. The OSI system management power should be maintained in order to support the efficient management of sophisticated future telecommunications infrastructures. In essence, this means transposing the OSI management model over the ODP-RM to become (part of) the latter's management framework. There have been fully object-oriented realisations of OSI management providing distribution and other transparencies [Pav94] [Pav95a], showing the power and benefits of the OSI management *cluster* or *ensemble* model and its applicability in management environments where engineering issues such as the amount of management traffic, timely response to events and early suppression of unnecessary notifications are of paramount importance.

One possibility for retaining the current TMN investment in advanced future long-term service architectures is to accept the existing TMN infrastructure as the means to manage network aspects of broadband infrastructures and provide adaptation to the ODP mechanisms used for service execution and management in those architectures. Recent research suggests this is possible and there are ongoing activities between the NMF and X/Open to define generic mappings between OMG CORBA and OSI management [XoJIDM]. In addition, the TMN itself may eventually migrate to the use of ODP-based technologies by retaining its information aspects but using ODP-based distribution and transport mechanisms as described above.

6. Discussion

As presented in the previous sections, IN and TMN are two important aspects of modern telecommunications networks and services, covering different aspects of their operation and management: IN is more closely related to the control plane, while TMN is concerned with the management plane. Although they are complimentary, there are areas of overlap, particularly from the point of view that IN networks need to be managed in the same way as any other network technology. The other major area of overlap is in the IN Service Creation and Management Functions. The SCEF in general lies outside of the scope of the TMN, but it provides an important input to the management functions in the form of service specifications. However, the IN SMFs directly overlap with the TMN. Whereas a significant amount of work has been performed in the TMN and management area, the exact capabilities of the IN SMFs are largely unspecified or untested through prototypes and experimentation. Because of the maturity of the TMN work, and the fact that the TMN approach to management has been proved through numerous research initiatives, prototypes, and now commercial developments, *we propose that the TMN is the best possible choice for implementing the IN SMF features.*

As argued in the previous sections, harmonising the approaches of IN and TMN is beneficial from a number of viewpoints: interactions between the two are needed anyway as the IN needs to be managed, and management decisions are based on data retrieved from the IN; data needs to be shared between the management and control planes; the dividing line between the latter is not completely fixed; and there is no need to provide a completely different framework for IN SMFs when the TMN framework exists.

Apart from the standardised INAP protocol specifications, IN platforms are in general proprietary and network operators have their own solutions for developing IN functionality. There are obvious benefits in moving towards a common technology with well defined generic protocols and APIs. In the TMN world, GDMO and CMIS/P provide the object-oriented framework while sophisticated software platforms have been developed and a number of initiatives are aiming at standardising high-level APIs for TMN developments. Furthermore, the TINA initiative has adopted a CORBA-based approach for its DPE, with well-defined APIs and IDL to specify object interfaces.

The TINA approach seems promising for the long term integration of IN and TMN. Though the main driving force behind it was the modernisation of IN, it does address, to a certain degree, the incorporation of network and service management. However, we believe that TINA still has some way to go before it can fully address the management issues which are currently resolved in the TMN. Currently, TINA incorporates the majority of what is known in the TMN world as network management into its "resource management" architecture. We feel that this is an over-simplification, and more importantly, many of the features identified in TINA as "management" (such as session management and connection coordination) are mainly involved in the on-line creation, maintenance and termination of user connections and calls, issues more related to the control rather than the management plane. However, it is recognised that the TINA work will progress, to address more directly issues related to the management plane such as fault, performance and QoS, routing, bandwidth management, etc.

Furthermore, the TMN has many essential features which are invaluable in the design and operation of management systems:

• *hierarchical layering* which allows for object-oriented abstractions at different levels, supporting system encapsulation and making possible to express different viewpoints and concerns (element, network, service, business management)

• *control and management plane functionality distinction*, which is closely related to the performance aspects of the TMN hierarchy

• *use of OSI systems management* to structure and cluster management information in an object-oriented fashion and to provide generic management features through the OSI SMFs

These features of the TMN have been proved indispensable, particularly when designing and implementing large TMN systems with sophisticated functionality [Gri95]. Although it is possible that TINA will adopt these facilities in the future, *we propose that the most efficient way of integrating TMN and IN in the medium term is through the TMN infrastructure and procedures.*

The medium term solution proposes that IN based networks are managed in the same way as other networks (ISDN, B-ISDN, etc.), by considering the IN physical entities as network elements providing management access through Qx or Q3 interfaces. The second aspect of the medium term solution is that the IN SMFs are implemented in the TMN itself as OSFs at the service management layer, using the existing TMN methodologies for management service decomposition [M.3020], design and implementation; using OSI systems management concepts for object-oriented information modelling; using the OSI SMFs for providing generic management capabilities; and using existing TMN platforms for APIs and other generic functionality.

As the IN SMFs will be implemented in the TMN framework, they can fully interact with other OSs in the TMN which provide the other TMN management services e.g. performance and QoS management, routing, bandwidth, fault, configuration management, etc. The relationship, dependencies and interactions required between the IN SMFs and the other management services of the TMN is an important consideration. IN SMFs cannot exist in isolation, and by bringing them into the TMN itself, such necessary interactions are more easily accommodated.

Finally it is proposed that the TMN takes on part of the control plane burden itself by implementing the SCFs and the SDFs in Control OSFs within the TMN[*]. To do this, it is proposed that a more lightweight version of Qx or Q3 is used, with CMIP mapped onto existing signalling protocols such as SS7. This is to ensure that the interactions between the SSFs in the network elements and the SCFs in the C-OSFs in the TMN[*] is as fast and efficient as possible. The advantages of this are numerous: there can be a common object oriented framework for both management and control; mapping between object oriented views is much easier than mapping between procedural and object-oriented approaches; the integration of the more intelligent part of the control plane with the TMN allows management and control to interact more easily within a common framework; the distinction between management and control no longer needs to have a clear dividing line; the capabilities of the control plane are no longer bounded by the capacities of the IN PEs as in the TMN framework, the SCFs and SDFs may be distributed (as opposed to centralised in traditional IN) and intelligence may be added as required through interaction with other components in the TMN*.

Although the TMN is in general not considered to be a real-time system, the TMN* can meet the constraints of both the control and management plane functions. By using lightweight versions of Qx or Q3 interfaces over existing signalling protocols, and by ensuring that strict engineering view constraints are considered in the design of the TMN*, these restrictions can be overcome. Signalling systems in the control plane of existing networks have been designed with performance in mind, signalling messages and decisions must be made quickly, within an acceptable time for the service user to wait between dialing the called number and receiving ringing tone. These performance considerations must be applied rigorously to the control parts of the TMN*, without unduly burdening the remainder of the TMN* (the original management functions and the new IN SMFs) when they are not necessary.

7. Conclusions

In this paper, we presented the issues behind the nature of operation of IN (service control) and TMN (service and network management) and explained their complimentary but also overlapping aspects. Given the fact that IN service management is largely as yet unspecified, we propose the use of the TMN for implementing the IN SMF features by modeling IN entities as TMN NEFs (Figure 1). Given also the distributed object-oriented nature of the TMN, the current understanding of relevant methodologies and modelling principles and the existing investment, we propose the integration of TMN and IN in the medium term through the TMN infrastructure and procedures (Figure 4). Finally, we envisage a target long-term architecture for their integration based on ODP principles, assuming the presented strengths of the current TMN approach are retained. An evolution of the relevant ODP-influenced base technologies such as OMG CORBA is necessary to support a powerful hybrid control and management DPE.

Acknowledgements

This paper describes work undertaken in the context of the RACE II Integrated Communications Management project (R2059). The RACE programme is partially funded by the Commission of the European Union.

References

[Q1200]   ITU-T Q.1200 series, Intelligent Networks

[M3010]   ITU-T M.3010, Principles for a Telecommunications Management Network

[M3020]   ITU-T M.3020, TMN Interface Specification Methodology

[I320]   ITU-T I.320, ISDN Protocol Reference Model

[I321]   ITU-T I.321, B-ISDN Protocol Reference Model and its Application

[INAP]   ETSI, Intelligent Network Application Protocol, Version 8, 1993

[X701]   ITU-T X.701, Information Technology - Open Systems Interconnection - Systems Management Overview

[X710]   ITU-T X.710 / X.711, Information Technology - Open Systems Interconnection - Common Management Information Service/Protocol, Version 2

[X500]   ITU-T X.500, Information Processing - Open Systems Interconnection - The Directory: Overview of Concepts, Models and Service, 1988

[X901]   ITU-T X.900, Information Processing - Open Distributed Processing - Basic Reference Model of ODP - Part 1: Overview and guide to use

[CORBA] Object Management Group, The Common Object Request Broker Architecture and Specification (CORBA), 1991

[TINA]   Mulder, H., J. Pavon, Telecommunications Information Networking Architecture, Tutorial presented at the 4th IFIP/IEEE International Symposium on Integrated Network Management, May 1995, Santa Barbara

[XoJIDM] XOpen / NM Forum Joint Inter Domain Management, Comparison and Interworking of OSI Management, OMG and Internet Management

[Mag95]   Magedanz, T., Modelling IN-based Service Control Capabilities as part of TMN-based Service Management, in Integrated Network Management IV, pp. 386-397, Chapman & Hall, 1995

[Pav94]   Pavlou, G., T. Tin, A. Carr, High-level APIs in the OSIMIS TMN Platform: Harnessing and Hiding, in Towards a Pan-European Telecommunication Service Infrastructure - IS&N '94, pp. 219-230, Springer-Verlag, 1994

[Pav95a]   Pavlou, G., K. McCarthy, S. Bhatti, G. Knight, The OSIMIS Platform: Making OSI Management Simple, in Integrated Network Management IV, pp. 480-493, Chapman & Hall, 1995

[Pav95b]   Pavlou, G., LCMIP: A Lightweight Protocol Architecture for Data and Telecommunication Network Management and Control, Research Note, Department of Computer Science, UCL, 1995

[Gri95]   Griffin, D., P. Georgatsos, A TMN System for VPC and Routing Management in ATM Networks, in Integrated Network Management IV, pp. 356-369, Chapman & Hall, 1995.

[Dic94]   Dickerson, K., J. Chester, Standards for Integrated Services and Networks, in Towards a Pan-European Telecommunication Service Infrastructure - IS&N '94, pp. 5-16, Springer-Verlag, 1994

Security

Session   2a   Introduction

Chairman - Alfred Ressenig

European Telecommunications Industrial Consortium (ETIC),
Boulevard du Souverain 165, B-1160 Brussels, Belgium
Tel : +322 674 8526
Fax : +322 674 8538
e-mail alr@ric.iihe.ac.be

The Sixth ETSI Strategic Review Committee (ETSI SRC6) had the mandate to look at the future European information superhighways in terms of requirements for specification, implementation and standardisation of the so called European Information Infrastructure (EII). In June 1995, the ETSI Technical Assembly accepted (in slightly modified form) the report and the recommendations of the SRC6. A dedicated group has already been set up to initiate and co-ordinate the necessary standardisation activities.

SRC6 concluded that security is one of the key areas which still require substantial efforts. The report contains a number of security related recommendations which emphasise:

- The need for specifying a comprehensive set of security features which must include cryptographic algorithms;
- The need that cryptographic techniques for authentication, integrity and non-repudiation may be specified and used anywhere in Europe;
- The need for confidentiality services;
- The need for standards for electronic signatures and for a framework in which such signatures may have the force of law.

It is clearly pointed out that an environment for commercial confidence has to be established. In the electronic information society "it is essential that service users and service providers have confidence in the security provided by the information infrastructure they use". Without adequate security features, the potential of the information superhighways will not be fully exploited. Business and residential users will be reluctant to use it e.g. for transmission of highly sensitive data.

There is undoubtedly the need for ensuring that a message is being received exactly as it was sent. Electronic signatures can satisfy this demand. The major representatives are based on the RSA (Rivest, Shamir, and Adleman) algorithm and the DSA (Digital Signature Algorithm).

There is another need that an intermediary cannot read nor modify nor substitute a message which is being transmitted from a sender to a receiver. This can be achieved by encrypting the plaintext message into a ciphertext at the sender site, transmitting the ciphertext, and decrypting the ciphertext into the original plaintext at the receiver site. Unfortunately there is a conflict with lawful interception which demands the ability to obtain transmitted messages as plaintext in order to combat organised crime. This demand is also the reason that frameworks for electronic signatures do not progress at present although the theoretical concepts exist. It is the wide applicability of algorithms such as the RSA which causes this disaster. Even when tailored for electronic signatures, one can adapt these algorithms for illicitly ensuring confidentiality.

In the U.S.A. the Clipper initiative tried to provide a solution to this conflict, however did not succeed for various reasons. An alternative approach is that of Trusted Third Parties (TTPs). The paper *A TTP-based Architecture for TMN Security and Privacy* sheds some light on this concept. It addresses also relationships between telecommunications services and their management services as well as implications on TMN interfaces.

The paper *TMN Security: An Evolutionary Approach* explains how the management of telecommunications systems changes. The new challenges include convergence with IT, multi-vendor environment, interconnection of TMNs and user access to management services and data. These changes pose new requirements with respect to the security of TMNs. The paper highlights the emerging scenario, presents security technologies and standards for TMNs, and pleads for a smooth, secure and economical migration from existing management solutions to TMN-based management.

Securing network management operations requires three main services: authentication, integrity/privacy and access control. The latter is the subject of the paper *Integrated Access Control Management*. In the past it was often believed to be sufficient just to protect the management applications with passwords. People who knew an application password had hence all access rights of the application. Because of the management changes stated above, new concepts had to be adopted e.g. from the IT world, such as the Access Control List (ACL). The paper discusses and compares the access control mechanisms of CMIP and SNMP v2, and presents an access control scheme which integrates the two systems.

Security Services for Telecommunications Users

Ralf Popp[1], Matthias Fröhlich[1], and Nigel Jefferies[2]

[1]Communication Networks
Aachen University of Technology, Germany
E-Mail: rp@dfv.rwth-aachen.de
[2]Vodafone, Great Britain

Abstract. The Pan-European RACE II project MOBILISE develops and demonstrates the *Personal Services Communication Space* (PSCS) as an approach for the deployment of personalised, mobile communication services based on *Intelligent Network* (IN) techniques.

This paper discusses - after an introduction to end-to-end security service requirements - the security architecture of the PSCS demonstrator. It gives examples how these requirements can be met by using existing demonstrator features.

A *Public Key Cryptosystem* (PKCS) using the RSA encryption algorithm is used for the mutual authentication between communication partners in the PSCS system. Available smart card technology is used for the secure storage of private and certified public keys, personal data, and for the processing of cryptographical functions needed by the PKCS.

A security architecture including a certification hierarchy and a suitable protocol for the mutual authentication has been implemented according to existing standards (X.509 [1], ASN.1 [2] [3], PKCS#1 [4]). The performance of the system has been evaluated in detail and examples for the provisioning of end-to-end security services are given.

1 Introduction to PSCS

The objective of MOBILISE is to define the *Personal Services Communication Space* (PSCS) concept, specify the PSCS architecture, build components for PSCS instances, and demonstrate PSCS systems and their use in applications in a project over four years.

PSCS offers to end-users, in their different roles, the ability to communicate, and to organise communication according to their own preferences, with respect to parameters such as: time, space, medium, cost, integrity, security, quality, accessibility and privacy. This user driven extension to the *Universal Personal Telecommunication* (UPT) approach will support end-user mobility over heterogeneous networks.

In order to demonstrate PSCS features, a PSCS demonstrator including advanced security features has been implemented.

This paper starts with a discussion of security requirements from the end-users and the service providers point of view. Furthermore, solutions are proposed by means of describing the security architecture of the PSCS demonstration system. The paper concludes by summarising the experiences made with the demonstration system and providing an outlook on ongoing research activities.

2 User Requirements for Security Services

Service providers, network operators, application service providers, subscribers, and end-users have security requirements of different significance and for different - sometimes contrary - reasons.

One approach to classify security requirements is to distinguish between the security features which support the secure operation of the underlying network and the end-to-end or value-added security services which can be provided by operators and/or service providers.

Some network security features enable the secure operation of a telecommunication network. Typical security features refer to user needs as for instance: correct billing, basic protection against eavesdropping[1], protection of personal user data, protection from being traced and so forth. These features also protect network operators and service providers and cannot be charged for, as they offer no services to the user beyond the basic telecommunication service.

It can be expected that these security features will be sufficient for a broad range of private telecommunication users, so that the demand for additional security services is limited.

However, there are groups of users - especially in the business area - who have a clear demand for additional end-to-end security services.

One example for the requirement of end-to-end security services is the context of *Virtual Private Network* (VPN) services:

Today, many companies manage their own private fixed networks in a physically closed environment. Corporation sites are usually interconnected via leased lines provided by public/private network operators. Here, new technologies like ATM offer the ability to change from usually not well utilised leased lines to a more cost effective communication between different company sites using bandwidth on demand. This is probably the first approach for a company to reduce the costs of transmission while preserving most of its telecommunication equipment. A further approach to reduce telecommunication and maintenance costs can be the change from use of private telecommunication equipment towards use of public ones. This change might also introduce mobile services in an area which traditionally focuses on fixed network services.

Keeping this development in mind, e.g. virtual private network services promise to be a new market for service providers and network operators. But these services and other services will only be accepted - and subscribed - if the customer can be convinced that his security needs are met.

The following sections provide an overview of fundamental end-to-end security services and how they can be introduced using available security features

---

[1] E.g. encryption of voice and user data during transmission via the air-interface in GSM. Basic protection against eavesdropping does not mean end-to-end encryption of information.

of the PSCS demonstrator. This is done with respect to the intended development from a private user oriented PSCS towards a PSCS system for corporate networks.

3 End-to-End Security Services for the PSCS System

The PSCS demonstrator incorporates two main security features: User-PIM (see below) authentication and PSCS authentication. User-PIM authentication merely ensures that the *Personal Identification Module* (PIM, mostly realised as a smart card) cannot be used by anyone besides the authorised user. The PIM also provides for the secure storage of authentication information and personal data. PSCS authentication authenticates the PSCS user towards the *PSCS Service Provider* (PSP) and vice versa. The latter authentication is implemented using a public key cryptosystem and a network of *Certification Authorities* (CAs).

The following end-to-end services are important when, e.g. VPN services are envisaged:

user-to-user authentication: This can be achieved by using the same mechanism as used for PSCS authentication, with the role of the service provider replaced by that of the second *end-user* (EU). This can also be used to authenticate a user to an *Application Service Provider* (ASP). Usually this can be done without the involvement of a third party, at least while roaming within an area *covered* by the certification authority of the home domain.

user-to-user confidentiality: The current implementation of the demonstrator does not support encryption of data, since encryption is not a normal PSCS feature. The proposal of a suitable encryption algorithm is beyond the scope of this paper. However, it can be presumed, that a user will not trust an encryption algorithm supplied by a service provider. Other legal aspects like the prohibition of encryption in some - even EC countries - will make it hard to cope with this user requirement.

non-repudiation of origin: This effectively means to sign a given message with a digital signature. This signature operation can be carried out in the secure environment of a smart card.

non-repudiation of delivery: A signed receipt message for the delivery of a previously sent message. Depending on the implementation, this service might require the assistance of a third party, e.g. the PSCS service provider.

message-integrity: This allows a recipient to detect any unauthorised modification to data transmitted. This service is independent whether the transmitted message is enciphered or not.

Other security services can be thought of which support joint working in corporate environments. For instance, group security services allow to set up and secure *Closed User Groups* (CUGs) .

4 PSCS Security Architecture

One of the main requirements for a PSCS security architecture as defined in the project [5] is to allow for the possibility of a mutual strong authentication between the users of the system, i.e. each user involved in a communication can obtain a secure confirmation of the identity of the communicating partner. For this purpose a security architecture consisting of a tree of certification authorities has been proposed (fig. 1) in agreement with recommendation X.509 of ITU-T [1].

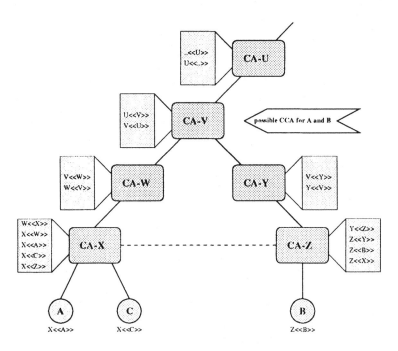

Fig. 1. Certification hierarchy of the PSCS security architecture.

The system makes use of a public key cryptosystem which enables the owner of a private key to electronically sign a message in a way that the correctness of the signature can be checked by everybody by means of the corresponding public key. Nevertheless is it not feasible to forge the signature without knowledge of the private key. The RSA [6] algorithm is used for the public key cryptosystem. Together with the *Secure Hash Algorithm* (SHA) [7], it is used to create digital signatures.

In such a way a certification authority V issues a certificate to the entities directly below it. This certificate (denoted as $V \ll E \gg$) confirms the relation between the identity of the certified entity (E) and the associated public key (E_p) by signing it with the *private key* of the certification authority (CA_s). A valid certificate means also that the corresponding private key of this entity (E_s) has not been compromised. For a mutual strong authentication between two users it is therefore important to find a *common point of trust*, i.e. a certification authority that both parties trust[2]. This CA is called *Common Certification Authority* (CCA, fig. 1). To reduce the processing of certificates direct links can be established between certification authorities.

5 Security Concept of the Demonstrator

The main features for the implementation of the PSCS security architecture in the demonstrator were identified to be:

- A strong authentication mechanism (mutual authentication) between the personal identification module of the end-user and the PSCS service provider. This mechanism relies on the exchange of three messages, each one containing a distinct digital signature.
- An authentication of the end-user towards the Personal Identification Module (realised here as a smart card) by using a *Personal Identification Number* (PIN). The PIM is blocked if the PIN has been incorrectly issued for three consecutive times. A blocked PIM can be unblocked by the end-user by using a longer so called *unblock*-PIN.
- A methodology for the generation, management and secure distribution of the public and private keys needed for the authentication mechanism.

Several other requirements regarding the security and consistency of data related to user interactions in the system were also considered for the design of the security architecture [5].

5.1 Certification Hierarchy of the Demonstrator

For the implementation of the demonstrator the certification hierarchy of fig. 1 was mapped to the demonstrator architecture.

Here the PSCS service provider acts as a common certification authority that issues certificates for all other entities in the PSCS enterprise model, e.g. an application service provider and several end-users. An extension of this system by integrating additional certification authorities is straight forward.

A certificate mainly consists of information, identifying the owner of the public/private key pair, the public key itself, the validity of the certificate, and management information. The whole set of data is signed by the issuing certification authority [1].

5.2 Authentication and Use of Digital Signatures

The processing of a mutual strong authentication is shown in fig. 2 where the originating party O (e.g. a PSCS end-user) initiates a communication with an-

---

[2] It is also possible to carry out a mutual strong authentication without a common certification authority [5].

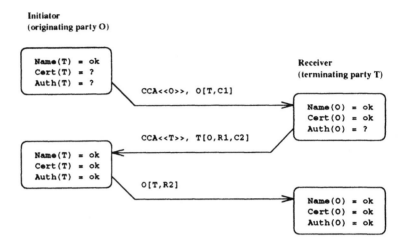

Fig. 2. Three message protocol used for strong mutual authentication.

other party (e.g. an ASP) by sending an authentication request message containing its certificate and a challenge C1 (e.g. a 160 bit random number) signed with its private key[3]. The terminating party T can check the validity of the certificate by checking the signature of the CA using the trusted public key of the CCA[4]. The public key of the originating party can be retrieved from the certificate after successful verification. This key is now used to check the signature of the previously received challenge. The authentication of the originating party is not yet complete, as an intruder could replay a previously recorded message. This thread is prevented by means of a second (random) challenge C2, that is sent to O together with the response R1 to the previous challenge C1. The correctness of the response R1 together with the certificate of T proves to O the correctness of the identity of T. Vice versa the correct answer R2 in response to the challenge C2 proves the claimed identity of the originating party.

An implementation of end-to-end authentication *services* is straight forward and can also be supported by the smart card.

6 Hardware Setup of the Demonstrator

A realistic implementation of the PSCS security architecture in the demonstrator has been considered important in order to allow for a first performance evaluation with respect to response times and delays imposed on the user by the security system. Therefore the implementation of the security system of the demonstrator

---

[3] A message M signed with the private key (E_s) of E is denoted as $E[M]$.

[4] The trusted public key of a CA/CCA is placed by itself on the smart card which is issued to the PSCS end-user. This operation is performed only once during the personalisation phase of the smart card, where the public key CA_p of the CA can be stored by its owner.

relies on available hardware. A PHILIPS PE 112 smart card system consisting of a reader/encoder and several smart cards has been used for its unique and advanced features.

The key features of the DX smart card are: 8051 CPU (8-bit) on smart card, clock speed ranging from 3.57–8.00 MHz, single 5 Volt power supply, 2 kB RAM (EEPROM) accessible to the user, 6 kB ROM containing the operating system of the smart card, and an RSA calculation unit capable of processing blocks of data of a size of up to 512 bit.

The storage capacity and clock speed of the smart card are at the upper limit of todays available technology, although a remarkable increase in performance is expected during the next few years, particularly regarding the size of the accessible memory.

Another outstanding feature of the card is the built-in RSA calculation unit. It allows the performance of public and private key operations directly inside the smart card. As it is therefore not necessary to read out the private key from the smart card, it can irreversibly be protected against read operations. This provides a maximum of protection against misuse of stolen or lost cards.

The time needed for a public or private key operation does not exceed $1.2\,s$. If the factors of the modulus of the key are also stored on the card, the operation can be accelerated to an average of $0.8\,s$ by using the *Chinese Remainder Theorem* [8].

7 Software Implementation

The main part of the security related software in the PSCS demonstrator deals with the integration of the required smart card system, the provision of cryptological functionality, and the system-independent representation of data structures. Most of this software is implemented in C++.

7.1 Smart Card Application Programming Interface

A *Smart Card Application Programming Interface* (SCAPI) has been designed, providing an interface for the communication between the security management of the PSCS demonstrator and the attached smart card hardware. The features supported include: simplified access to the smart card system, virtualisation of smart card functions, access to PSCS information, PIN-management, invalidation of individual smart card data, RSA-calculation using private/public keys of the end-user, verification of certificates using the public key of the CCA, management of telephony-credit features, and acceleration of smart card operations by using caching-techniques.

7.2 Cryptology – Fundamental Methods and Protocols

There are two views on the security mechanisms in the PSCS demonstrator. The first view focuses on the protocols which provide the distributed security functionality for the PSCS system. This mainly concerns the communication between the security managers of the PSCS users and service providers. These functional entities are specified using the *Specification and Description Language* (SDL) [9].

The second view focuses on the underlying methods of cryptology which are the building blocks of every security architecture. While the exchange of security-related PDUs and the realisation of security-services are implemented in SDL, the cryptographical operations are performed using these elementary functions.

The cryptological library was implemented so as to reflect the relations between different methods of cryptology in a hierarchical and object oriented way.

The main tasks of the security library can be identified as[5]:

- En- and deciphering of given data (*EncryptionAlgorithm*: RSA_DX, RSA_Emu),
- Calculation and verification of digital signatures (*SignatureAlgorithm*: DSA, RSA_Signature),
- Calculation of digital digests (*HashAlgorithm*: Com_SHA, MDigest2, ...),
- Support of session key agreement (*KeyAgreement*: DHKeyAgreement),
- Mutual authentication with challenge/response protocols (*CR_Authentication*: PSCS_Authentication).

These classes allow the creation of cryptological applications in a fast and easy way. They also offer the ability to compare the influence of different cryptological classes for a given problem. E.g. the creation of digital signatures requires a hash function which reduces a given plaintext to a digest value of a fixed size. Thus, a signature algorithm can make use of different digesting algorithms derived from the base class *HashAlgorithm* with respect to requirements such as speed and level of security.

In the demonstrator implementation, the *Secure Hash Algorithm* [7] has been chosen to perform the task of calculating digital digests with a fixed size of 160 bit.

The digest is encrypted by using an RSA private key operation, resulting in a digital signature for the given message. Fig. 3 reflects this procedure.

Generation and verification of digital signatures offers also the possibility for the implementation of a wide range of security services. Services like non repudiation of delivery, non repudiation of origin, data integrity and so forth can be realised using digital signatures. In an intelligent network for instance, service requests are triggered by an *Initial Detection Point* (IDP) within the *Basic Call Process* (BCP). Even these IDPs could be protected by means of methods of non repudiation of origin, giving protection against fraudulent service requests.

The generation and verification of messages used for the strong mutual authentication procedure (section 5) is encapsulated in one single class (PSCS_Authentication). In addition, this class also performs the validation of certificates via the help of a revocation list and a successive check of certification paths. An

---

[5] Abstract base-classes are marked using *slanted* text . Examples of derived classes are given using `typewriter`.

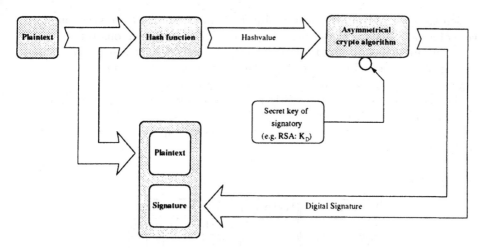

Fig. 3. Creation of a digital signature.

internal state machine tracks the order of calls to member functions and thus provides an additional fail-safe mechanism. The communication between this class and the SDL specification is established via an *Abstract Data Type* (ADT).

Other utilities were developed to provide tools for the following tasks: generation of RSA private and public keys (512 bit modulus for the demonstrator), secure storage of the RSA key-pairs (password secured triple DES encryption with participation of up to 16 so called *key masters*), certification of keys by the CCA, and personalisation of smart cards.

8 Evaluation of Performance Aspects and Outlook

A critical issue in the development of a security system for the PSCS consists in the response times and delays imposed on the end-user. It is essential that the end-user is not seriously hindered in performing his or her communication tasks. The performance of the presented implementation is therefore considered in detail. Possible improvements and issues for further studies are outlined in the last section.

8.1 Performance Aspects of the Implementation

The performance of cryptological functions generally requires the processing of integer numbers consisting of several hundreds of bits. Arithmetic operations on such data are not supported in general programming languages. Several C++ libraries supporting large integer numbers were evaluated for their performance. As a result of this evaluation the INTEGER library delivered with the GNU C++ set of libraries (Free Software Foundation, Inc.) was selected. Although it was slightly slower than another implementation programmed in C, it has been chosen since the advantages of the object oriented design of the C++ language were of great value for the security implementation of the demonstrator.

A fast implementation of a hash function is essential for a security system, since it allows the reduction of a large amount of data into a small unique bit-sequence which can be used for the creation of digital signatures.

The chosen hash algorithm (SHA) has been implemented for two different hardware platforms (SUN-1PX (30 MIPS), PC (486/50 MHz)).

Table 1 shows the results of the performance comparison between the two implementations.

System	Rate (byte/s)	normalised
SUN	771,604	100.0 %
PC	208,333	27.0 %
PC (assemb.)	284,900	36.9 %

Table 1. Comparison of hashing rates on different architectures.

A remarkable improvement of the performance could be achieved by assembler optimisation of the most time consuming functions for the PC implementation (last row of table 1) .

The total time to complete a full authentication cycle as outlined in section 5.1 strongly depends on the number of certificates to be processed. For the certification hierarchy of the demonstrator this time averages to a total of 7 seconds.

Although the time might appear satisfactory in most cases it has to be considered that an authentication may occur several times during a communication session, e.g. when switching between services or between different service providers.

8.2 Outlook

Setup times for a communication task exceeding a few seconds are in general not considered acceptable towards end-users. A reduction of the time can be achieved in different ways.

First, it is possible to increase the performance of the RSA calculation on the smart card. This will more or less automatically happen with the expected progress in smart card technology, but without weakening the strength of the underlying cryptological algorithm. Nevertheless it has to be considered, that the data transfer rate to the card will most probably be limited by the use of a serial I/O connection.

A second approach to decrease the delays caused by security functions consists of the use of different security concepts or the redesign of the protocols.

In a PSCS environment, a typical session consists of an initial mutual authentication between the end-user and the service provider while the end-user registers him- or herself to the PSCS system. This *first* registration process can be used to exchange a secretly shared session key in parallel, i.e. without any additional signals required. The shared session key can be used in all further required authentications during this session. Of course it is possible to agree upon a new session key, if one of the involved parties demands it.

The use of a shared session key speeds up authentication, since it can be used as a parameter for the generation of symmetrical digital signatures[6]. A significant speed up of the authentication process can be expected, since first measurements indicate times for signing/verification of less than $100\,ms$ each.

However, the time needed for the first authentication/registration is prolonged by approx. 3-4 seconds, due to the additional computations required for the key-agreement.

The first authentication can be accelerated, if the receiver of an authentication request (usually the PSCS service provider) is willing to offer services for a short period of time while the authentication has not been completed. To enable a minimal authentication, the end-user (initiator) can offer e.g. a *registration ticket* to the service provider. If the ticket is accepted both authentication and key agreement can take place in parallel while limited services are provided. After a successful authentication has taken place, the service provider can calculate a new ticket and deliver it to the end-user. The end-user stores the registration ticket (in his PIM) which can be used for the next registration (key-agreement). This results in delays imposed to the end-user of less than $2\,s$.

Appropriate protocols and security mechanisms for the handling of registration tickets are currently under evaluation.

Work is ongoing for the implementation of end-to-end security services. Within the MOBILISE consortium, a migration towards PSCS for corporate networks is continuing. Especially the introduction of end-to-end security services in IN based telecommunication systems, based on results given in [5], is under study.

9 Conclusion

After identification of security services required by end-users - especially in a business environment - the security architecture of the PSCS demonstrator has been discussed in detail. Suggestions were made how security services can be realised with existing security features of the PSCS demonstrator.

A first performance evaluation of the security functions of the demonstrator has been carried out and suggestions for their improvement have been made.

One serious problem could be identified while looking at the user requirements. The provision of end-to-end confidentiality services are not a technical

---

[6] These symmetrical signatures only allow to prove that two or more entities share the same session key. Therefore they cannot be used for certain services, like e.g. non repudiation of delivery. In this case digital signatures using the private RSA key would have to be used.

but a political problem. The user requirement for confidentiality[7] of information interferes with the legitimate requirement of national governments for authorised access to telecommunication systems in order to prevent crime or to protect national security. One of the difficulties - especially in Europe - is the differing nature of these requirements for different countries. Ideally, there would be a single set of criteria, which would then be met by a European solution.

References

1. ITU-T. *Recommendation X.509* The Directory - Authentication Framework. The International Telegraph and Telephone Consultative Committee, Melbourne, November 1988.
2. ITU-T. *Recommendation X.208* Specification of Abstract Syntax Notation One (ASN.1). The International Telegraph and Telephone Consultative Committee, Melbourne, November 1988.
3. Burton S. Kaliski. A Layman's Guide to a Subset of ASN.1, BER, and DER. RSA Laboratories Technical Note (URL: ftp://ftp.rsa.com/pub/pkcs), RSA Data Security, Inc., November 1993.
4. RSA Laboratories. PKCS#1: RSA Encryption Standard. RSA Laboratories Technical Note Version 1.5, RSA Data Security, Inc., November 1993.
5. Mobilise. PSCS Security, Network and Access Aspects, volume I. Deliverable 23, Race Project R2003, December 1993.
6. Ronald L. Rivest, A. Shamir, and L. Adleman. A Method for Obtaining Digital Signatures and Public-Key Cryptosystems. *Communication of the ACM*, 31(2):120–126, February 1978.
7. FIPS PUB 180-1. Secure Hash Standard. Federal Information Processing Standards Publication (FIPS PUB) Category: Computer Security, U.S. Department of Commerce/National Institute of Standards and Technology (NIST), April 1995.
8. Donald E. Knuth. *Seminumerical Algorithms*, volume II of *The Art of Computer Programming*. Addison-Wesley, Reading, Mas., 2nd edition, 1981.
9. CCITT. *Functional Specification and Description Language (SDL). Criteria for using Formal Description Techniques (FDTs). Recommendation Z.100*, volume X – Fascicle X.1 of *Blue Book*. The International Telegraph and Telephone Consultative Committee (CCITT), Geneva, 1989.
10. D. Maillot, J. Ølnes, and P. Spilling. In service security and service management security and their relationships – using upt as a case study. In *Towards a Pan-European Telecommunication Service Infrastructure - IS&N'94*, pages 513–524. Springer, 1994.

---

[7] End-to-end authentication can be achieved with cryptological methods which cannot be used for encryption of information. Use of these methods is usually permitted in most countries.

TMN Security: An Evolutionary Approach

G. Endersz[1], T. Gabrielsson[1], F. Morast[2]
A. Bertsch[3], H. Bunz[3], M. Jurečič[3]

[1] Telia Research AB, Rudsjoeterassen 2, S-13680 Haninge
[2] Telia AB, Network Services, S-12386 Farsta
[3] IBM European Networking Center, Vangerowstraße 18, D-69115 Heidelberg
bunz at vnet.ibm.com

Abstract. This paper focuses on the practical aspects of the integration of security services into telecommunication management applications as well as the building blocks needed to achieve a company wide security policy and ensure the consistent usage throughout the Telecommunication Management Network. The OSI security architecture [7] provides the flexibility necessary for the adaption to environments that deal with highly sensitive data like security management in a distributed telecommunication management network. However, to achieve interworking between implementations of different vendors it is of crucial importance to define functional security profiles and hence define an implementable subset of the security services suited to TMN. Prototype implementation and testing in the RACE project R2058 SAMSON has demonstrated that the security services and architectures currently under standardisation are capable to provide sufficient security mechanisms.

1 Introduction

The purpose of this contribution is threefold; to show the need of security in telecommunication management networks, to present available and adequate security technologies and standards from a TMN point of view and last to review possible ways to introduce security into the complex and heterogeneous management environment.

2 Scenario: Telecom Management in Change

Integration is driven by the need for uniform availability of services independent of geographic place, operator domain or networking technology. Effective and flexible management is essential for the set-up and maintenance of a global service environment. Among the functional requirements for management applications are interoperability, scalability, easy access and mobility of operators. These requirements are fulfilled by open architectures, standardized interfaces and protocols which also increase the exposure and vulnerability of the management systems. Risks are further elevated by the availability of increasingly advanced cracking and intrusion tools and resources, at almost no cost, to criminals and hackers.

Increased exposure calls for protection. Multiple actors in management require selective access to resources and information with no access to competitors or other actors information. Business relationships will interact with security policies and services. Steady changes in roles and service/network configuration require fast and flexible response of set-up and reconfiguration of security services protecting management. Equally important is the effective handling of intrusions and other security incidents.

The present low level of integration and harmonization of security services, policies, management and monitoring makes those services and their management inefficient and will slow down the deployment of powerful European and global Telecom-based services.

A multitude of actors are present in this evolving scenario: service providers, equipment/system vendors, users and Public Network Operators (PNO) share management resources and information necessary for their corresponding roles. Below are some examples illustrating the need for future security solutions.

Office Information Systems A network operator's IT-systems and data networks can be described in many ways. Figure 1 shows the Network Elements (NE) in the center – like Exchanges, Transmission equipment and Digital Cross Connect (DXC). These are supported by Operational Support Systems (OSS), which in turn are connected with Workstations at Management Centers.

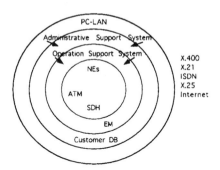

Fig. 1. IT-systems and data networks of a network operator

At these centers the personnel does not only work towards the NE or administrative support systems and databases, sometimes they also need access to word processors and other office programs. Therefore they are also connected to the office LAN and servers. Office workers on the same LAN and Servers have other needs for example external connections to Internet and modems. Existing physical connection between NE and public network outside the company makes security extremely important. Security services such as; authentication and access controls in systems, correct use of access control lists in the data network routers and introduction of firewalls between interconnected networks are needed.

Vendor Access The equipment/system vendor who may claim access to the TMN of a PNO. The reason is advanced support of supplied equipment which implies new hazards to the TMN and puts requirements on security measures. Access will be made via telephone line and modems or data communication networks and will comprise assistance by remote interrogation to make immediate corrective action. The internal – private – data communication network carrying TMN traffic is in this case interconnected with a public network. The security problem is twofold: firstly, all other, not authorized users of the outside public network must be kept outside the TMN. Secondly, all actions of the outside operators have to be monitored and/or controlled in order to offer interoperability in a safe way.

Interacting TMNs In the case of interconnected TMNs different organizations may access resources across security domain boundaries. One example is when network operators cooperating in path provisioning for SDH. An originating PNO has the right to allocate bandwidth on a link owned by another PNO. Another example is when a Network Layer OSS at one Network Operator receives an alarm of a fault on a cable connecting to another Network Operator. The OSS analyses the fault and then sends restoration commands, both to his own DXC and to the other Network Operators OSS which, after access control, relays the commands to his corresponding DXC. Before it is possible to connect the TMNs of two network operator's, by using the X-interface, a contract must be signed stating a Security Policy on how the TMN OSS should identify each other and what they may access in the other TMN.

Customer Access Customers who require access to "their" resources and information in the network can be granted access only if the granularity of the access control system can guarantee that the customer only can see and manipulate his "own" data.

Multiple Sign-On A PNO's management center is a central point from which the Network or the Network Elements are remotely maintained. This means that the personnel must know and be able to use a number of different OSS's from different vendors, each of which have its own login procedure (if any). Therefore they must remember a lot of personal user identifications and corresponding passwords. IDs and passwords are often noted at public accessible locations (e.g. under the keyboard). It is also understandable that being forced to login to each OSS slows down the speed of reaction in critical situations considerably. The situation is also a burden for the administrator, keeping track of all user identities. Especially when a user changes work within the company or quits.

3   Security Requirements

The generic security objectives of any PNO are to ensure that stored and communicated data and information in systems and networks, owned by any legitimate stake-holder, are treated according to the appropriate policies stating correct users, form, time, place, assurance level and cost. Stake-holders may be operators, VASPs, customers, VPNs, vendors, etc. These objectives raise requirements on several security services to ensure security of management, like identification, authentication, access control, integrity and confidentiality but also on functions to manage security like intrusion detection and security audits. Extensive risk and threat analysis shows that Telecom management is in need of the entire set of security services and security management capabilities.

However, since control is a key word in operating large networks it is important that individuals taking actions are accountable for these actions (accountability) and that the course of events may be reviewed (traceability). Therefore priority should be given to the services that support accountability and traceability. An important tool for traceability is the security audit trail which is a journal of security-related events collected for potential use in intrusion detection but also for example in gathering evidence for use in prosecuting an attacker. Log records of operations performed by the authenticated operator complete the set of information needed to achieve accountability. The integrity of all log and audit records is essential for confidence in the system.

Furthermore availability of TMN functions and resources is very important. Public Operators are prepared to spend large amount of money to keep their management systems and networks operable. Management applications are often put on expensive fault-tolerant platforms. This wish to guarantee availability also applies to security matters. Security services to support this are above all integrity of communicated data and access control to management information bases. Accountability as well as access control requires authentication which is the security service that gives assurance of identity. Without the support of a reliable authentication service, the other measures are without effect.

As a conclusion, efforts to secure management of telecommunications shall initially concentrate on authentication of users and inanimate-d entities, access control to stored management information, integrity of communicated data and on tools for accountability purposes. The relative priority between these services is however dependent on the characteristics of a specific TMN and the chosen migration path towards secure management.

4 Standardization

In the evolving competitive open business environment Telecom networks and the corresponding management systems are supplied by a large number of different vendors. The need for standardized security solutions is as obvious as the need for standardized management protocols and information models. Telecommunication Management Network (TMN) concepts and standards have been worked on for a long period of time and new network management products include a growing share of standardized facilities and interfaces. However, adequate security features to protect management operations and information are still missing in the standards, and consequently, in the emerging products. A main prerequisite for the provision of European-wide and global Telecom services and for Open Network Provision (ONP) is that all involved parties can interact in a secure and safe way at the management level.

A network operator acting in a deregulated market, is very much aware of the pressure from his competitors. The PNO depends on the ability to quickly buy equipment in order to provide new services at a speed comparable to that of his competitors. The time period between pre-study and installation is very short, much shorter than 5 years ago. The consequences of a delay or a slip-up are much more severe, counted in money and market shares. In this context it is impossible to stay with only one vendor. PNOs have to buy from those that can provide the adequate products in the right time. Therefore the number of different network element types are increasing faster than before. Each network element normally has its own local operation support system. The process of provisioning focuses primarily on the network element, secondarily on the operation support system. The process of provisioning is focused on functionality not on security. The time between stated requirements and delivery is short, often too short to introduce major additional development.

ETSI NA4, in cooperation with ETSI STAG, has initiated standardization work on TMN security. The first results of the new rapporteur group will be available during the autumn of 1995. The major objective is to develop security profiles which can be implemented by vendors ensuring secure inter-operation of TMN systems and components.

5 Available Security Technologies

The proposed security architecture is to be aligned to [7]. Security profiles for TMN will be based on existing standards for security services, mechanisms and interfaces. Some of the security services standards under consideration for profiling are: authentication – X.500 [1], X.509 [2], access control of MIB – X.741 [9], security alarms – X.736 [10], security audit trail – X.721 [11], X.740 [12], integrity, confidentiality in OSI – NLSP, TLSP and Key Management. The placement of the security services in the communication infrastructure is shown in figure 2.

Firewalls are security gateways between two subsystems. They can perform authentication, access control and log/audit of relevant events. Firewalls offer means to implement security services, standard or non-standard, without the need for major changes of the interacting architectures. Hence, firewalls are expected to be useful components, especially during the migration period towards TMN systems with integrated, standard based security.

6 Functional Security Profiles

The idea of grouping security services into functional classes, which are described in Race CFS H211 [3] needs to be validated and exemplified. A functional class in this context is defined as a set of security services and possibly mechanism that together provide a system with a certain security level. A functional sub-profile on the other hand is a specification in which a functional class is realized using standard components. The idea is to define a set of security functional classes

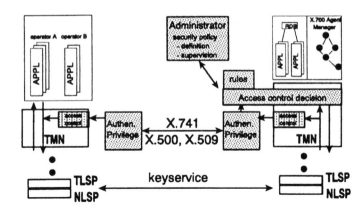

Fig. 2. Security services for telecommunication management networks

for the TMN environment and to suggest their implementation by pointing at existing security standard components.

A basic question is the number of classes to be defined? Practical reasons imply a low number of FCs. It is also more likely that a low number of FCs will be accepted and implemented by vendors. On the other hand if the classes are to few, many network operators will find it hard to express their specific requirements on security by means of a small set of FCs. Thus the concept will not be practicable in this case. A middle way is to suggest four levels that is; no security, minimal security, basic security and advanced security.

No Security Class The no-security class will of course comprise no security mechanisms or only very insufficient ones like simple passwords without replay protection or encryption over public networks, screen locks without boot-protection, etc. In short measures that can be circumvented without expertise and may be misleading regarding the security state of the system or network.

Therefore the security of the TMN is achieved by organizational processes and legal measures, physical separation of the network and the connected hardware, and last but not least by trust in network operators.

Minimal Functional Class The minimal functional class focuses on correctness of stored data and provides minimal accountability for management activities. It provides secure handling of management operations within a single security domain, offering the integration of existing telecommunication management systems and protects against unintentional misuse within one organization.

Based on simple authentication, access control shall be performed for the association between manager and agent and also for the management information base. The weakest form of integrity protection during data transfer is included as well as redundant audit tools like security alarms and security audit trail. The minimal FC could be used in two ways. The first is as an intermediate step towards more advanced security mechanisms. The secured way is when the target system is not very distributed. Typically it should be physically protected as a whole and the purpose of the security mechanisms is mainly to protect against accidental misuse and insiders attacks. Confidentiality is not very important in comparison with not allowing unauthorized personal to perform operations dangerous to the well being of the system and the services it provides. Therefore cryptographic functions for integrity and confidentiality protection are not considered.

Basic Functional Class This FC adds additional security services to counter the threats introduced by using public untrusted networks – namely integrity and confidentiality of transferred data. Especially strong authentication mechanisms are included to protect the transmitted authentication information. This profile allows to open a TMN to a small external trusted group under special relations (contracts) like: operators on stand-by at home, maintenance staff of manufacturer and selected cooperating telcos and customers.

The access control of associations and the MIB is still there but the simple authentication has been replaced by strong authentication using cryptographic means. Confidentiality and data integrity should be possible to apply both connection oriented and on selected fields. The basic level could also be regarded as an intermediate step towards a more secured environment or as a solution to intra-domain security in TMNs where requirements for security are of low or medium level. It is suited for distributed environments and outside threats. Considering the focus of this class it is suggested that access control of notifications should be added.

Advanced Functional Class This FC is specially intended for very sensitive areas and allows for management actions across organizational and jurisdictional domains. In order to open the TMN for external access of large external group – like end users, service provides and direct competitors – strong accountability of all management actions is required. This FC adds security services for mutual strong authentication, non-repudiation and means for detecting denial of service attacks. It aims at a very high level of security and is specifically suitable for the inter-domain case. Special arrangements, including the use of the advanced FC may also be required for highly sensitive functional areas like security management and accounting management.

It is of course not economic to apply e.g. the Advanced Functional Class to the entire TMN - although this would be the secure solution. Therefore a optimized mix of Functional Classes is needed. Several alternatives exist. Functional classes may be applied to types of applications or functional areas. This approach is indicated in H211 [3]. A functional class should then be valid for security management and/or accounting management and another one for performance monitoring. The disadvantage of this approach is that it does not help operating a TMN and makes the implementation even more expensive. Storing and communicating data will use the same infrastructure regardless of functional area.

Naturally there is a difference in the inter- and intra-domain case. Therefore different FCs should at least be applied to these two cases. Best is to apply FCs to generic functional relationships, that is to relationships between functional blocks. Note however, that these relationships do not consider distribution aspects which have to be added. The following matrix is suggested for use when specifying functional classes for TMN:

	Not distributed	Secure Distribution	Unsecured Distribution
Human-WSF	impossible	Secured room	Office Landscape
WSF - OSF	Application integrated in workstation	Remote login but within the same "bunker"	Remote login
OSF - OSF	Integrated in the same server	Physically protection of link. Dedicated network	Geographically separated. Communication over "public" network
OSF - ext. OSF	impossible	Does it exist?	Over the X interface
OSF- MF	etc......		
MF-NEF			
OSF- NEF			
MF- MF			

The figure above shows the different relationships possible between function blocks within and between TMNs (vertical) and how a physical implementation of these reference points may be distributed (horizontal). Within the matrix, examples are given on how the combination of reference points and distribution may be reflected in reality – impossible combinations are not considered. For a TMN implementation, this table could be used together with the functional classes to define security levels in the following way;

	Not distributed	Secure Distribution	Unsecured Distribution
Human-WSF	–	minimal	basic
WSF - OSF	none	minimal	basic
OSF - OSF	minimal	minimal	basic
OSF - ext. OSF	–	advanced	advanced
OSF- MF	none	minimal	basic
MF- NEF	none	basic	basic
OSF - NEF	minimal	basic	basic
MF- MF	none	minimal	basic

The advantages with specifying functional classes for TMN are several. They create a common base of terms that may be used and referenced in polices and strategies related to security in TMN. This may later form a base also for automated implementation of security measures according to specified policies which otherwise would be quite impossible. It might be possible to write contracts regarding co-operation in such a way that it is interpretable directly by the systems involved. With functional classes it is easier to negotiate security issues between TMNs, both directly over the common interface but also in terms of TMN internal security which might be an interesting issue when deciding what should or should not be allowed over the X interface. The profile defined allow to release the current needed physical separation and level of trust in a controlled manner. Another advantage is the facilitation of migration from today's situation where security is low and irregular towards a more homogeneous picture.

The information given in the functional classes may be made more concrete by specifying functional sub-profiles. Such a profile is a specification in which a functional class is implemented using standard components. Suggested examples of baseline security profiles for authentication, access control, integrity and confidentiality are given below.

7   Prototype Implementation

This section describes a practical realization of the concepts outlined above within the RACE project R2058 SAMSON. The integration of the security services is based on the standards suitable for the TMN Q3 interface. The prototype setup addresses the connection of networks with different degrees of security functionality – namely the minimal and basic functional class. The complete spectrum – especially the migration aspect – is covered by the combination of the prototype implementations for the connection integrity and confidentiality provided by Telia with the application layer security service elements – authentication, access control and audit – developed by IBM.

In order to achieve a cost effective operation of a TMN it is necessary to apply strong security services only when and where needed. The minimal FC allows an organization to define and enforce a security policy within the physical protected, internal TMN. Strong security services are necessary at the connecting point between an PNO internal TMN and an external network. They have to control the management operations and restrict them to those allowed for external users.

Authentication: A way to include security services into the TMN is by using the Access Control Parameter field of the CMIP protocol data unit. Within SAMSON the credentials defined in [2] are used. Due to this definition simple and strong credential can be transmitted and implementations for directory authentication can be reused. It is possible to use external defined tokens, like the privilege attribute certificate (see [9], [5]) allowing the transmission of additional privileges. The external defined PAC is not used in the prototype due to the added complexity. Within the internal network a simple authentication mechanism is sufficient to provide the authenticated identity as the base for access control.

Access control: [9] defines the objects and attributes for a rule based access control service. The implementation provides outgoing access control on the management center using definable rules supporting a flexible granularity from domains over roles, operations down to attribute values. For incoming operations the access control decision function is additionally checking the strong credentials provided by the initiator. The operations demonstrated in SAMSON were the management of strong authentication users of the X.500 [1] directory.

Audit: All security relevant management operations can be logged in the security audit trail [12]. Alarms [10] are displayed in real time on the monitoring application of the security manager in the management center. All external management requests are evaluated by the access control decision function, therefore logging and auditing is placed in this module. This will also ease the future implementation of trusted third party notary services.

Integrity and confidentiality: This section describes a practical realization of connection integrity and confidentiality services based to large extent on standards and suitable for the TMN Q3 interface. The design addresses requirements in the "Basic Functional Class" of CFS H211 [3] and has been implemented and demonstrated as part of the Telia prototype in Race II SAMSON. The security services are provided by the ISO Transport Layer Security Protocol (TLSP),

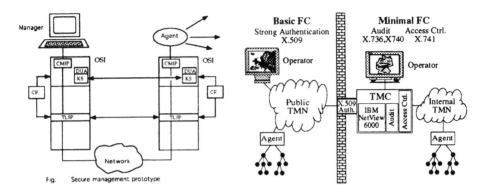

Fig. Secure management prototype

Fig. 3. Prototype: Integration of TLSP (left) and external access to a protected TMN (right)

integrated as an encapsulation protocol in the lower part of the ISO Transport Layer [8]. The TLSP protects the data carried by the transport protocol providing secure communications between manager and agent in the TMN environment or over the X-interface if a similar protocol suite is applied there. Figure 3 shows the security architecture applied to an OSI management environment. The TLSP needs support by two additional entities in order to perform its functions: a key service and a crypto facility.

Key Services: First, cryptographic functions and parameters are to be agreed between the communicating systems. This includes agreement on cryptographic algorithms to be used, establishment of secret session keys, etc. The process is called the creation of a secure association between the two parties. A generic application layer Key Service defined by IEEE 802.10 [13] and also submitted to ISO SC 21 was chosen to handle secure associations and provide key material for the TLSP in the SAMSON prototype. The handshake between the Key Services of two systems starts with mutual strong entity authentication, satisfying by coincidence also the corresponding requirement of the Advanced Functional Class in CFS H211. Both the TLSP and the KS Protocol [6] are algorithm-independent. In our case, symmetric encryption was chosen for data encryption (DES), while authentication and key exchange are based on RSA. Trusted third party services for the provision of public key certificates can be accessed by the X.500-conformant Directory User Agent.

Crypto Facility: Second, the functions of both the TLSP and the KS rely on cryptographic calculations performed by the Crypto Facility. No standard was available at the time of the definition of this part of the SAMSON prototype. For the purpose the software package SECUDE was chosen in the SAMSON project. Although there are present efforts to define standards for cryptographic calls and functions see [6], it remains to elaborate on this task until independence of specific implementations of cryptographic functions can be achieved.

A significant advantage with this approach to the Basic FC is that the TLSP and the IEEE KS provide application independent, generic services, i.e. their services can be invoked by different applications and lower layer security protocols in a simple way, e.g. by using Quality of Service (QS) parameters. No additional adaptation of applications or other entities is required in order to use the above security services.

In the Telia SAMSON prototype connection integrity, confidentiality and entity authentication were applied to protect communication between security manager and agent and also as targets of security management, the latter having been that time the main activity of the SAMSON project. A more detailed presentation of the prototype design and validation is given in [4].

The experiences gained from the prototype can be summarized as follows:

- it was possible to build the prototype mainly based on stable standards and standard proposals;
- the implementation and integration was straightforward;
- it performs well and is manageable using OSI-based management concepts (SAMSON);
- it is generic with no need for additional adaptation in the system.

Therefore, it can be concluded that the presented realization of the integrity, confidentiality and authentication services qualifies as one possible basis to be developed further into a security profile for TMN.

8 Migration

Migration to TMN, understood as a smooth, secure and economical transition from existing management solutions to TMN-based management, is considered as one of the keys for the success of TMN. TMN standardization is a long process which progresses in steps. Practical TMN deployment will therefore face a partially standardized environment with a great number of conventional "pre-TMN" components. Furthermore TMN will need to incorporate new and enhanced features not included in standards that suppliers want to introduce in the innovative and highly competitive scene of Telecom management.

From the security point of view network operators in the past relied on physically protected local control of network resources and on the visibility and accountability of the operating staff in such a scenario.

As awareness of security threats is increasing, vendors and PNOs implement security facilities to protect the most sensitive information and resources. The proliferation of vendor-specific security solutions will cause, in addition to difficulties in interworking, also a heavy burden on security and system administrators. During this period of evolution add-on security solutions, such as firewalls, which are able to protect heterogeneous subsystems, will play an elevated role. Standards for TMN security should allow modular implementation and integration into existing architectures.

During the evolution phase towards a standard based TMN management environment, including also security profiles, some of the most urgent security needs are to be met. One of these is the requirement for secure information exchange between TMN systems (X-interface), often belonging to different security/operator domains and separated by public data or telephone networks. A pragmatic, but still standard based solution is to introduce a security gateway (referred also as firewall) at the network entry point to each of the involved TMN systems. Connection integrity and confidentiality and authentication can be provided between the gateway points by introducing the ISO Network Layer Security Protocol (NLSP). NLSP provides similar services to the TLSP but it has the option to exchange session keys without the support of a separate key service entity. This concept does not provide any intra-TMN protection and is more demanding considering administration and management but it offers the convenience of add-on solutions. Some further improvements are possible by the addition of application-level firewall components for authentication and access control of external operators. However, there are no standards available in this area so far.

One major task for security work in the evolution of management systems and networks is the integration of security in the short- and long term systems development achieved by an increased cooperation between vendors and PNOs in the security area.

References

1. CCITT, Melbourne. *Recommendation X.500 – The Directory: Overview of Concepts, Models, and Services,* 1988.
2. CCITT, Geneva. *The Directory - Part 8: Authentication Framework, Recommendation X.509,* 1992.
3. Commission of the European Communities, *RACE Common Functional Specifications: H211: Security of Service Management,* Aug 1994.
4. G. Endersz and R. Zamparo. Key management and the security of management in open systems: the SAMSON prototype. In *Information Security - the Next Decade.* Chapman & Hall, 1995.
5. European Computer Manufacturers Association, Geneva. *ECMA-138, Security in Open Systems Data Elements and Service Definitions,* 1989.
6. FIPS. *Security Requirements for Cryptographic Modules,* 1992.
7. International Standards Organization, Geneva. *ISO 7498-2-1988(E), Information processing systems – Open System Interconnection – Basic Reference Model – Part 2: Security Architecture,* 1988.
8. International Standards Organization. *ISO/IEC 10736: Information Processing Systems, Open System Interconnection, Transport Layer Security Protocol,* 1992.
9. CCITT, Geneva. *Systems Management: Objects and Attributes for Access Control, Recommendation X.741,* 1992.
10. CCITT, Geneva. *Systems Management; Security Alarm Reporting Function, Recommendation X.736,* 1992.
11. CCITT, Geneva. *Definition of Management Information, Recommendation X.721,* 1992.
12. CCITT, Geneva. *Security Audit Trail Function, Recommendation X.740,* 1992.
13. IEEE 802.10D. *Key Management Protocol,.*

A TTP-based Architecture for TMN Security and Privacy

[1]Dominique Maillot, [2]Jon Ølnes, [3]Pål Spilling

[1]Télis, Dominique.Maillot@synergie.fr
[2]Norsk Regnesentral, Jon.Olnes@nr.no
[3]Telenor Forskning, Pal.Spilling@tf.telenor.no

Abstract. It has been recently stressed [1] that the rapid development of new telecommunication services could be undermined by the lack of commercial security for the service providers and the lack of privacy for their users. Security solutions to cater for those needs will almost necessarily employ encryption techniques which are still, and perhaps more and more, subject to severe control by government authorities with respect to the protection of the public order and the national security. Security architectures which might satisfy those apparently conflicting requirements have to be proposed. Another aspect is that those security architectures should permit both the security of the telecommunications services and the security of their management services, given that the final users will more and more have access to those management services. The objective of this contribution is to show how a security architecture based on Trusted Third Parties can bring an answer to the above problem, and what would be the consequences of using TTP services to secure management systems, in particular, the implications on the specification of TMN interfaces. This work was carried out in the framework of the RACE project 2041 PRISM.

1   Introduction

A large concern is currently arising among international organisations regarding the security and privacy of telecommunication services to be deployed on a trans-national scale; the Bangemann Report *"Europe and the global information society"* drew the attention to the need for privacy in telecommunications services in order to encourage consumers in using such services. It also recommended that the security of international telecommunications and the legal protections of remote transactions be rapidly given a solution. The INFOSEC Business Advisory Group (IBAG) had previously stated [2] that the conflicts between the requirements of users for privacy, the requirements of providers for commercial security and the legitimate demand by national authorities for decryption capabilities in order to ensure public order and national security could possibly be reconciled by using security techniques based on TTPs.

European research has concomitantly developed (ECMA) and validated solutions to secure distributed services by use of TTP techniques: the RACE SESAME architecture enables the provision of most of the security services. The objective of this contribution is to illustrate the use of such a technology for providing the security of network and service management taking into account the requirements of the various actors (users, providers, national authorities and regulator). Then some implications of such a technology on the service management configurations and the TMN standardisation are presented.

2   TMN Security Requirements in Brief

2.1   Network Integrity and Service Providers Commercial Security Requirements

A TMN system will have security requirements with respect to information transfers and actions on its internal F and Qx/Q3 interfaces, with respect to the information stored in the system, and concerning the integrity of the management system itself. In a pan-European setting, external communication is added to this set-up, and correspondingly security requirements concerning information transfers and actions on external interfaces are also added (all storage is assumed to be internal to the individual systems involved).

Access to a TMN is via reference points of type `f' or `x'. In the case of external access, these will almost definitely be realised as physical interfaces of type[1] `F' or `X'. If access is from another TMN system, or from a non-standard management system through an X Adapter Function (XAF), an X interface must be used, otherwise access from a Work Station Function (WSF) through an F interface is the only possible solution. The security requirements for a given role in the pan-European set-up can thus be determined as the requirements the role has with respect to the relevant external F or X interfaces, plus internal requirements.

The internal security level of a system should correspond to the level at the external interfaces. Threats may be different for internal and external security, but note that the mere presence of external interfaces constitutes a threat, and in many systems even the internal interfaces may be exposed to external attacks. Thus, the security requirements for the internal F/Qx/Q3 interfaces may be derived from the requirements for external interfaces.

Security is often characterised by availability, integrity and confidentiality [3]. In this paper, integrity is split into correctness and accountability, since the security services needed will be different. Wherever appropriate, these requirements must be met either by specific security functions or by general functionality in the management system.

- Availability requirements may strongly impact the configuration of the management system. High availability is achieved by duplication of functionality, and backup solutions. In some cases, denial of service attacks must be considered, e.g. intentional flooding of a network or a computer to reduce performance or completely block authorised access.
- Correctness implies a certain guarantee that information in the system is not deleted, added to or modified in an unauthorised way, neither during transfer, nor when being stored. It also implies that functions in the system behave correctly and that they cannot be accessed without proper authorisation. A variety of security functions exist to achieve correctness, the most important being access control to functions and data, and integrity mechanisms like checksums and digital signatures. For communication, data origin authentication services may be desired. Other necessary functionality may be proper software distribution and version control systems, and virus detection programs.
- Accountability is the property that some action may be traced to a given entity. Often auditing and simple logs will be sufficient. In other cases, data origin

---

[1]   A F-interface is internal to a TMN according to M.3010. However, there is a need for external access to a TMN from actors without their own TMN system. PRISM suggests external F-interfaces for this purpose.

authentication will be requested. Ultimately, strong services for non-repudiation with proof of origin and/or non-repudiation with proof of delivery may be needed. Accountability requirements will often be persistent, e.g. it must be possible to construct a proof, with some strength, that some data were requested by or delivered to a certain entity, even an "arbitrarily" long time after the actual data transfer took place. Strong mechanisms require exchange and logging of cryptographically secured messages and/or utilisation of services from a notary server.

- Confidentiality means protection against unauthorised disclosure of information. For stored data, access control is the most important function. During data transfer, in general encryption is needed, and data may be stored in an encrypted representation as well.

A few requirements cannot be placed directly under any of these parameters. Examples are logging and alarms to detect and trace security breaches, and authentication and authorisation within the management system, essential to meet any requirement in any parameter category.

2.2 Regulatory Requirements for the Organisation of Fair Competition

The ONP Framework Directive [4] adopted by the European Council of Ministers establishes the principles to be adopted for the provision of access to public networks and, where applicable, also to services, within the EU. It identifies three measures:
- technical interfaces for terminals and network interconnection;
- harmonised usage and supply conditions;
- common tariff principles.

Conditions for access to public networks must be based on objective criteria, be transparent and published, and guarantee equal and non-discriminatory access. However, the ONP directive introduces the concept of "essential requirements", representing the only circumstances where a national regulatory authority of a member state may restrict access to the public network. These are all concerned with security and integrity, as follows:
- security of network operations;
- maintenance of network integrity;
- inter-operability of services;
- protection of data.

Therefore providers, possibly at the same time competitors, will have to co-operate in order to enable pan-European services, and to fulfil the ONP requirements for non-discriminatory access. The TMN functional architecture identifies one open interface type, the X-interface, for the purpose of management operations between TMN domains of different service providers. This interface will be the preferred option for secure exchange of management information between service providers [5], and is thus essential for the application of ONP to TMN. The co-operative TMN model, presented in 3.3, aims at fulfilling ONP requirements for non-discriminatory access via open, well-defined and well-secured X-interfaces.

The National Regulatory Authorities (NRA) - or an hypothetical EU regulatory authority - should then be able to monitor the enforcement of the ONP regulation, in terms of non-denial of management services requested by VASPs from a BSP and, to a finer granularity, in terms of access rights to those services. It may also be its role to provide the certification of the trustworthiness of the various providers with regards to the above mentioned *essential requirements* of ONP. Those roles are typically trusted party roles that can be played by TTP, accredited by the NRAs, acting as certification authorities, authentication servers or authorisation servers.

2.3 National Security and Public Order Requirements

For national security reasons or law enforcement purposes, severe restrictions exist for use of cryptography[1] in some countries. Provision of services involving encryption is not too controversial in some countries (Norway), while in other cases the use or trade of cryptographic devices must be declared, or even explicit permission must be granted (France).

The regulators will not encourage use of open, published algorithms, and agreement on which encryption algorithms to use (if allowed at all) for international service provision is a wide open issue. Some countries (Norway, Sweden) specify national encryption algorithms which are suitable for national services, but cannot in general be used abroad. Awareness of this problem pushed the International Chamber of Commerce (ICC) to issue a position paper [6] where the governments are called " to remove barriers to trade and impediments to the use of commercially available encryption methods and to remove unnecessary export and import controls, usage restrictions on encryption methods ".

In some cases, agreement on a specific algorithm for a specific service may be obtained, as was done for GSM mobile telephony. This will probably be the solution for pan-European management. ETSI has started work on a dedicated algorithm specifically for this purpose [7]. Equipment for this algorithm will be made available only to licensed service providers, and only for management purposes.

Whether or not the use of cryptographic techniques will require prior authorisation, the mechanisms employed should provide an interface permitting « tapping » by authorities in the legal framework of court decisions. This could be achieved by using accredited key distribution centres (TTPs) acting as key-escrow where the encryption keys would be available to law enforcement agencies.

2.4 Users' Privacy and Protection of Personal Data Requirements

The data protection acts of European countries differ. For example, a particular piece of information may be regarded as highly confidential in one country, but not in another. Attempts have been made [8] to harmonise legislation in order to achieve free movement of personal data between the EU-countries. However, at present it seems like personal data protection will be permanently left to the individual states. National regulations may contain rather detailed rule-sets defining under which conditions personal data may be requested, recorded, stored, transferred, processed and made available (and to whom). In many countries, privacy and protection of personal data is monitored by an independent Data Inspectorate. However, not all countries have determined an authority for monitoring and control of such issues.

A draft CEC directive [9] intends to set the framework for privacy and related issues in the telecommunications area. It defines what should be understood as personal data,

---

[1]   Different restrictions will apply depending on whether cryptography is used for confidentiality purposes and with respect to key management, or used for cryptographic authentication and integrity protection.

and certain obligations of the telecommunications organisations with respect to such data:

- directions for use and provision of call forwarding and calling line identification services;
- contents of calls may only be accessible to third parties if all parties concerned have agreed (e.g. conference calls);
- any collection, processing or storage of personal data:
  - is restricted to be used only directly for provision of the service;
  - is restricted to the shortest time range possible (e.g. user information may only be stored during the time of transmission);
  - must be known and permitted by the individual party concerned;
  - must be kept confidential, and must not be given to other parties (e.g. other service providers) without the individual's prior consent.
- collection or filtering of information (temporary location, personal and business circumstances etc.) into electronic profiles of subscribers is not permitted.

Discrepancies between countries will still be a problem, since the definitions of [9] will leave open issues with respect to personal data not directly covered by the proposed directive. For management, especially logging, auditing and accounting must be in accordance with the (national) rules.

3   A TTP-Based Security Architecture

3.1   Establishing Trust Within and Between Service Management Domains

A likely great number of entities, human agents and application entities realising the various TMN Functional Blocks, will need to interact in the operation of Service Management. Unless a given TMN domain is regarded as totally secure because it is physically well protected, the identity and the associated rights of an entity should be authenticated before a management interaction can take place. In the context of the IBC network, where a same DCN could link the management systems of various service providers, the assumption of physically protected networks cannot be made. Therefore, whenever two entities have to interact they should have the means to trust each other; that trust can be established:

- either on a unilateral basis, that is, each entity, which needs to authenticate other entities, must know a specific information for authenticating any of those entities,
- or, when the authentication must be mutual, the trust is established on a bilateral basis (mutual trust), that is any entity must know information related to any other entity,
- or by trusting a third party, able to assure one entity of the trustworthiness of the other one.

Such TTPs can be categorised according to their communication relationships with the management entities they serve: TTPs can provide on-line services (interacting in real time with one or both of the management systems, e.g. an authentication server) or in-line services (positioned in the path between the two management systems). As for off-line TTPs they do not have interactions with the management systems in the process of the given security services, but may have to interface with the management systems in order to manage the private keys used by those systems. Following those characteristics, various kinds of TTPs can be used to secure the various TMN interface types.

3.2 Trust Within a Service Management Domain

An on-line TTP architecture is characterised by the ability to revoke or update certificates of security information within a short delay; this is suitable to secure interactions of management applications with human agents whose roles, and thus capabilities, may rapidly change. Such an architecture could be used to secure the F interface between WSFs and the service OSF-S; it should support security services such as authentication, access control and integrity of the exchanges, perhaps also the confidentiality of the data transfers if the transport network is not trusted. Non-repudiation is not a strong requirement here, since the service provider agents are assumed to be trusted.

Most of the interactions between agents and the OSF-Ss will result in operations performed on Network Elements (NE). Those operations can be secured by using an off-line TTP, i.e. the Certification Authority (CA) of the domain, since the certificates granted to the management applications are not subject to frequent update, in contrast with those of human agents. The control of the access to the managed resources will be based on the Privilege Attribute Certificate (PAC) of the initiating agents, propagated by the OSF-S to the NE through the Q3 interface. Those PACs are certified by means of the private keys of the domain PAS, so that they can be validated by both the OSF-Ss and the NEs.

Figure 1 depicts a scenario where the service management system of a provider is distributed over a LAN. From the WSFs, management operators authenticate against the Authentication Server (AS), then obtain the PAC for accessing any of the OSFs of the SP (single log-on). Management functions operating on the Q3 interfaces use the same mechanisms to get access control information that will be presented to the Access Decision Function located in the NE agents. In the depicted scenario, the LAN is regarded as trusted, and thus the confidentiality and the integrity of the transmitted data have only to be protected on the TMN DCN; this can be achieved by using security services at the network level (NLSP).

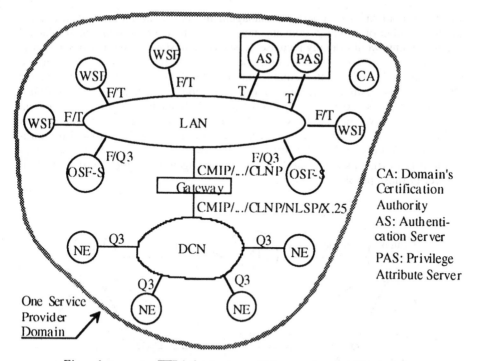

Fig. 1. TTP infrastructure within a service provider domain

CA: Domain's Certification Authority
AS: Authentication Server
PAS: Privilege Attribute Server

3.3 Trust Between Service Management Domains

The use of TTPs within a given provider's domain has been illustrated in the previous section. A similar question may be raised when interactions take place between entities located in different domains, as it is especially implied by the capability given to telecommunication service customers to perform some management operations, or by the liberalisation of the provision of services by service providers separated from the network operators needing access to the operators' TMN systems.

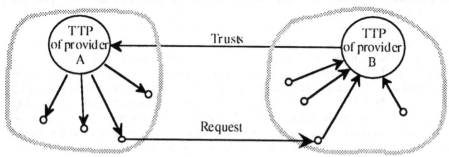

Fig. 2. Unilateral trust between domains

At this level of aggregation, trust could also be established on a peer-to-peer basis between the TTPs of each actor's domain, as shown in figure 2. As trust is transitive, an entity of a domain B can trust an entity of domain A, if the TTP of domain B trusts the TTP of domain A. Such a solution would be suitable for interactions occurring in a joint-venture management scenario.

But, in the context of co-operative management, as for intradomain interactions, when the number of providers domains increases, the number of unilateral or bilateral trust agreements would rapidly increase. In such circumstances, another possibility is the use of trilateral trust agreements between the different domains, involving an interdomain TTP. When interactions seldom occur between two domains - or for policy decisions - a direct agreement may not be appropriate, and the two authorities may have recourse to a TTP domain with which each domain agrees on an interdomain policy. Figure 3 illustrates that situation, where an entity of domain A must provide with its request some information to prove that it is trusted by intra-domain TTP A and that TTP A is trusted by interdomain TTP C. The target entity in domain B must trust its interdomain B TTP, which again trusts the interdomain TTP C, to verify those proofs.

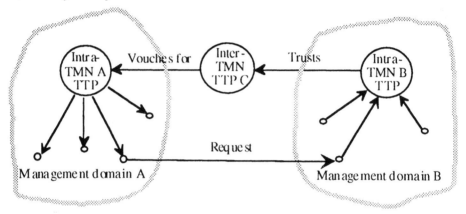

Fig. 3.   Interdomain Trusted Third Party

As for the Q3 interface, interactions on the X interfaces do not require that the validity of security certificates is tightly controlled, and then long-term off-line certificates can be used to establish trust between Service OSFs pertaining to different domains. In such a situation, the target OSF-S interacting through a X interface will have to know - or be able to obtain - the public key of their initiating correspondent - either transmitted in the security exchange or retrieved from a directory - certified by the CA of their respective TMN domain. The given certificates will be checked by use of the public key of the inter-TMN TTP that also can be retrieved from a directory. As most of the interactions will be initiated by human users operating on a F interface in the initiating TMN, the access privileges of those users will have to be propagated into the remote OSF-S; this requires that the remote system should be able to check the PACs certified by the Privilege Attribute authority of the initiating domain, i.e. be able to obtain the public key certificate signed by the interdomain Certification Authority. The privileges of the initiating agent should be combined with those of the initiating OSF-S before the access decision is made by an Access Decision Function of the target NE. Here the assumption is made that the syntax and semantics of the access privileges are the same in both domain, otherwise an inter-domain function should be involved in order to perform the mapping of the syntax and semantics.

Although most of the needed security services can be provided with the use of an off-line TTP, some forms of non-repudiation may require the use of on-line TTP services when the level of trust is especially low between the two management domains.

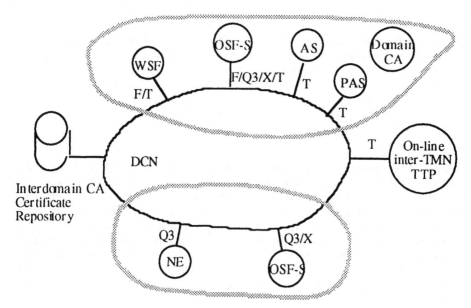

Fig. 4.   Interdomain Certification Authorities between Management Domains

3.4   Access by End Users to Service Management

For access by end-users, two particular aspects should be investigated: first, from the user point of view, the access to service execution and control, and the access to service management must be controlled in a consistent and even seamless way (i.e., one would say that the user could require single log-on to service and service management during the life time of a connection); second, the user may not be known by the telecommunication service provider if the use of the service has been subscribed to through a one-stop shopping (OSS) provider.

That may lead to consider that the authentication of the end user should be made in the domain where he is known, i.e. his OSS provider's domain, and the privilege attributes retrieved from the target domain, either the service execution domain or the service management domain. A by-product of such a mechanism is that it provides for the privacy of the user, since the identity provided by the authentication server of the OSS provider may contain an access control identity (perhaps together with an accounting identifier) different from the "real" identity of the user.

The user's AUC will then be either presented to the service control PAS, or used by the management system to obtain the user's PAC from the service management domain PAS.

If we took the alternative view that access to the management system by end users should be made through an external interface X, then securing that sort of access would be similar to securing interactions between management domains. The user's PAC would be obtained from the originating domain PAS and forwarded through the X interface.

3.5   Implication on TMN Standardisation

Considering first the internal interfaces of a TMN, the use of centralised authentication servers and privilege attributes servers to secure the access of WSFs to the service OSF(s) will require the addition of a security interface to the WSF function block. Such a "T" - for TTP - interface would likely be supported by OSI security protocols SESE for performing the security exchanges between the WSF and the security server(s). The F interface itself should then offer the capability of transferring the users' PACs to the target management application. This also should normally be made by use of SESE.

Assuming that public key cryptography is used to secure the Q3/Qx interfaces, the standardisation of security on those interfaces would consist in the incorporation, in the manager and agent applications, of an authentication application service preferably employing "standardised" public key mechanisms. Access control to the managed objects would require the transfer of the initiating user's and manager application ACIs to the agent. The exchanges required by those authentication and access control mechanisms should rely on SESE, rather than employing the facilities provided respectively by ACSE and CMISE. Such a solution would avoid the need of defining an additional interface for the OSF functional block, with regard to the intra-TMN security.

On the X-interface, the basic services required by network integrity and commercial security can be satisfied by means of off-line certificates and security mechanisms not involving on-line TTPs, provided that the integrity and confidentiality of the management associations are ensured by lower layers security mechanisms (NLSP). The use of on-line TTPs acting in the roles of key distribution server - playing a role in any legal request for decryption - and notary servers required for strong non-repudiation services. The definition of an additional interface for the OSF-S would then be necessary to interact with the involved TTP(s).

With OSI techniques, management Application Service Objects (ASOs) security is provided by a Security Support (SS) ASO; that Security Support ASO interacts with its peer SS ASO by means of the Security Exchange protocol; those exchanges can be made on the association already established for the management application protocol exchanges, or a dedicated one established by means of ACSE.

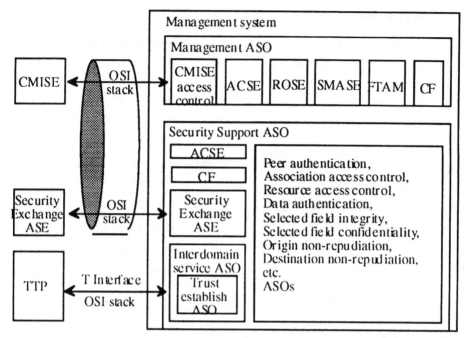

Fig. 5.   Realisation of management application security with OSI technique

4   Summary and Conclusions

The combined use of on-line and off-line Trusted Third Parties, both based on public key technology, should be suitable to provide for the various security and privacy requirements of Service Management Systems. The combination of such solutions can lead to a consistent and harmonious security architecture where the amount of security information and its management are minimised. While off-line TTPs can satisfy the security needs on the Q3/Qx interfaces and most of those of X interfaces, on-line TTPs are necessary to fulfil the security requirements of the F and enhanced security requirements of X interfaces, such as non-repudiation.

While standardisation of the F interface is not regarded as a priority, thus allowing the use of proprietary distributed security solutions, it goes differently for Q3 and X-interfaces in a multi-provider and multi-vendor context. The most urgent actions for standardising TMN security in a first stage consist in the selection of the appropriate security mechanisms - and also algorithms - to be used at the management application level to perform authentication and access control, being assumed that the integrity and confidentiality of the management exchanges are ensured at the network layer level. In a second stage, the standardisation of on-line TTP interfaces to ensure key management or non-repudiation services appears as a less urgent action. This will finally lead to the definition of various security profiles suitable to the specific requirements of the various management applications, some of those profiles implying the use of on-line TTPs, others not. The consistency of the security solutions adopted for the various TMN interfaces (F, Q3, X) relies on the standardisation of the security information objects, such as the ACIs.

5 Acronyms

AS	Authentication Server	ONP	Open Network Provisioning
ACI	Access Control Information	OSF-S	Operations System Function-Service
AUC	Authentication Certificate		
BSP	Basic Service Provider	OSS	One-Stop Shopping
CA	Certification Authority	PAC	Privilege Attribute Certificate
DCN	Data Communication Network	PAS	Privilege Attribute Server
ECMA	European Computer Manufacturers Association	PRISM	Pan-european Reference configuration for IBC Service Management
ETSI	European Telecommunication Standardisation Institute	TMN	Telecommunication Management Network
GSM	Global System for Mobile	TTP	Trusted Third Party
IBC	Integrated Broadband Communication	SESE	Security Exchanges Service Elements
NE	Network Element	VASP	Value Added Service Provider
NLSP	Network Layer Security Protocol	WSF	Work Station Function
NRA	National Regulatory Authority		

6 References

[1] Recommendations to the European Council, Europe and the global information society, "The Bangemann Report", 26 May 1994

[2] IBAG Statement on the Availability of Commercial Cryptography, March 94

[3] CEC: Green Book on the Security of Information Systems, Draft 3.6 (July 14, 1993)

[4] CEC: "On the Establishment of the Internal Market for Telecommunications Services through the Implementation of Open Network Provision (ONP Directive)", Council Directive DIR(90) 387 ECC (1990)

[5] NERA and MITA: "Study of the Application of ONP to Network Management", report prepared for CEC DG XIII (1992)

[6] ICC Position Paper on International Encryption Policy, International Chamber of Commerce, 1994

[7] ETSI: "Requirements Specification for an Encryption Algorithm for Operators of European Public Telecommunications Networks", ETSI TC-TR NA/STAG 5 (93) 123 rev. 4 (1993)

[8] CEC: "Amended Proposal for a Council Directive on the Protection of Individuals with Regard to the Processing of Personal Data and on the Free Movement of such Data", Commission communication to the Council COM(92) 422 final (1992)

[9] CEC: "Proposal for a Council Directive Concerning the Protection of Personal Data and Privacy in the Context of Public Digital Telecommunications Networks, in Particular the Integrated Services Digital Networks (ISDN) and Public Digital Mobile Networks", Commission communication to the Council COM(90) (1990)

Integrated Access Control Management

Günter Karjoth*

IBM Research Division, Zurich Research Laboratory,
8803 Rüschlikon, Switzerland

Abstract. In this paper, we discuss access control for objects distributed over heterogeneous network management systems. In particular, we compare the access control mechanisms of SNMP version 2 against those proposed for CMIP. We employ the Typed Access Matrix model as a framework to study the two approaches in detail. Apart from the differences due to their management model, both schemes use an identity-based ACL scheme on object groups defined as collections of subtrees. We present an access control scheme integrating both systems.

1 Introduction

Network users and operators are becoming increasingly concerned about the security of information accessible through networks, and about the security of the network components and resources. In particular, the management of computer networks depends on the ability to exchange network control information with confidence. This implies, among other things, that access to the Management Information Bases (MIB), a structured collection of objects belonging to network management agents, must be adequately protected. Securing network management operations requires three main services: authentication, privacy/integrity, and access control. The first two are a mandatory requisite for access control. They are not discussed further here.

Access control provides the ability to determine precisely who is authorized to access which objects in what ways. Advanced communication services can employ complex access models, which must be both economically implemented and securely managed. The large number of users and objects, the varying working relationships among users in such environments, and the frequent changes of access control information pose challenges to the design of the authorization system.

There are many standards in the area of access control, e.g. POSIX 1003.6, X.501, X.741, and RFC 1445, but they are not coherent and differ with respect to their expressiveness. Furthermore, security technology for network management must be able to interoperate with security mechanisms provided by other systems, be it the local security infrastructure of a client management application or the security services of a foreign private or public network. To achieve

---

* This work was supported in part by the European Communities under RACE II project no. R2058, Security and Management Services in Open Networks (SAMSON).

seamless, end-to-end management in such an environment, it is recognized that the interworking of CMIP with SNMP is essential. A formal integration of the security technologies involved is therefore necessary.

Authorization is an independent semantic concept that should be separated from its implementation in system-specific mechanisms. For its representation, we need a language that is expressive enough to specify commonly encountered authorization requirements. The language must have a formal semantics so that the meaning of an authorization requirement stated in that language can be precisely determined. This way, the security administrator is able to reconcile easily what should be authorized with what is actually authorized [WL92].

In this paper, we use Sandhu's Typed Access Matrix (TAM) model [San92, AS92] as a framework for studying and constructing access control policies while taking the diversity of the above-mentioned systems into account. A TAM model consists of an access matrix accompanied by a finite set of commands that control the update of the matrix. Thus, it constitutes a language with a semantics that is independent of specific implementation mechanisms and allows easy expression of a variety of access control policies.

We compare the access control mechanisms of CMIP to those proposed for SNMPv2 by developing TAM specifications of both systems. We show that the TAM specifications of X.741 and RFC 1445 identify a common subset of expressible access control policies. This intersection can clearly be managed, thus covering administrative systems with different access control services. The TAM specifications presented also define a set of operations to manage the common access control information and the functions performing access control decisions. In the next section, the TAM model is reviewed in some detail. Section 3 contains TAM specifications of the access control systems investigated: CMIP and SNMPv2. A generic access control scheme integrating them is presented in Section 4. Finally, conclusions are drawn in Section 5. This paper assumes some familiarity with network management concepts; further details may be found in [Sta93]. A list of the acronyms used can be found in Section 6.

2   A Formal Model of Authorization

The access matrix of a TAM model represents the protection state of the system in which *subjects* are granted *rights* to perform operations on a set of *objects*. Formally speaking, a protection state is a triple $\langle S, O, P \rangle$, where S is the set of subjects, O is the set of objects and P is an access matrix with a row for every subject in S and a column for every object in O. The contents of a cell, denoted $[s, o]$, is the subset of rights to object o possessed by subject s. The access matrix can be pictured as in Figure 1. Note that the set of subjects is a subset of the set of objects. Further, row s of the matrix in Figure 1 is like a "capability list" for subject s and column o is like an "access control list" for object o. The intersection of the row and the column is the cell $[s, o]$.

A TAM model further includes a set R of rights and a collection of commands of the form:

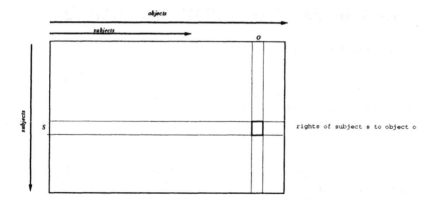

Fig. 1. Access control matrix P

> **command** $\alpha(X_1, X_2, \ldots, X_k)$
>   **if** $r_1 \in [X_{s1}, X_{o1}] \wedge r_2 \notin [X_{s2}, X_{o2}] \wedge \ldots$
>   **then** $op_1, op_2, \ldots$
> **end**

Here, α is a name and $X_1, \ldots, X_k$ are formal parameters, each X_i being either a subject or an object. Subscripts s_i and o_i are integers in the range $1 \ldots k$ indicating whether a parameter is a subject or an object. Each $r_i \in R$ is a right and each op_i is one of the primitive operations

enter r **into** $[X_s, X_o]$	**delete** r **from** $[X_s, X_o]$
create subject X_s	**destroy subject** X_s
create object X_o	**destroy object** X_o

whereby each s and o is an integer in the range $1 \ldots k$, k being the number of parameters in the command; X_s denoting a subject and X_o denoting an object. The execution of a TAM command is controlled by a test for the presence or absence of certain rights in certain positions of the matrix verifying that the action to be performed by the command is authorized. Each operation defines some modification to be made to the access matrix.

Here we are concerned only with the condition part of commands. However, we extend the TAM language to be able to express implicit rights. For this let us include special symbols in the set of rights to denote relations, e.g. membership is expressed by the symbol $\sqrt{}$. Further, the logical formula of a condition may also include functions on object names and bit strings. ϵ is the empty string and for two names x and y, $x \preceq y$ holds if x is a prefix of y. & denotes the pairwise logical and of bits. Finally, we interpret 0 as false and 1 as true.

3 TAM Specifications of CMIP and SNMPv2

3.1 Access Control in CMIP

Access control for OSI management is based on the model for access control defined in [X.812]. We describe hereafter the ACL scheme of [X.741], in which access control information itself is specified as managed objects (MOs). The `accessControlRules` MO defines the conditions under which an access may take place. It maps ⟨initiator, target⟩-pairs to the access permissions.

Within the managed object containment tree, collections of objects sharing the same protection properties are grouped into targets. A **target** is uniquely identified by its base object and scope. Targets may be nested. Figure 2 shows a containment tree with targets A, B, and C; C is contained in B. The base object of a target is identified both on the basis of individual instances and on the basis of classes. A class is treated as the set of all its instances, i.e. the right to an object class implies this right to all objects of this class. Using the base object as the root, a subtree of the MIB tree is obtained. Its scope is defined by four mechanisms (see Figure 2). A filter may be used to select individual objects within the scope of the subtree.

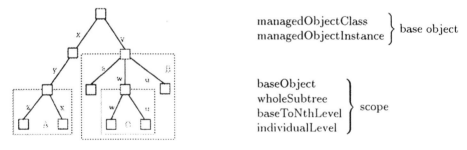

Fig. 2. CMIP containment tree with targets A, B, and C

The CMIP operations **M-Get**, **M-Set**, **M-Create**, **M-Delete**, and **M-Action** are mapped on operation types listed on the right side of Table 1 to include access to individual attributes, access to more than one object within one request, or to select objects (or their attributes). The **operations** MO defines the admissible operation type together with constraints upon the parameters of the CMIP operation[2]. Table 1 defines the admissible constraints for each operation type. By using the **attributeIdentifierList**, access granularity is on individual attributes; by using the **attributeFilterList**, access control granularity is on values of individual attributes, the finest granularity achievable.

Unless the **targets** MO contains **operations** MOs, a target identifies all operations upon the selected managed objects. When a target contains one or

---

[2] For the purpose of access control, multiple object selection and filtering are regarded as operations upon managed objects.

constraint	operation types
attrIdList	get, replaceWithDefault, filter
attrFilterList	create, delete, replace, addValue, removeValue
actionFilterList	action
scopeFilter, syncFilter	multipleObjectSelection

Table 1. Operations and associated constraints

more operations, an access request must match the operation type for one of the operations and must satisfy the constraints for this operation type. If the constraint-defining attribute, e.g. the attrIdList attribute for operation type get, is empty, no constraints are imposed, i.e. all attributes are part of the target.

Finally, **rules** map pairs of **initiators lists** and **targets lists** to a set of enforcement actions: denyWithResponse, denyWithoutResponse, abortAssociation, denyWithFalseResponse, and allow. A rule with an empty initiators list applies to all initiators; a rule with an empty targets list applies to all targets in the security domain. A rule with an empty targets list is called a **global rule**. The other rules, called **item rules**, grant or deny specific initiators access to specific targets in the domain. In case none of the global or item rules match, the value of the defaultAccess attribute of the accessControlRules object defines whether the management operation request is to be accepted or rejected. The access decision function (ADF) evaluates these rules in a predefined order, favoring denials and global rules.

Mapping

The set of initiators lists and targets lists are denoted by iL and tL respectively. There are two distinguished symbols All and GR denoting all initiators (empty initiators list) and representing global rules (empty targets list) respectively. We require that $All \notin iL$ but $GR \in tL$. The set of initiators is denoted by $\mathcal{I}$, the set of classes is denoted by $\mathcal{C}$, and the set of object instances by $\mathcal{O}$. Let $\mathcal{A}$ be the set of attributes, and let $S = tL \cup \mathcal{I} \cup \mathcal{C} \cup \mathcal{O} \cup \mathcal{A}$ be the set of subjects. The object dimension $O = \{All\} \cup iL \cup T$ of the protection matrix consists of the set of initiators lists, all initiators, and the set T of targets. Operation types constitute the rights set $R = \{a, c, d, g, r, aV, rV, rD, m, f\} \cup \{+, -, 0, 1, \sqrt{}\}$. The additional rights (symbols) are used in the following sense: $+$ gives unrestricted access to all objects (and their attributes) within the targets, $-$ denies any kind of access, and $\sqrt{}$ denotes membership. Zeros and ones define the scope of a target.

Figure 3 shows the components of the CMIP access control matrix. The matrix defines that the defaultAccess attribute rejects access if there is not any rule that specifically grants or denies access to the object. Initiator i_2 is included in the initiators list iL_2, which gives (general) access to the targets listed in targets list tL_2 but denies access to all targets of targets list tL_1. Target t_2, included in targets list tL_2, is defined by base object moc3 and moi1 with a scope of

Objects Subjects	All	iL1	iL2	t1	t2	t3	t4
GR	-			+	+	+	+
tL1		+	-	+			11
tL2			+		001	+g	
i1	√	√					
i2	√		√				
i3	√		√				
moc1				√			
moc2				√			√
moc3					√	√	√
moi1				√	√		
moi2						√	
a1					g,r,c		g,r,c
a2				a,c,d			
a3					c,d		

+ – wholeSubtree
1 – baseObject
$1^N 1$ – baseToNthLevel
$0^N 1$ – individualLevel

g – get
rD – replaceWithDefault
f – filter
c – create
d – delete
r – replace
aV – addValue
rV – removeValue
a – action
m – multipleObjectSelection

Fig. 3. CMIP access control matrix P

individual level 2 (001). For operations of type **get**, attribute $a1$ is included in target t_2; for operations of type **create**, target t_2 includes attributes a_1 and a_3. Note that in our modelling we collapse the default access value onto a global rule with an empty initiators list. To preserve the original semantics, we must ensure that $[GR, All]$ is evaluated last.

If $x \preceq y$ then $y \backslash x$ returns a string of ones such that the number of ones is larger by one than the number of levels between x and y:

$$y \backslash x = \begin{cases} 1 & \text{if } y = x; \\ 0.(x.x' \backslash x) & \text{if } y = x.n.x'. \end{cases}$$

Thus, $[B,t] \& o \backslash B$ is true if the object o is within the scope $[B,t]$ of target t defined by base object B. A request of initiator i to access attribute a through the execution of a CMIP operation p of type r on object o is allowed if it holds that:

$B \preceq o \wedge \sqrt{} \in [B,t]$ /* target */
$\wedge$
$\sqrt{} \in [i, iL]$ /* initiatorsList */
$\wedge$
$(+ \in [tL, t] \vee [tL, t] \& o \backslash B)$ /* targetsList */
$\wedge$
$- \notin [tL, iL]$ /* global/item rule deny */

$$\land$$
$$(+ \in [tL, iL] \lor + \in [GR, All]) \quad \text{/* global/item rule grant, default access */}$$
$$\land$$
$$r \in [tl, t] \lor r \in [a, t] \quad\quad\quad \text{/* operationsList */}$$

For each target t containing object o, the above ADF checks the absence of any global or item rule that denies access. If a global rule, item rule, or the default access value $[GR, All]$ permits access, the requester i can access attribute a if the operation type of the CMIP operation is contained in $[a, t]$.

3.2 Access Control in SNMPv2

SNMP also uses a tree to store managed information in the MIB. The structure of the MIB is static and is determined at its design time. Figure 4 shows the structure of a MIB. As managed objects consist of one scalar value only, they are only at the leaves of the tree. SNMP object names are unique only within a system; to provide for global naming, object names must be qualified with globally unique system names. Each instance of a managed object type can be identified only within some scope or "context", e.g. a physical device.

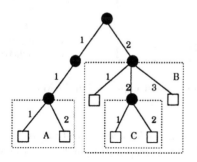

(A)	1.1	include
(B)	2	exclude
(C)	2.2	include

Fig. 4. SNMP Naming Tree

The access control model for SNMPv2 is defined in [RFC1445], introducing the following elements: A *party* is an execution environment residing in an agent or management application. Each management request includes a *context* which refers to a collection of managed objects defined by a *MIB view*[3]. A MIB view, a subset of all the managed objects held by an entity, is defined in terms of one or more *subtree families*. A subtree family is determined by an object identifier in combination with a bit mask, where zero serves as a wild card.

An *access policy* defines the operations that may be performed when a source party communicates with a destination party and makes reference to a particular context. Operations are Get, GetNext, Response, Set, GetBulk, Inform,

---

[3] For simplicity, we assume that the objects identified by a context are local.

and **SNMPv2-Trap**. A SNMPv2 message contains fields to identify the sender (**srcParty**), the receiver (**dstParty**), the managed objects visible to the operation (**context**), and the requested operations (**PDUs**). The receiving entity must have an access policy entry that relates **dstParty**, **srcParty**, and **context**. Implicitly defined, access is denied by default.

Mapping

In contrast to the access control of CMIP discussed above, SNMPv2's access decision function is based on pairs of subjects, i.e. principals. Furthermore, the operation requester has to provide a context in which the accessed objects have to be. The context resembles initiator-bound access control information such as group membership or role.

	AA	LH
Chico,Groucho	g,gN,gB	g,r
Groucho,Chico	t,r	i
internet	+111111	
ifEntry	−	
system		+
partyMIB		−1111111101

g – Get
gN – GetNext
gB – getBulk
r – Response
s – Set
i – Inform
t – SNMPv2-Trap

Fig. 5. SNMPv2 authorization expressed in TAM

Figure 5 shows the components of a TAM access control matrix. The bit mask value that, in combination with the corresponding instance of viewSubtree, defines a family of view subtrees, is used as an access right; preceded by a plus symbol the view is included, whereas preceded by a minus symbol the view is excluded. For example, $+111111 \in [\text{internet}, \text{AA}]$ defines that subtree **internet** is an included MIB view in context **AA**. However, MIB view **ifEntry** is excluded as $- \in [\text{ifEntry}, \text{AA}]$.

For each specific object instance either no MIB view entry matches or a family can be uniquely determined by taking the lexicographically largest family name to avoid conflicts. Let $x = x_1.x_2.x_3 \ldots x_n$ be an object instance, let v be an entry in the viewTable with a viewSubtree value of $v_1.v_2.v_3 \ldots v_k$, and let m be a viewMask value of $m_1.m_2.m_3 \ldots m_k$. Object instance x is contained in the family of subtrees defined by an entry in the viewTable if

$$F(x,v,m) = \begin{cases} true & \text{if } m = \varepsilon \wedge v \preceq x; \\ F(x',v',m') & \text{if } m_1 = 0 \vee x_1 = v_1 \text{ for } x = x_1.x', \ m = m_1.m', \ v = v_1.v'; \\ false & \text{otherwise.} \end{cases}$$

A request of party *src* to party *dst* to execute operation p on object o within context *ctxt* is allowed if it holds that:

$$B \preceq o \wedge \neg(B \preceq B' \wedge B' \preceq o) \qquad \text{/* dominating family name */}$$
$$\wedge$$
$$(+ \in [B, ctxt] \vee F(o, B, [B, ctxt])) \qquad \text{/* viewTable */}$$
$$\wedge$$
$$p \in [\langle dst, src \rangle, ctxt] \qquad \text{/* aclTable */}$$

For each viewSubtree B that is a prefix of object o, the ADF tests whether it is the (lexicographically) longest prefix. If so, the object must be included in the context, taking wildcards into account. Access is granted if this context also contains the required right.

4 Integrated Access Control

The TAM specifications in the previous section show that there are several differences between the access control schemes of CMIP and SNMPv2; often due to the richer model of OSI network management, e.g. access granularity on values of individual attributes, but also due to diverging designs. Although both schemes support object groups (targets/views) and negative rights (inclusion/exclusion of object groups), the tuples are sorted differently: $(T_+ \cup T_-) \times Op$ in SNMPv2 versus $(T_+ \times Op) \cup (T_- \times Op)$ in CMIP. Therefore the ADFs have to differ in the way they derive access rights from these access data stored in lists/tables by relating subjects to (included/excluded) subtrees and a set of associated operations. This illustrates the need for a rich "reference" model to serve as the basis for integrated access control management, providing a representation of access control that can be mapped on managed objects defined under different proprietary paradigms.

Using the above comparison we identify the common subset of expressible access control policies as described in the access control matrix of Figure 6. Targets are defined only by MOC values as there is no structural difference between object classes and object instances in SNMP. The level of access granularity is on object instances; thus filtering on attributes cannot be supported. Owing to different orderings the access data tuples are normalized. This means that a context is split into two targets lists, one containing all included subtrees and the other containing all excluded subtrees. In SNMPv2, a subject has the same set of rights to all objects it can access, whereas in CMIP operation rights are associated with each target. The latter provides greater expressibility but may cause redundancy in the storage of the access data. This is the case in our mapping of the SNMPv2 access control scheme onto the common access matrix (Figure 6).

Taking the access data of the SNMPv2 TAM example of Figure 5, we see that each context is split into four targets lists, e.g. for context LH targets list LH+64 defines all included subtrees where the subject has the right to execute operation Inform. In the ADF we include functions to abstract from particular

Objects / Subjects	iL1	iL2	t1	t2	t3	t4	t5	t6	t7	t8
AA+35	+		111111 g.gN.gB							
AA−35	−			+						
AA+132	+						111111 t.r			
AA−132	−							+		
LII+5	+				+ g.r					
LII−5	−					1111111101				
LII+64	+								+ i	
LII−64	−									1111111101
Chico,Groucho	√									
Groucho,Chico		√								
internet			√				√			
ifEntry				√				√		
system					√				√	
partyMIB						√				√

Fig. 6. Common subset of expressible access control policies

encodings. The ADF of this matrix is defined as

$$\sqrt{} \in [i, iL] \wedge o \mapsto [tL, t] \wedge - \notin [tL, iL] \wedge + \in [tL, iL] \wedge r \in [tl, t] .$$

Whether an object is contained in a target (MIB subtree) is determined by function $\mapsto$. The SNMP source/destination party pair and the context value might be determined by an out-of-bound mechanism, e.g. on the basis of an authenticated CMIP manager identity.

The Network Management Forum's OSI and Internet management coexistence (IIMC) activity defines an application level gateway (proxy) between CMIP and SNMP. The above TAM specification provides the integration of both access control systems, filling the missing part for access control in the end-to-end security architecture of [IIMC].

5 Conclusion

It has been shown that the access control schemes of CMIP and SNMPv2, although both using an identity-based ACL scheme on object groups, bear contrary design decisions that makes it difficult to integrate the management of objects protected by these access control policies. The TAM model proved to be a good language to abstract the specific access control semantics. Its value will become even more visible when monitoring the changes, as neither of these access control schemes is stable yet. Furthermore, the number of security standards for access control is still increasing. Currently, the Object Management

Group solicits proposals from vendors in response to the Object Services Task Force RFP 3 on security and time [OMG]. Our TAM models will serve as a basis to judge the interrelationships with the access control schemes presented in this paper. Eventually it should be possible to integrate the numerous concepts of access control information into a common concept that is suitable to provide a structure of proficiency. Furthermore, definitions of advanced access control policies in the TAM model will ease the task of human security managers. These TAM specifications will provide the security administrator with a common view of the managed resources, independent of the employed management protocol.

6 Acronyms

ACI   Access Control Information
ADF   Access Decision Function
CMIP   Common Management Information Protocol
MIB   Management Information Base
MO   Managed Object
MOC   Managed Object Class
MOI   Managed Object Instance
SNMP   Simple Network Management Protocol
TAM   Typed Access Matrix

References

[AS92]   P.E. Amman and R.S. Sandhu. Implementing transaction control expressions by checking the absence of rights. In *Eighth Annual Computer Security Applications Conference*, pages 131–140, 1992.

[IIMC]   L. LaBarre (Editor). Forum 027 – ISO/CCITT to Internet Management Security. Issue 1.0, Network Management Forum, October 1993.

[OMG]   Object Management Group. Object Services RFP 3. TC Document 94-7-1, 1994.

[RFC1445] K. McCloghrie and J. Galvin. Administrative Model for version 2 of the Simple Network Management Protocol (SNMPv2), RFC 1445, Hughes LAN Systems, Trusted Information Systems, April 1993.

[San92]   R.S. Sandhu. The typed access matrix model. In *1992 IEEE Computer Society Symposium on Research in Security and Privacy*, pages 122–136, Computer Society Press, 1992.

[Sta93]   W. Stallings. *SNMP, SNMPv2 and CMIP*. Addison-Wesley, 1993.

[WL92]   T.Y.C. Woo and S.S. Lam. Authorization in distributed systems: A formal approach. In *1992 IEEE Computer Society Symposium on Research in Security and Privacy*, pages 33–50, Computer Society Press, 1992.

[X.741]   Information technology – open systems interconnection – systems management: Objects and attributes for access control. DIS 10164-9, ISO/IEC, 1994. ITU-T Rec. X.741.

[X.812]   Information technology – open systems interconnection – security frameworks in open systems – part 3: Access control. DIS 10181-3, ISO/IEC JTC1, 1994. ITU-T Rec. X.812.

User Design Issues

Session 2b Introduction

Chairman - Nick Hine

Applied Computer Studies Division, MicroCentre, University of Dundee, Park Wynd,
Dundee DD1 4HN, Scotland
Tel: +44 (0)1382 344711
Fax: +44 (0)1382 345509
e-mail: nhine@mic.dundee.ac.uk

As broadband telecommunications services become available, interpersonal communication and access to information will become possible in ways that have formerly been completely impractical. It is impossible to underestimate the impact that this will have on society and the behaviour of people. In the process, however, users will be confronted with interactions with technical systems that are far more powerful than anything they have encountered before. Services will provide, users (perhaps untrained), with the ability to handled a variety of media, to manipulate that media and to retrieve, store and forward it to countless locations around the world. New behaviour patterns will result in new concepts of personal and national identity, space and time.

The challenge for the designers of services and terminal equipment is to present the user with an interface to this infrastructure which enables these users to utilise it and to take advantage of the opportunities it presents.

This challenge is best understood by considering the potential barriers that a user has to negotiate if they are to successfully use a telecommunications service. In the first place, the telecommunications terminal should be accessible and usable. It is important that the physical design and placement of the terminal accurately reflects the abilities and expectations of the user. This is a far from trivial issue, as a terminal may have the capability to handle the complex mix of media available currently using a television, computer and telephone handset and yet may need to be easier to use than a video recorder.

Secondly, the fact that the user is interacting with a complex and distributed network infrastructure could be a source of confusion and difficulty. The user will have the facility to connect to other people or information sources located in an infinite number of sites around the world. Sometimes some simple numerical facts will be required for a bank account. On other occasions, the user will want to show a picture to a relative during a videophone call. In each case, the system will need to minimise the tasks associated with negotiating network specific properties, and the costs associated with the use of the system should be readily understood and accepted by the users.

Thirdly, the interaction with other people and the access to information will be mediated by a telecommunications service. Information will be conveyed in a variety of media, and the user will control the flow of information through an interface to the service. Face to face interactions have complex rules that have evolved over centuries and are handed down from one generation to the next. We are only now beginning to understand the richness of these interactions, and the potential for misunderstanding that arise when some cues are emphasised more than others. Multimedia communication has the potential for at last providing much of this rich interaction at a distance, but incorrectly implemented, it could lead to serious examples of misunderstanding. Not only will it be necessary to provide information in the appropriate media, but it will be important to ensure that the media exchange can be properly and readily controlled by all involved in the interaction.

The magnitude of the challenge is only becoming evident as researchers and system developers seek to make these facilities available to users. In short, this activity is aiming to put into the hands of all people, a system that is more complex than anything they have ever encountered before, and is a platform that integrates and makes accessible everything that mankind has learnt and experienced in all domains of knowledge, in every type of media, throughout history.

Much work has taken place to model the problem and to uncover the user interface and interaction issues involved. This activity began as an exercise involving HCI experts who were convinced of it's importance. As data emerges, and concrete applications have started to be demonstrated, both the scale of the challenge and the necessity for work in this domain has become evident to the wider telecommunications research and provider community. The papers in this session build on the early foundation work, but at the same time demonstrate technical progress in applying sound HCI principles that address the complex environment that will be encountered by the users. Whilst basic user interaction and modelling work must continue, these papers reflect the results of the applied research that has taken place over the last 5-10 years, particularly within the framework of the CEC RACE and RACE II programmes. They offer design and engineering specialists concrete solutions to some aspects of the complex task of ensuring that multimedia telecommunications services and systems are usable.

An Experimental Evaluation of a Normative User Interface Design for the Configuration of Telecommunication Services

Ian Denley, Becky Hill & Andy Whitefield
Ergonomics & HCI Unit
University College London
26 Bedford Way
London WC1H 0AP
England
tel: 0171-387-7050 x5342
fax: 0171-580-1100
e-mail: i.denley@ucl.ac.uk

Abstract. This paper reports an experimental evaluation of a prototype normative user interface designed to support end-users in the dynamic (real-time) configuration of future telecommunication services. The normative user interface supports configuration tasks which are common to typical broadband services, and is claimed to be generally applicable across such services. The paper also illustrates the value of user interface consistency to skill transfer between services. The prototype user interface performed well with respect to criteria of learnability and the transfer of skills between services. The experiment demonstrates the feasibility and value of a normative user interface for the dynamic configuration of services by end-users.

1 Introduction

The design of new telecommunication services is a rapidly growing field of technical research and development. New services such as videoconferencing, multimedia electronic mail and 'video on demand' are becoming possible as the digital broadband networks (the so called "information superhighways") which can carry such services are being implemented at local, national and international levels.

End-users are likely to face a bewildering choice of different telecommunication services, offering different features and capabilities at different costs. Faced with such variety, end-users will benefit from software support in configuring services, i.e. in identifying the most appropriate services for their needs, and in specifying the service parameters or options most suited to their situation. The ASCOT project (see acknowledgements) has specified and developed a set of prototype software tools to support end-user service configuration tasks. This toolset aims to increase the usability of services and hence promote usage of both services and networks.

This paper examines one aspect of service configuration by end-users: the dynamic configuration of services[1]. Dynamic configuration concerns some basic tasks that an

---

[1]This paper concentrates only upon the software tools to support dynamic configuration tasks. Other parts of the toolset support other types of configuration tasks; for instance, the static configuration of a service prior to use (e.g. specifying the participants of a videoconference in advance, or setting cost/duration limits).

end-user carries out during the real-time use of a telecommunication service (for instance, adding new users to a videoconference already in progress, or changing the bandwidth allocated to a videoconference during a call to increase the picture quality). It is asserted here that many dynamic configuration tasks are common across different services, and can be supported, therefore, by a normative user interface. By the term "normative user interface" we mean a user interface which can be implemented similarly, or in a standard form, in all or most future telecommunication services. This paper reports an experiment which examines the benefits to end-users of a normative user interface of this kind.

Section 2 describes the background to the study, the rationale for a normative user interface, and the aims of the experiment reported here. Section 3 describes the experimental design and data collection techniques. Section 4 presents the results of the study, whilst section 5 concludes the paper with a discussion of the implications of the study for the design of new telecommunication services and user interface design.

2 The Context of the Experiment

2.1 The Rationale for a Normative User Interface

The ASCOT project has identified a number of dynamic configuration tasks that are common to most telecommunication services (Hill, Denley and Whitefield, 1993). The most important tasks of this kind include:
- changing the connections of a service (e.g. users, or databases)
- changing the media used by a service (e.g. from audio only to audio and video)
- changing the quality of media used by a service (e.g. from low resolution images to high resolution images)
- transferring the management of a call to other parties
- being informed of the costs of a call in real time

The idea of a normative user interface for the dynamic configuration of telecommunication services is based on the notion that it will be beneficial to end-users if these common tasks are supported by a single user interface with appropriate functionality to support these tasks in all services.

Such a normative user interface exploits the concept of user interface *consistency* which is often cited as an important principle of user interface design (Shneiderman, 1987; Barnard et al, 1981; Payne and Green, 1986; Kellogg, 1987). Consistency of application functionality and interface representations typically increases the ease of use of a software application for end-users.

It was an important assumption in the design of the ASCOT toolset that user interface consistency would be beneficial to end-users given the multiplicity of

---

the participants of a videoconference in advance, or setting cost/duration limits).

potential services as described in Section 1. The experiment reported here was designed to test this claim with respect to the dynamic configuration of services, and used a prototype of the user interface designed for the dynamic configuration tools.

2.2   Aims of the Study

This experiment had a number of aims. The primary aim was to assess if a normative user interface for dynamic configuration tasks common across different services will be beneficial to users.

The secondary aims were: (a) to provide diagnostic feedback on the user interface design to support redesign before the final implementation (this experiment used a laboratory simulation of the toolset, which was still being developed at the time the experiment was performed); and (b) to illustrate what such common dynamic configuration tasks might be in two particular services.

3   Method

This section describes the hardware and software, the experimental design and the data collection techniques used in this experiment.

3.1   Hardware and Software

Platform
A laboratory simulation of the user interface shown in Figure 1 was implemented, and ran on a Macintosh™ Quadra 950. The simulated interface had two parts: the normative user interface, and the simulated services. To support the validity of assessing the normative user interface with simulated services it was important that these simulated services were as realistic as possible i.e. the videoconferencing and surveillance services should contain some type of moving image, either real time or stored. Authorware Professional™, an object oriented authoring environment, was used as a prototyping tool as it supported the integration of both stored video clips and real time video into the simulation.

User Interface
The user interface used in this study is illustrated in Figure 1. The interface is made up of three major components:

- The *Service Finder* (on the right of Figure 1) - this is the basic toolset user interface, present at all times when the toolset is in operation, it coordinates access to the basic information which is common across the configuration dialogues (i.e. to information about services, users, terminals and other communicating entities).

- The *Dynamic dialogue* (in the lower left corner of Figure 1) - This provides a single dialogue for those dynamic configuration tasks which are common across all services; it was implemented here as a menu with most selections leading to individual dialogue boxes.

- The content of the (simulated) service (the large upper central area in Figure 1). Figure 1 replaces the video clips used in the actual experiment with head and shoulder line drawings.

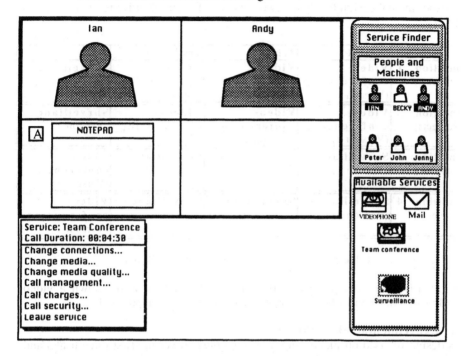

Fig.1 The Experimental User Interface

Note on the User Interface Presentation and Representation The representations and screen designs used in the experiment (see Figure 1) were intended primarily to illustrate the roles and functionality of the various components of the user interface, and the relationships between these components. The experiment was not concerned with presentation level issues (e.g. the design of icons and buttons), but concentrated instead on *task* level concerns. The user interface used for the experiment, then, was not intended to be a definitive statement as to how the final user interface was to be presented; this was to be determined as the toolset's user interface manager was developed further by other project partners.

3.2 Experimental Design

Experimental Scenarios

Two experimental scenarios were used: *surveillance* and *videoconferencing*. The surveillance scenario involves the remote monitoring by a security guard of a number of locations within a particular building. The videoconferencing scenario involves the management of a workgroup videoconference in an office setting. These scenarios were taken to be representative of the types of services likely to be supported by broadband telecommunication networks. Two versions of each scenario were produced. These versions used different video clips, and consequently differently named locations and workgroup participants, but were otherwise identical.

The experimental design (using the two versions of each scenario) is summarised in Table 1:

	Trial 1	Trial 2	Trial 3	Trial 4
Condition 1 (4 subjects)	Video-conferencing (Version 1)	Video-conferencing (Version 1)	Video-conferencing (Version 2)	Surveillance (Version 1)
Condition 2 (4 subjects)	Video-conferencing (Version 2)	Video-conferencing (Version 2)	Video-conferencing (Version 1)	Surveillance (Version 1)
Condition 3 (4 subjects)	Surveillance (Version 1)	Surveillance (Version 1)	Surveillance (Version 2)	Video-conferencing (Version 1)
Condition 4 (4 subjects)	Surveillance (Version 2)	Surveillance (Version 2)	Surveillance (Version 1)	Video-conferencing (Version 1)

Table 1: A Summary of the Experimental Design

Additionally, after completion of the final trial, subjects were shown a video recording of Trial 4 and asked to provide a retrospective verbal protocol.

The Experimental Hypotheses

This design focused on identifying *learning effects* (over the first three trials) and identifying *transfer effects* between the two scenarios (i.e. between the third and fourth trials). Learning effects were taken to be improvements to task performance over the 3 repeated trials of a particular scenario, and transfer effects were taken to be changes to task performance between the two types of scenario. The two versions of each scenario were produced to ensure that subjects were not simply rote learning the task actions to be performed on particular screen representations. No training was given before Trial 1; this allows the two learning components of guessability and learnability to be treated separately (Jordan et al, 1991 - see also Section 5.1)

Given the design described above the experimental hypotheses can be summarised as predicting:

- A significant improvement in task performance between Trial 1 and Trial 3
- No significant difference in task performance between Trial 3 and Trial 4
- No significant difference in task performance between the videoconferencing and surveillance scenarios

Subjects

There were sixteen subjects - three males and one female in each of the four conditions. Subjects were all Psychology Department students from University College London. It was a requirement of the experiment that subjects had at least some experience of use of a graphical user interface (e.g. Macintosh™ or Windows™). Subjects were paid £5 for taking part in the experiment.

Originally, twenty subjects were tested. However, four exclusions were made because the subjects managed to 'crash' the software by opening too many windows, and leaving them open, leading to shortages in computer memory. A system crash required a restart of that trial, and so the subjects were excluded on the grounds of increased exposure and hence increased learning opportunities.

Tasks and Scenario Details

There were two experimental scenarios: surveillance and videoconferencing. The tasks to be performed in each scenario were at least conceptually equivalent and often identical, and are believed to be representative of the larger set of dynamic configuration tasks described in Section 2.1. The tasks required exactly the same menu and dialogue box selections to be made in the two scenarios. The instructions concerning what tasks to perform were presented on screen. The desire to create a meaningful 'storyline' from the combination of these instructions meant that the tasks were presented in the same order in each trial. The tasks and the computer presented instructions for one version of the surveillance scenario are shown in Table 2 below.

Surveillance	Computer Presented Instructions	Menu Commands and Dialogues
Task 1 **Start service** Users start the service by clicking on the icon, and are presented with a window consisting of three panes showing security videos; a fourth pane is empty.	"Please click on the *Surveillance* icon to start the service"	Service finder
Task 2 **Add new site** Users are asked to add a new site for surveillance (i.e. the empty window pane).	"You need to watch the door - please add this site to the surveillance service. "	'Change connections' + dialogue
Task 3 **Increase video quality** Users monitor the four sites until an intruder is identified and then increase the quality of the image for that site.	"The quality of the video image from the door is very poor - please increase the picture quality."	'Change media quality'+ dialogue
Task 4 **Add new service component (audio from site)**	"You might want to hear what is going on at the door - please add an audio capability."	'Change media' + dialogue
Task 5 **Hand over control of service (to Head Office)**	"Things are looking serious, hand-over control of the surveillance service to head-office."	'Call management' + dialogue
Task 6 **Find out total cost of service use (for records)**	"Please calculate the total costs of the work you have done."	'Call Charges' + dialogue
Task 7 **Leave Service**	"Please leave the service."	'Leave service'

Table 2: The Task Scenarios and Computer Presented Instructions

Procedure

The subjects were given a short general demographic questionnaire (age, sex, education, first language and computer experience). The results of the questionnaire are described here for convenience.

The ratio of male to female subjects was 3:1; the age range was 20-50 years, and the average age was 24. English was the first language of twelve of the subjects, and the four remaining subjects spoke Swedish, Norwegian, Hebrew and Icelandic as a first language. All subjects were students educated at least to A- level standard (or equivalent), and 30% of the subjects also had degrees. All subjects were computer literate, and all had experience with graphical user interfaces for software applications such as word processing, graphics, spreadsheets and databases.

After completing the questionnaire the subjects were provided with a written general introduction to the experiment. The introduction briefly described the type of tasks they were to be asked to perform, and the number of times they would be asked to do each task. The introduction also informed the subjects of the data collection techniques to be used. It was emphasised that all data were completely confidential. The subjects were then given specific information concerning the relevant task scenarios to be carried out.

3.3   Data Collection

Two measures of task performance were used:
  • Task performance times (see Section 3.3.2)
  • User errors and difficulties (see Section 3.3.3)

Data Collection Techniques

The subjects were observed during task performance, and notes were taken by the experimenter. Additionally, the screen image was captured using a VideoLogic Mediator™ and recorded to S-VHS video tape. The video recording was time stamped. Audio was recorded direct to the video cassette recorder (VCR).

After completion of the final trial, subjects were shown the video recording of that trial and gave a verbal protocol. The protocol was recorded on a separate channel direct to the VCR.

Task Performance Times

Task performance times were taken for the seven individual tasks comprising each trial. Because the prototyping software showed minor variation between trials in the time taken to present dialogue boxes, it was necessary to control for this variation. To this end, times were taken for the start of the task, menu selection, dialogue box presentation, and task completion. The elapsed time between dialogue box presentation and menu selection was subtracted from the task completion time. No dialogue boxes were presented for starting and quitting the service as no other actions were necessary. Table 3 shows an example data collection table.

Subject 1   Trial 1

Tasks	Start (OK)	Menu Bar	Dialogue Box	Finish (OK)	Time to Menu	Time to Finish
T1	15.4			21.4		6.0
T2	46.5	141.9	146.2	150.3	95.4	99.5
T3	171.3	174.6	177.8	184.4	3.3	9.9
T4	199.8	210.1	212.5	242.3	10.3	40.1
T5	256.1	276.1	279.4	284.6	20.0	25.2
T6	298.9	305.2	305.7	315.3	6.3	15.9
T7	323.1			327.1		4.0
						200.6

Table 3: An example data collection table

User Errors and Difficulties

As mentioned earlier, the experiment did not address presentation level issues (e.g. icon design, button placement). Any errors, therefore that were clearly caused by presentation level issues were simply noted for the purposes of redesign of the interface, but ignored for the purpose of the experiment. Also, as the experimental design allows for three repetitions of the first scenario it was expected that by the third repetition users would have learned to overcome most presentation level problems with respect to user-service dialogue. This expectation was confirmed during the experiment.

Of particular interest were those errors which are related to the consistency of user interface commands between different services. It was predicted that the main influencing factor on this type of error would be the task semantics of the normative dialogue. For example, a user may recognise that a task (such as adding a new entity) is common to different services, but may find that the normative user interface dialogue is not equally meaningful for all possible different types of entity (where an entity might be a human user or a machine, say).

Data were also collected on the number of times subjects asked for help, and non-optimal task performance (e.g. choosing the wrong menu command, adding the wrong user).

4   Results

4.1   Task Performance Times: Overall Data

As an indication of the trends in task performance, Figure 2 plots the average trial completion time for the two scenario groups over the four trials.

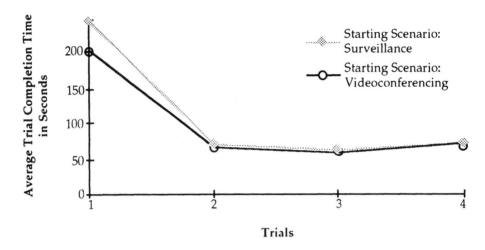

Fig. 2: Average Trial Completion Times

Because the variances in the times for each trial were unequal the data were transformed using a log transform to make the use of parametric tests appropriate. The transformed data, both overall and for each individual task, were then tested with:

- Two Way Anova
- Planned Pairwise Comparisons

The results of the Anova for the overall task data are summarised as follows:

- No difference was observed between the scenarios (videoconferencing and surveillance): $F = 0.04$, $p = 0.8444$.
- A strong difference was observed between the trials: $F = 56.48$, $p < 0.0001$
- No interaction was observed: $F = 0.3$, $p = 0.9917$

The results of the planned pairwise comparisons for the overall task data are summarised in Table 4 below.

Trial 1 vs Trial 2	$F = 106.55$	$p < 0.0001$
Trial 1 vs Trial 3	$F = 128.6$	$p < 0.0001$
Trial 1 vs Trial 4	$F = 100.92$	$p < 0.0001$
Trial 2 vs Trial 3	$F = 1.04$	$p = 3.146$
Trial 2 vs Trial 4	$F = 0.08$	$p = 0.7835$
Trial 3 vs Trial 4	$F = 1.68$	$p = 0.2026$

Table 4: Planned Pairwise Comparisons for the Overall Task Data

The results clearly show a significant difference between the task performance times for Trial 1 and those of all other trials, whilst no difference in task performance was observed between Trials 2, 3 and 4. Comparisons within each scenario (videoconferencing and surveillance) show exactly the same pattern.

4.2 Task Performance Times: Individual Task Data

The Anova results for the individual tasks show the same pattern as the results for the overall times: strong differences between the trials, with individual tasks in Trial 1 having significantly longer performance times than individual tasks in subsequent trials; no differences between trials 2, 3 and 4; and no interactions. This pattern is repeated for each scenario type as well as for both scenarios considered together. The between trial figures for the individual task data (and their comparison with the overall task data) are summarised in Table 5 below.

	F Value	Probability
Overall Task Data	56.48	<0.0001
Task 1 - Start Service	8.89	0.0001
Task 2 - Change Connections	44.46	<0.0001
Task 3 - Change Media Quality	15.73	<0.0001
Task 4 - Change Media	25.89	<0.0001
Task 5 - Call Management	20.67	<0.0001
Task 6 - Call Charges	29.06	<0.0001
Task 7 - Leave Service	3.04	0.0392

Table 5: Individual Task Results

4.3 Some Minor Anomalies

While the data as described above evidently show a very strong and consistent pattern, there were also three minor anomalies in the data which need to be explained.

i) An interaction effect was observed on Task 3 (Change Media Quality). On Trial 4 both the videoconferencing group and the surveillance group were equally fast, whereas on Trial 1 the videoconferencing group are significantly faster than the surveillance group. A possible explanation for this effect is that moving images are used in the videoconferencing scenario whilst the images used in the surveillance scenario appear virtually static, and the need for quality improvement may be more apparent with the moving images.

ii) As can be seen from Figure 2 the surveillance group are slightly slower (but not significantly slower) on Trial 1 than the videoconferencing group. This results in the learning effect for surveillance from Trial 1 to Trial 2 being greater than that for videoconferencing (although this is not a significant difference). A possible explanation for this difference is that subjects are more familiar with the concept of videoconferencing than remote surveillance, and therefore performed slightly better on first using a videoconferencing service.

iii) Relatively weaker effects were observed for Task 1 (Start Service) and Task 7 P(Leave Service) than for tasks 2-6. A possible explanation for this effect is that subjects sometimes appeared to daydream on these two tasks after clicking 'OK' on the instruction box. The subjects wrongly believed that they had completed the tasks as instructed.

4.4   Interpretation of User Errors and Difficulties

Table 6 summarises the most frequently occuring task level user errors. These errors typically concern the selection of an inappropriate command for the task under consideration. Most of these errors were caused by confusion over the semantics of the menu items with respect to the task instructions.

The total number of task level command selection errors was only 22 over the 448 separate tasks performed by the 16 subjects. Two particular confusions accounted for 16 of the 22 errors. The interpretation of these errors is based on the experimenter's observations and the subjects' retrospective verbal protocols.

Task	Appropriate Command	Selected Command							
		Click on Icon	Change Connections	Change Media Quality	Change Media	Call Management	Call Security	Call Charges	Leave Service
Start service	Click on icon								
Add new user/site	Change Connections				▓	▓			
Increase picture quality	Change Media Quality								
Add new media (audio or graphics)	Change Media		▓	▓					
Hand-over control of service	Call Management		▓	▓			░		
Calculate total cost	Call Charges								
Leave service	Leave Service								

▓ = Errors occurring in both scenarios

░ = Errors occurring in surveillance scenario only

Table 6: User Errors in Command Selection

The most frequently occurring error (10/22) was the selection of the *change media quality* command instead of the *change media* command when subjects were instructed to add a new media. This error most frequently occurred when subjects were instructed to add audio in the surveillance scenario. This confusion is probably due to differences in the salience of information presentation in different dialogue boxes. Changing media quality is performed using radio buttons in a readily identifiable graphic display representing only audio and video as media with variable quality. Typically subjects recalled seeing audio as a major component of this graphic display and subsequently selected this most salient location rather than the less obvious list box containing audio as one of many media also including text, graphics and video.

The second most common error (6/22) was the selection of the *change connections* command instead of the *call management* command when subjects were instructed to hand-over control of the service. The dialogue box presented by selecting *change connections* provides a list of connected entities, and since subjects are asked to hand-over control of the service to a particular entity this error is likely to have resulted from the expectation that all actions associated with particular entities might be located within this dialogue box.

An interesting error made by two of the subjects with English as their second language occurred only in the surveillance scenario. When asked to hand-over control of the service to head office these subjects selected the *call security* command. This is perfectly understandable if *call* is interpreted as a verb and not a noun.

Of course, other errors occurred in the performance of the experiment, but these errors were identified as presentation level issues and therefore not of interest to this experiment. Examples of these errors include interaction style errors such as attempting to drag icons where this was not possible due to the limitations of the prototype, and poor positioning of interface elements (e.g. buttons) by the designers.

5 Discussion and Conclusions

The primary aim of the experiment was to assess if a normative user interface for dynamic configuration tasks which are themselves common across different services will be beneficial to users. An important component of the experimental design focused, therefore, on identifying learning effects over the repeated trials, and transfer effects between the two scenarios.

5.1 Learning Effects

The results clearly show an highly significant difference between task performance for Trial 1 and for all other trials, whilst no difference in task performance was observed between Trials 2, 3 and 4. This learning effect was demonstrated within each scenario (videoconferencing and surveillance).

Jordan et al (1991) describe three components which they consider influence the ease of use of an interface. *Guessability* is a measure of the user costs (in terms of time and effort) associated with the first-time use of a system. Jordan et al suggest that guessability is an important property of an interface in situations where a user must perform a task correctly at the first attempt. *Learnability* is the amount of time and effort required to reach a user's peak level of performance with a system. *Experienced user performance* (EUP) corresponds to the asymptotic level of a user's performance with a system over time. When they have reached this level of performance users can be said to have learned how to use the system.

In this experiment it appears that guessability is fairly high. Although they have never seen this interface before, the subjects can complete the seven experimental tasks, and can complete them in an average total time of 226 seconds, or less than 35 seconds per task. Against this is the fact that two of the original 20 subjects were excluded from the analysis on the grounds that they needed explicit help from the experimenter to complete the tasks. While there must, therefore, be room for improvement, there is support for the view that the interface is fairly easy for new users to use.

In terms of learnability, the interface must rate very highly. In general, there is a large and significant performance improvement between Trial 1 and Trial 2, with little significant improvement thereafter. Learning to use the system is therefore completed rapidly, and largely through the experience of the first trial. Such a high level of learnability would be an important property of an interface for the dynamic configuration of a large number of different services, many of which will often have been previously unseen.

Because there are only four trials, it is difficult to be confident about comments on experienced user performance. But there is no real improvement on Trials 3 and 4 over Trial 2, indicating that experienced user performance might be reached fairly quickly. There were signs that the fastest subjects on some tasks were approaching a performance ceiling imposed by the prototype software, since they were occasionally waiting for the software to finish presenting something before entering their next input. This suggests that experienced user performance might not have been reached during the experiment, but also that it could be reached quickly.

In terms of learning effects, therefore, the dynamic dialogue leads to generally very good performance on the tasks used in this experiment.

5.2 Transfer of Skill

The results show no significant difference in task performance between Trials 3 and 4. This indicates that there is no difference in task performance between the last training trial in a given scenario (e.g. videoconferencing) and the transfer trial to the new scenario (e.g. surveillance).

Skill transfer between services can be positive (helpful) or negative (harmful); (Johnson, 1992; Pollock, 1988). The design of the dynamic configuration dialogue aimed to exploit positive transfer effects brought about by the use of a normative (and consistent) user interface. The results indicate that the dynamic dialogue successfully exploits the subjects' learning and understanding of the tasks to be performed and supports the transfer of this knowledge to a new and previously unseen service. To some extent this approach to the transfer of skill trades off the consistency of the interface against the *compatibility* of services with users' expectations and knowledge gained from outside the system. This approach is sensible in the first instance since it attempts to exploit domain independent (or application generic) knowledge with respect to the largest number of possible future services. With respect to the design of domain dependent (or application specific) components of a service then evidently the additional use of the compatibility principle would be beneficial.

5.3 Implications for Re-Design of the Toolset

The experiment reported here is a component of a sequence of evaluations aimed more particularly at providing diagonostic feedback to support the redesign of the toolset. However, a number of design recommendations were made on the basis of the results of this study. These included changes to: menu format and layout; menu item labelling; the presentation and content of a number of specific dialogues; permissable interaction styles; and the inclusion of additional functionality in the Service Finder.

5.4 Implications for Service Design

The Dynamic Dialogue described above gives users access to information and functionality that allows them to configure during service operation those aspects of a service which are common across services. Such an approach has implications for service designers and service providers who will need to design their services to be compatible with the ASCOT toolset (or some other similar software tool). The benefit of such an approach for end-users is that it ensures consistency of essential configuration functions (and their representation) between different services. This will require a standardised service-terminal interface. The analysis of configuration tasks was in part based on a proposed usage reference model for broadband communication services (Byerley and Bruins, 1992) which is itself included in the Common Functional Specifications of the RACE programme and is hence a candidate for the standardisation of services. There is, therefore, a possible route for the ASCOT toolset to relate to future service standards.

5.5 Conclusions

In summary, this experiment demonstrates (albeit in a limited context) the feasibility and value of a normative user interface for the dynamic configuration of telecommunication services by end-users. The prototype user interface performed well with respect to the criteria of learnability and the transfer of skills between services. These factors will be important in the design of software to support the dynamic configuration of services since it is likely that the number of services will be very large and that the turnover of services will be high. It is proposed, therefore, that a normative user interface which supports a basic set of dynamic configuration tasks will be of considerable support to an end-user faced with such a plethora of rapidly evolving services. Future work should aim to confirm the applicability of these dynamic configuration tasks across a wider range of service types and to increase the range of tasks supported by the toolset.

6 Acknowledgements

This work was carried out as part of the R2089 ASCOT project funded by the Commission of the European Communities under the RACE programme. The project partners are: MARI Computer Systems Ltd, UK; Ergonomics and HCI Unit, University College London, UK; Saritel S.p.A., Italy; Alcatel-SEL-AG, Germany; Deutsche Bundesposte Telekom, Germany; and Octacon Ltd, UK.

7 References

ASCOT (1993). Analysis of Configuration Management Enabling Tasks. CEC Ref.no. R2089/UCL/ERG/DS/P/004/b1

Barnard, P.J., Hammond, N.V., Morton, J., and Long, J.B. (1981). Consistency and Compatibility in Human-Computer Dialogue. *Int. J. Man-Machine Studies, 15, pp 87-134.*

Byerley, P., and Bruins, R.W. (1992). Conceptual Framework for the Usage of Telecommunication Services. In: P. Byerley and S. Connel (Eds). *Integrated Broadband Communications: Views from RACE.* North-Holland Studies in Telecommunication Volume 18. Elsevier.

Hill, B., Denley, I. and Whitefield, A. (1993). User Requirements for End-User Service Configuration. In: *RACE IS&N Conference. International Conference on Intelligence in Broadband Services and Networks, 23rd-25th November, 1993,* Paris, France.

Johnson, P. (1992). *Human Computer Interaction: Psychology, Task Analysis and Software Engineering.* McGraw-Hill. Section X/5 pp 1-9

Jordan, P.W.; Draper, S.W.; MacFarlane, K.K., and McNulty, S. (1991). Guessability, Learnability and Experienced User Performance. In: D. Diaper and N. Hammond (Eds). *People and Computers VI, Proceedings of the HCI '91 Conference, 20-23 August 1991.* Cambridge University Press.

Kellogg, W.A. (1991). Conceptual Consistency in the User Interface: Effects on User Performance. In: H.J. Bullinger and B. Shackel (Eds). *Human-Computer-Interaction - INTERACT '87.* Elsevier Science Publishers B.V., North Holland.

Payne, S.J., and Green, T.R.G. (1986). Task-Action Grammar: A Model of the Mental Representation of Task Languages. *Human Computer Interaction, 2, pp 93-133.*

Pollock, C. (1988). Training for Optimising Between Word Processors. In: D.M. Jones and R. Winder (Eds). *People and Computers IV, Proceedings of the HCI '88 Conference, 5-9 September 1988.* Cambridge University Press.

Shneiderman, B. (1987). *Designing the User Interface: Strategies for Effective Human-Computer Interaction.* Reading, MA. Addison-Wesley.

Providing Future Telecommunication Services to Naive Users

T.J. Hewson, G. Allison and A.M. Clarke
HUSAT Research Institute, The Elms, Elms Grove
Loughborough, Leicestershire, England
Tel: +44 (0)1509 611088   Fax: +44 (0)1509 234651
e-mail: t.j.hewson@lut.ac.uk

Abstract. The RACE II LUSI (Likeable and Usable Service Interfaces) project has investigated the use of existing and future telecommunications services by the general public. The project has an underlying premise that the requirements of business users in the telecommunications marketplace have been extensively dealt with, however, the particular needs of members of the general public are often overlooked.

The paper will present case studies of two such RACE experiments and compare these evaluations with the experiences of the LUSI project evaluations. The case studies are:

The provision of an expert system and PC integrated videoconferencing facilities for rural farming in Bari in the South of Italy. Making use of CD ROM, the World Wide Web, and the TELES.VISION videoconferencing system. The study was conducted as part of the RACE AREA project.

The installation of a public access multimedia shopping and information kiosk in Basel, Switzerland. This study was conducted as part of the RACE ESSAI project

Recently a great deal has been written about the role of human factors and usability issues in the future of telecommunications, and some commentators have even gone so far as to suggest that usability is not a primary factor in determining the success of a new system. However the reported studies lend strong support to the case for usability in services.

The overriding factor affecting the uptake of new services studied by the LUSI project is that most of the target customers are not aware that they need the service, in fact, some may even be resistant to the service. This is to some extent due to the fact that they are not technology literate, and in some circumstances they are technophobic.

The most startling empirical finding from the evaluations was that the first impressions of the ease of use of a new service is absolutely crucial in the acceptability of that service.

1   Introduction

The key focus of the design of advanced telecommunications services and products is the business user domain. Most services have seen their origins in the business arena with a 'trickle down' effect into the domestic market. This has been true of the PC, fax machine, and videotelephony services which all have, or intend to have, a foothold in industry and commerce, but long term aims of establishing a market for the domestic user. The result of this is that products and services are primarily designed for the business user, and the domestic user is faced with products that can either fail to meet exactly their requirements, or are often unusable. In addition, business users can often pay for installation, training and technical support, whilst domestic users are increasingly left to work many problems out for themselves. It is clear that domestic users of systems have different needs and requirements, but designers often over estimate people's abilities to adapt to new technology.

The RACE[1] II LUSI (Likeable and Usable Service Interfaces) project has investigated the use of existing and future telecommunications services by the general public. The project has an underlying premise that the requirements of business users in the telecommunications marketplace have been extensively dealt with, however, the particular needs of members of the general public are often overlooked (Hewson and Clarke 1993). The project is conducting research which will lead to the production of a set of guidelines for designers of future systems, taking into account the particular requirements of the more naive and inexperienced users.

The research conducted by the LUSI project has been augmented by studies from other RACE pilot projects, which are also addressing the design of systems for inexperienced users. All of the research so far has suggested that individuals are less inclined to embrace a new technology if the usability of the product or service has not been properly addressed.

2 Background

Within the RACE programme there are a number of pilot experiments which are providing services for naive users and they have all encountered the same kinds of problems. The LUSI project now has nearly three years experience of evaluating future telecommunications services with the general public, and as part of the project, has used this experience to assist in the evaluation of these other services in different domains. All of these projects have one thing in common; that they are providing services to customers with no experience of advanced telecommunications and in some cases no knowledge of computers.

One of the key battles being fought by all projects is in widening the audience of new technology by concentrating on the issues that affect acceptability by inexperienced users. That is, those people who come from technophobic backgrounds. Many technological advances will not succeed in proliferating into our everyday lives, because the requirements of inexperienced users have not been considered.

Rubin (1994) identified four reasons why unusable systems are designed;

- During product development the emphasis and focus have been on the machine or system, not on the person who is the ultimate end user.

---

[1] Research and Technology Development in Advanced Communications Technologies in Europe

- As technology has penetrated the mainstream consumer market, the target audience has changed and continues to change dramatically. Development organisations have been slow to react to this change.

- The design of usable systems is a difficult, unpredictable endeavour, yet many organisations treat it as if it were just "common sense."

- Organisations employ very specialised teams and approaches to product and system development, yet fail to integrate them with each other.

The LUSI project has witnessed all of these problems with a variety of manufacturers; after evaluation of fourteen existing telecommunications services the project concluded that the vast majority of usability problems stemmed from a disregard of well establish human factors design guidelines (Hewson et al, 1995). After discussions with representatives of the AREA and ESSAI projects, the problems identified by Rubin were evident within other RACE projects.

Many models of usability have been published over the years, the best of these, when addressing utility, account for accessibility issues as well as an individual's need for a service. Nielsen proposed the following model for system acceptability.

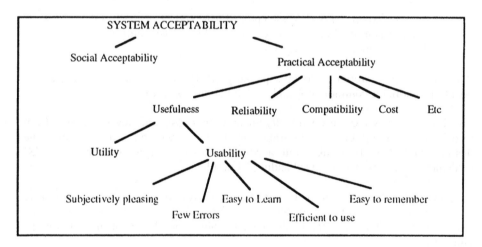

Fig. 1 Categories of System Acceptability (after Nielsen, 1990)

It is the factors that are listed under practical acceptability that are of importance in providing services to the general public, particularly with these two case studies. The most important category within practical acceptability being that of usefulness. Usefulness is the degree to which a system can be used to achieve a desired goal. It can be broken down into two categories of utility and usability, where utility is concerned with whether the functionality of the system can do what is needed, and usability is the question of how well users can use that functionality.

3 Overview of Projects

3.1 The ESSAI Service

The ESSAI project has developed a touchscreen multimedia shopping kiosk, offering tour operator ticket sales with credit card transactions. For the testing period the kiosk was placed in a shopping centre in Switzerland, outside a major Swiss store. The system is designed to be used intuitively, as a 'walk up and use' service. The system included a presence detector, which triggered the start screen, as well as starting an audio sequence in order to attract users. There was also a print facility, which allowed details of tickets and information to be printed and taken away by the user.

3.2 The AREA Service

The AREA project is developing advanced telematic services to support agricultural professionals in the rural areas around the Mediterranean. There are fundamentally three services which make up the AREA system.

- Remote Expert Service (Made up of SEMM and TELES.VISION)
- Remote Training Services (Made up of CBT and Teletutoring)
- AREA Information Services (ARIS)

All of the services were developed separately, presenting the AREA project with a goal of providing a 'common look and feel' to the three systems.

The design of the screens is continually undergoing redesign in view of findings from evaluations. A trial in Greece identified many problems with the layout and dialogue design, which have been taken into account. Trials in Bari, attended by the LUSI evaluator, were expected to reveal additional problems.

The services all run on a PC, but use different applications including TELES.VISION, Mosaic (using the World Wide Web) and SEMM, a dedicated expert system providing a knowledge base for phytopathologies of olive and grape cultivation.

TELES.VISION is a videoconferencing application running on a PC. It runs under Windows™ and as such conforms to Windows™ conventions. The TELES system has been the focus of some research on the LUSI project, where it was evaluated by members of the general public. In summary these results suggested that members of the general public with no experience of using Windows™ and other similar operating systems struggled to use the system without a considerable amount of training and regular usage. The LUSI project was intrigued to find out how the system would perform within the AREA project as they also focused on inexperienced computer users as their subject population, in this case farmers.

Mosaic is an application which enables access to the Internet (or more specifically the World Wide Web). It is relatively simple to create and amend pages on this system, each page is then given a Uniform (or Universal) Resource Locator (URL). This is a unique address point, which can be navigated to from other pages using hypertext links. The World Wide Web is an ideal medium for presenting new information which needs updating on a regular basis.

4 Data Collection

Different data collection methods were employed by the three different projects;

The LUSI project conducted a number of usability evaluations in a laboratory with members of the general public in order to assess which particular design features of a system might contribute to a problematic interface for naive users. The trials collected objective performance data, but most importantly in this case, the trials elicited the subjective data assessing the naive users' first impressions of the service.

The trials in Italy with the rural farming community used the SUMI (Software Usability Measurement Inventory) questionnaire developed by the ESPRIT MUSiC project. This questionnaire gives subjective data in categories of likeability, learnability, support, control, and global satisfaction. These are the indicators which give a guide to the acceptance of this service. The questionnaire asks questions such as 'would you recommend this system to your colleagues?' and 'did you enjoy your sessions with the system?.' The results using this type of subjective metric were not encouraging to the developers, and suggested that because of the poor usability of the system, the would be a high chance of rejection of the service.

The ESSAI trials in Switzerland measured subjective data using questionnaires. Again the results suggested that usability of the system would be a primary factor in determining whether the system would be widely used.

5 The Field Trials

5.1 The ESSAI trials

The evaluations took place between April and May, 1995, in Basel in Switzerland. User sessions with the kiosk were filmed, and a log of all button presses was recorded. Users who explored the system beyond the introductory screen were asked to complete a questionnaire, aimed as establishing the level of acceptance of the service. In all 23 people completed the questionnaire, covering a fairly even spread of age and experience with public information kiosks of this nature.

May (1993) identified a number of unique characteristics of interactive public information systems. Although they can be technologically simple, they have unique characteristics which require special consideration.

- Public information kiosks are walk up and use systems, and so there can be no initial training of potential users.
- No selection of users is possible as the entire population are potential users.
- There is no compulsion to use a system, so there must be a perceived benefit
- Motivation to use may not be high
- Users may require specific goal oriented information or more general information
- The surrounding environment may be noisy and public
- Local experts are not available

The ESSAI trials had mixed success. There were a few usability problems with the system, e.g. glare on the screen at some hours of the day which made parts of the screen unreadable. Some users had a disturbing feeling of being watched whilst using the terminal. However, the user interface performed well, with all of the screens being readable and easy to understand, and all of the auditory outputs were heard clearly.

There was a fairly positive response to the level of help given by the system as well as the ease of navigation through the system, particularly when compared to older iterations from the ESSAI project. Most importantly there was a very positive response to the question regarding future use of the system whereby over 90% of users said that they would use the system again in the future. Crucial to people's impressions of the system were the levels of concurrence with the following two questions;

'I think that the ESSAI system is a good idea'

'I enjoyed using the ESSAI system today'

These two questions reflect the level of utility and usability that contribute to the overall system acceptability discussed earlier.

5.2 The AREA trials

The evaluations took place in October 1994, in Bari, Italy. Approximately 8-10 subjects were used which were drawn from farms in the surrounding area. All of the subjects were viniculturists, and most had only limited experience with computer technology.

It was clear from the evaluations that the test subjects were indeed very inexperienced users, some had never used a mouse before, and a few had never used any kind of computer equipment. It was little surprise therefore that the evaluations took more the form of a guided walk through the system, and there were very few recommendations regarding interface design.

There is a danger when exposing inexperienced users to a complex system that they will regard the system in a favourable light, as the evaluations show services which appear to provide them with impressive new features, which they feel would help them considerably. However one must guard against taking too much heart from a trial of this nature because it will often be the case that although the system looks impressive when demonstrated, the learning required for the user to realise the system's full potential is often inhibitive.

The disappointment experienced in not being able to make full use of the potential of the system can lead to dissatisfaction and ultimately rejection of the service. It is therefore imperative to acknowledge the interface design and training issues in the design process, before the launch of the service.

6 Usability Issues

The overriding factor affecting the uptake of the two services is that most of the target customers are not aware that they need the service, in fact, some may even be resistant to the service. This is to some extent due to the fact that they are not technology literate, and in some circumstances they are technophobic.

The most startling finding from these evaluations was that the first impressions of the ease of use of a new service is absolutely crucial in the acceptability of that service.

The provision of advanced telecommunications systems for members of the general public will face many of the problems well documented with providing IT to naive users. Experiences of past IT implementations into the home have led to an all too common technophobia. The proliferation of PC's into the home is still met with some scepticism as the genuine usefulness of PC's in the domestic market is still not clear. The only true successes of IT proliferation in the home is with products that are either very easy to use (in the case of the television set) or provide a useful service (as with the VCR).

It is now difficult to avoid using advanced technology products and services; many products of the IT age have successfully become a part of our everyday lives without encountering the resistance to new technology by naive users. This is because most people would not consciously identify themselves as users of IT. They have a latent consumption of IT whilst still regarding computers and microchips as part of a completely different culture. Most people do recognise that common electronic products do have microchips in them, and are not intimidated by them, because they are inherent within an ostensibly simple machine.

A way of understanding why people differ in their degree to which they are attracted to IT systems is to analyse the choice in terms of 'value.' People will use IT systems as long as they get, or anticipate, value from the systems. The 'value' attributed to a system can be said to result from an implicit process of cost-benefit analysis. If the benefits outweigh the costs, then the system will be perceived as having a positive value.

Three types of conscious users of IT have been identified (Frude 1987); experts, workers and hobbyists. 'Experts' would be expected to place a high positive value on IT because they primarily have an interest in the products and services and also gain numerous intellectual, professional and financial rewards from working with advanced IT systems. Allied to this their effort costs are likely to be low.

Many workers may not share the fundamental enthusiasm for IT as the experts and do not operate IT systems through choice, but anticipate that use of such systems may bring financial and professional benefits, not least a enhanced career progression. IT may be seen as a useful means of achieving a worthwhile end.

Hobbyists interaction with IT systems can be an end in itself. They may develop uses for systems which do not necessarily maximise efficiency but are seen as rewarding in their own right e.g. building a database of recipes.

All of these individuals are prepared to invest some effort in interacting with IT systems, however these people are still in the minority. Most people are prepared to reap the benefits of IT systems as long as it does not require any effort on their part. When special skills are required to operate a system this brings a cost which, in the absence of apparent benefits, leads them to evaluate the system negatively.

It is often useful to look at failures in providing IT services to the general public, one of the most startling of which was Prestel in the UK. The service simply did not offer anything that was cost effective over other potential sources of the same information. The uptake of Prestel by members of the general public has been minimal. To some extent, this could be attributed to the financial costs of the system, and the skills necessary to access data, but it was its lack of usefulness and usability that caused the demise of the service. Future systems, if they are to learn from the mistakes of Prestel, will need to be easier to operate and more useful.

7   Designing Systems for Inexperienced Users

It was clear from the trials that the users were not in any way experienced with computer equipment. This raises a number of questions, and points for consideration by the designers of this system. These questions also raise matters of concern for the LUSI project.

Many systems designed for inexperienced users are developed with the aim of being a 'walk up and use' system. For this to work successfully there are a number of special factors that need to be considered. Maguire (1981) identified the following;
  • Consider first impressions

  • Devise simple, concise and unambiguous output messages

  • Provide quick access

  • Build in all desired retrieval paths

  • Ask appropriate questions

  • Avoid unnecessary clarification

  • Provide facilities for memory refreshment

  • Suggest the causes of errors

A complementary approach to ensuring the use of a system by inexperienced users is to implement training procedures. The relationship of training to interface design is represented in the following diagram.

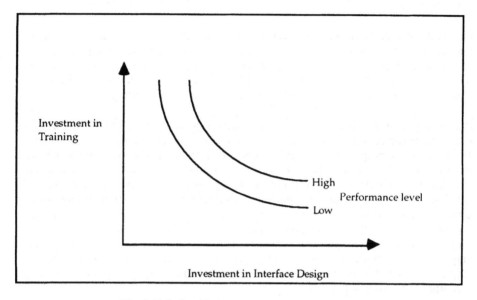

Fig. 2 Relationship between training and usability

The graph shows that in order to attain a required level of performance a level of user training is required or an increased effort in user interface design is necessary. For some systems it is likely that however intuitive the interface design is, a certain level of training will always be required. For services such as the AREA system this training may be as basic as mouse usage, or an introduction to Windows™, as long as the appropriate investment is made in the development of the user interface.

What the graph also shows is that for optimum performance investment is required in both interface design and training. One has to conclude that for very complex systems an element of training will always be required.

8 Guidelines for Service Design

The issue of what end users need from a new service is rarely dealt with. It is very difficult to be prescriptive about specific usability issues whilst a interface is still in its conceptual stage, however many general guidelines exists for designers of systems, to facilitate good design practice. These usability aspects of an interface are well documented, publications from Shneiderman, Laurel, Nielsen, Smith and Mosier etc. go into great detail about how a user interface should be presented, and provide guidelines for good user interface design.

A frequently heralded set of design principles are the 9 usability heuristics of Nielsen and Molich; use a simple and natural dialogue, speak the user's language, minimise the user's memory load, be consistent, provide feedback, provide clearly marked exits, provide shortcuts, good error messages, prevent errors (or have generous error tolerance). Further to these should be added make the service suitable for the task, make it self descriptive, allow user control, conform with user expectations, enable individualisation.

Guidelines can be more specific, but this leads to hundreds of pages of design advice like that offered by Smith and Mosier, which goes into details about appropriate use of colour, text, navigation issues etc. With this plethora of usability design advice, with more being published almost by the day, the interface design issues seem now to have become so well established, that there must be more impending issues.

Recently a great deal has been written about the role of human factors and usability issues in the future of telecommunications In a paper delivered at the HFT conference 1995, Schwartz proposed that usability has now become a secondary issue and that the most important issue to be addressed now is that of accessibility. He stated that "We know of no evidence that usability plays a primary role in most cases of product selection...We have concluded that easy accessibility to a particular product or service seems often to outweigh usability as a choice factor." The paper then proposes a shift in the role of human factors to concentrate on those other factors such as accessibility and aesthetics, rather than the traditional ease of use issues. This may have some truth in it, but when providing a service to naive users who are sceptical as to whether they need a service at all, usability is crucial in creating favourable first impressions.

9   The Future of Telecommunications Services

The ease of use of a system, for naive users, will be a major influence in forming first impressions of likeability of a service. Whilst participating in the three different sets of trials it was clear that designers have very different priorities when developing systems. They are often faced with demanding technical problems, and feel that the human factors practitioner is simply a critic of the system. Usability is often overlooked by system developers with a common reaction being 'We'll concentrated on releasing version 1.0, and worry about usability in the next version." This is a very damaging misconception, as a person's first impressions of ease of use is the single most critical factor in the acceptance of many advanced telecommunications systems. This attitude will serve to alienate consumers from the product, and contribute to the phenomenon on technophobia. It is of particular importance when providing a service when the consumers need of the service is under question. All prospective users will conduct their own cost benefit analysis of a new system, and if the benefits are not clear, then the cost of frustration, anxiety, and depression caused by poorly deigned systems will be enough to cause a rejection of the service.

It is very common for the development team to be excited about the capabilities of a new service, and in appreciating the functionality and the potential of the service themselves assume that the general public will respond favourably to it. If however the target customers are technologically naive, a poorly designed system will only serve to exacerbate the negative feelings towards new technology, and that after one failed attempt at trying to achieve a goal with the system, the service will be rejected.

10 Acknowledgements

Particular thanks in the preparation of this paper must go to the members of the AREA project team, Revital Marom and Mirco Bolzoni, and to Daniel Felix of the ESSAI project.

11 References

[Brown,88]   Brown, C.M. "Human Computer Interface Design Guidelines" Norwood, NJ: Ablex 1988

[Frude,87]   Frude, N. "Information Technology in the Home: Promises as yet unrealised" in 'Information Technology and People' ed Blackler, F and Oborne, D. British Psychological Society, 1987.

[Hewson,95]   Hewson, T.J., Allison, G., Clarke, A.M., "Evaluation of Telecommunications Services and Terminals: A Heuristic Questionnaire Protocol versus Empirical User Testing," Proceedings of the 15th International Symposium on Human Factors in Telecommunications, 1995.

[Hewson,93]   Hewson, T.J. and Clarke, A.M., "Overview of the RACE II LUSI Project", Proceedings of the 2nd International Conference, Spaniol, O. and Williams, F. (eds), North-Holland, Broadband Islands: Towards Integration, Athens, Greece, 14-16 June, 1993, pp 267-270, ISBN 0 444 81710 7.

[Laurel,90]   Laurel, B. "The art of Human Computer Interface Design" Addison Wesley, 1990.

[LUSI,93]   LUSI Project Deliverable No. 11 " Report on Results of the User Trials Across the User Group Panels." CEC Del. No. R2092/HUS/-/DS/P/011/b1.

[Maguire,81]   Maguire, M.C., "A Study of the Problems of Man-Computer Dialogues for Naive Users" PhD Thesis, Leicester Polytechnic, 1981.

[May,93]   May, A.J. "Development of Generic User Requirements for Interactive Public Information Systems," MSc Ergonomics Thesis, University of London, 1993.

[Nielsen,90]   Nielsen, J., "Hypertext and Hypermedia" Academic Press Inc, 1990.

[Nielsen,90]   Nielsen, J., and Molich, R. "Heuristic evaluation of user interfaces" In proceedings of CHI '90 ed. Chew, J.C., and Whiteside, J., Addison Wesley.

[Rubin,94]   Rubin, J. "Handbook of Usability Testing" Wiley and Sons. 1994.

[Rubin,88]   Rubin, T. "User Interface Design for Computer Systems", Chichester, England, Ellis Horwood, 1988

[Schwartz,95]   Schwartz, B., Salasoo, A., Egan D., McConkie, A. "Usability Vs Accessibility: "Easy to Get" is more important than "Easy to Use", Proceedings of the 15th International Symposium on Human Factors in Telecommunications, 1995.

[Shneiderman,87] Shneiderman, B. "Designing the User Interface" Addison-Wesley. 1987.

[Smith,86]   Smith, S. & Mosier, J. "Guidelines for Designing User Interface Software" Technical report ESDE-TR-86-278, 1986, Hanscom AFB, MA: USAF Electronic Systems Division.

Usability an Effective Methodology for Designing Services in the Agricultural Sector

R. Marom (Theseus Institute, Nice - France; marom@theseus.fr)

M. Bolzoni (University of Padova, Italy; psbolzo@ipdunivx.unipd.it)

M. Goldberg (Theseus Institute, Nice - France; goldberg@theseus.fr)

D. Rotondi (Tecnopolis CSATA N.O., Bari - Italy; dom@mvx36.csata.it)

Abstract. The adoption of technology in co-operative working environments deeply affects social and cultural practices. This is particularly the case in environments which traditionally show much resistance to innovation. These are issues which can influence systems appreciation and evaluation of performance in an important way. The AREA project has been concerned with the design and testing of a multimedia distributed service able to provide remote access to expertise (experts systems or human experts), training facilities and distributed information in Mediterranean agricultural communities. To anticipate potential drawbacks, a combination of usability, usability- in-context, and scenario building has been used to ensure the compatibility of the system with social and cultural working practices and activities of an environment characteristic of rural agricultural regions of southern Europe. This paper demonstrates that such an approach can yield important benefits in design and development.

1 Introduction

In order to react effectively and immediately to market needs, the agricultural sector requires a closer co-operation between the different players: farmers' co-operatives, research institutes, and educational agencies. By focusing on improving the management of information through a system of efficient knowledge updates, such co-operation could be strengthened. The study of plant pathology (Phythopathology) is an agricultural field which is particularly influenced by the rapid changes in chemical products, crop protection procedures, agriculture policies and health/ environment legislation, and is thus prone to having outdated information sources. The use of advanced telecommunication and IT technologies can help facilitate access to current information and knowledge resources, yielding the following overall benefits: improved crop quality and increased level of expertise. The objective of the AREA project is to design and test a multimedia distributed service able to provide remote access to expertise (experts systems or human experts), training facilities and distributed information for Mediterranean agricultural communities.

Advanced communication technology, such as the AREA services, cannot simply be inserted in the existing agricultural environment without taking into consideration the specific aspects of the context. For instance, the potential impact of the AREA services on the everyday life of the users, e.g., farmers, technicians and experts from remote agricultural areas in Greece and Italy, is considerable. Agriculture provides a particularly interesting case study as, it is uniquely characterised by being based in rural and traditional societies, while, at the same time needing to be oriented towards the highly competitive markets which require constant technological innovation. Existing social and cultural practices in rural agriculture has the potential to be significantly influenced by the introduction of new technology, although such a strongly homogenous environment will certainly display important levels of resistance to innovation and change. It became clear from the beginning of the development of the AREA project that the services provided would attempt to both complement an existing social and cultural reality by supporting established relationships while also creating new forms of exchanges between participants.

A particular issue in the development of the AREA services was its advanced typology which offered little reference points in the everyday experiences of the users. As a consequence it has been difficult to expect users to make useful references to previous experiences as an aid for using the new system. The AREA services also makes use of various applications utilising quite different user interfaces. This entails added effort on the part of the users who may already have little sophisticated knowledge and experience using computers and application programs.

To ensure that these services are efficiently and effectively developed, implemented and deployed we have been systematically conducting evaluation studies, and this from the beginning of the project. The formal model we use for representing the development and deployment of technology is illustrated in Figure 1 below.

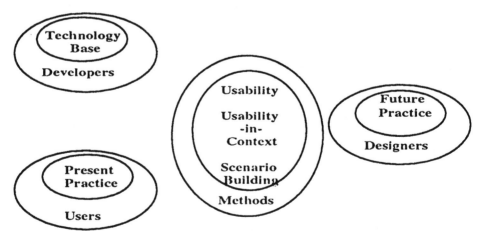

Fig. 1 The four components of the model for design and development

There are four components to this model: present day practice- i.e. user real world experiences; the technology base available for deployment as perceived by the engineers; the future landscape conceived by the system designers where technology has changed current practice; and finally a formal methodology which combines these three components. Our formal methodology is based upon a combination of three complementary tools: usability, usability in context, and scenario building[Goldberg]. Usability is concerned with such questions as how easy is it to use the technology and focuses on the design of the user interface [ISSUES 94]. Usability-in-context focuses on issues related to the use of the technology as part of the work process and addresses issues such as workflow and efficiency [Bevan,94]. Scenario building concentrates on defining new services and processes which alter the character and organisation of work and learning. In practice this formal methodology helps to create a bridge between the designers of the system, the developers of the technology and the users. This bridge can ensure that the technology deployed not only, complies to usability criteria and solves present day problems, but also supports new organisational practices.

The paper is organised as follows: we first present a succinct overview of the AREA project, describing the technology, the services and the partners. The next section concentrates on the different prototypes used and the user-based and expert-based evaluation tools employed during the design and development process. Finally, the six studies we carried out are then described and the main results are detailed and demonstrate how our formal methodology improved the overall design and development process.

2 Overview Of The Area Services

The aim of AREA was to develop a communications/information network to provide three classes of services: *Local/Remote Expertise Access, Specialised Training Services,* a *Distributed Information service (ARIS)*. These are built upon Multimedia, Videoconferencing and Workspace Sharing Facilities.

The AREA service communication infrastructure is based on a star-shaped model with a central site, called the *AREA HUB,* and a set of peripheral sites, called *AREA Service Access Points.* The Hub is the centre where the AREA services are maintained and provided. The Access Points are the sites, typically far from the Hub, where the AREA services are provided to the end-users (farmers, technicians and experts). The Hub and Access Points are integrated through a communication infrastructure which allows the connectivity and the transportation of information between the users' workstation and the computer systems which host the applications. There are 7 sites (Hubs and Access Points) across rural areas in the south of Italy and Greece.

The *Local/Remote Expertise Access* consists of the following services:

- A Multimedia Expert System (SEMM), able to provide descriptions of phytopathologies cases. The tool is used to perform or diagnose pest management procedures.
- Off-site human expert consultation, and additional resort for phytopathology diagnosis and appropriate actions identification.

The service consists of a two-way, real time video/voice and data communication facility using ISDN and satellite links as well as support services for booking experts, agenda manager and report exchange offered by the distributed information services.

The Specialised Training Services includes:

- Usage of ad-hoc courseware packages. The Computer Based Training packages (CBTs) are made available (via network), so that end-users are free to select and use them according to their needs.
- Real time remote training based on multi-point video telephony, screen sharing, and conversation management facilities.

Other services such as checking availability of CBTs, teleseminars or teleconsultation, ordering CBT, registering to teleteaching courses are provided via the Area information distribution service.

The *ARIS Information Service* is based on the World Wide Web and distributes specialised information such as newsletters, announcements, weather condition, EU regulations, seminars announcements, and other useful information in agriculture. The information services supports both the diffusion and collection of information across all sites.

3 Methodology

Our objective when designing the AREA system was to take into account the potential social and cultural impact on the users without neglecting technical performance. Our strategy has been the adoption of a usability -based approach, requiring the development of a set of services geared as much as possible to the needs of a particular agricultural environment, thus allowing for ease of cognitive effort in its use and producing high performance results. The AREA system has thus been designed to be easily understandable, easy to use, affordable, approachable and somehow referable to the previous experience of its users. By doing so we would like to achieve the following:

- increase the acceptance of the AREA services by end-user (reduce resistance),
- maximise the effectiveness of the user's performance (reduce the cognitive effort),

- reduce training and support costs (error prevention),
- reduce costly system revision following implementation.

3.1 Dynamic System Design

We used an iterative design methodology for the development of the AREA services. According to this perspective the design process is seen as a dynamic procedure whereby different modules are revised, redesigned and modified according to the feedback received in the usability trials. We focused on two types of data: social (the impact of the system on the actual practices of users) and cognitive ergonomics (i.e., screen layout, button positions, interface structure etc.). Both types of information informed the system designers as to the development status of the system from the users' perspective.

3.2 Combination of Usability Tools Techniques

To meet our objectives in terms of social and cognitive compatibility of the AREA system, a "combination" of usability tools have been adopted and applied interchangeably according to the development status of the prototype and according to the results of the previous usability tests conducted by the AREA evaluation team.

In order to preserve the ecological validity of the testing we put an emphasis on the elements that would approximate the trials' scene to an operative real environment. Instead of considering the contextual influences as a threat for system functionality, they have been conceptualised as useful and valid resources for feeding designers with pertinent information. On-site testing provides us with a credible real context of work embedded in the proper social and cultural environment. The users are immersed in the everyday local experience and a realistic operational setting is created.

To meet the complexities of the system and the study, different types of evaluation tools were utilised. Tools were used in accordance with their intrinsic characteristics (i.e. qualitative or quantitative) and the needs stated in the specific development phase of the project. The tools captured data on two different levels: "Pre-Post interaction" (user's accounts) and "while interacting" (direct observation). Pre-Post interaction information is gathered on the attitudes, expectations, previous experiences with computers, and related aspects of their everyday work practices prior to the usage of the AREA system. While interacting, refers to the use of the AREA services in the work context of the users involving issues such as task accomplishment and users' satisfaction.

Different prototypes were used throughout the design cycle to elicit this information: *PMU (Paper Mock-Ups), Scripted Work Simulation and Formal Operation.*

Paper mock-ups

This kind of prototype involves creating a paper and pencil mock-up of different parts of the product as a basis for discussion. The prototype represents information screens, windows, message structures etc. We used this approach early during the formative stage of the design for brain storming and discussion on broad design features. We selected this prototype because it is less threatening for the users, very fast to construct, and even faster to make the changes. New ideas are not inhibited by the use of the technology thus, less mistakes are prone to be made.

Work simulation

This kind of prototype consist of a computer-based representation of the different AREA services possessing sufficient functionality to allow users to 'play with' and explore its facilities in a flexible way. We used this approach for simple testing and for product specification. We selected this technique as it is very cost effective and can give significant advantages for comparatively little investment of resources. This prototype also allows the users to interact with the 'system' and obtain a feel for how the intended product will operate. It also allows us to discover some of the difficulties which the end-users might encounter in a real working situation.

Operational service

This kind of prototype is used for a fully developed operational service in the users' work context. This form of prototyping is used once the overall design has taken shape, in order to test the product specification and usability. It placed in the real work context of the users.

3.3   Usability Evaluation

Two approaches to usability evaluation were adapted by the AREA project: a user-based approach and an expert-based approach. The user-based approach consists of assessing user performance, behaviour, knowledge, attitude, stress, physical and mental effort. It uses a variety of methods and measures to elicit information which relates directly to the users' current interaction with the system or service under evaluation. The expert-based approach involves the system matching predefined design criteria.

Methods for expert based approach

The LUSI project [LUSI,92] is involved in the expert evaluation for the AREA services using the QUEST tool (Questionnaire for the Evaluation of Services and Terminal). A human factors practitioner from the LUSI project conducted a "walkthrough' of the AREA services in the various stages of development. Aspects such as control, consistency, feedback were identified for potential usability problems[LUSI,90] .

Methods for user-based evaluation

Objective and subjective data were collected. Objective data refers to performance where the feeling and beliefs of the user have no observable influence. Thus, Task Completion, User Error and Usage of Facilities are quantitatively measured.

Subjective Measures. We used subjective measures to detect the user's understanding of the system, satisfaction and the user's subjective opinion while employing the services. Qualitative data was collected using this approach. The following are the different tools used for the user-based evaluation;

- **On line evaluation tool:** The on-line evaluation tool was developed by the evaluation team of the AREA project. The tool is made of a software package for deriving the metrics and statistical information on the usage pattern of the AREA services and their effectiveness.

- **Non-participant observation:** Users were observed in real time noting pertinent aspects of the interaction. Recording of user comments when interacting with the system provided us with information on issues concerning the users.

- **Thinking aloud:** We used this technique to encourage the users to explain what and why they are performing on a specific action. This provides us with an insight into the users' strategies while interacting with the AREA services.

- **SUMI (Software Usability Measurement Inventory):** The SUMI questionnaire[SUMI] is a tool which identifies performance of a system through a formal evaluation of user's satisfaction over the artefact. SUMI was modified to the specification of the AREA system. It provides us with a quantitative measurement of a user's subjective opinion.

- **Questionnaires and interviews:** These tools were used to collect data on the users understanding and satisfaction with the system. These provided us with the information necessary to interpret quantitative measures such as error rates or task completion times. Two questionnaire were developed by the AREA team and used in the form of semistructured interviews. One set focused on the videoconferencing-related services and the others on the rest. The former focused on the interaction aspect and ease of use of the service while the latter is based on the QUEST questionnaire and focused on the degree of user support (navigation, interface feature and overall impression of the system.

Evaluation Tools	Information Target	Data Type	Informed Design
Think-Aloud (TAT)	Insight on user's cognitive activity	Qualitative	System image, errors, misleading invitations
Non Participant Observation (NPO)	Behavioural evaluation of user's activity	Qualitative	Design rationale, navigation, errors
Walk-Through (WT) (Participant Observation)	General overview of system achievement and interacting skills	Qualitative	Procedures, action sequences, coherence and consistency
User Comments in the interaction (UCI)	Underlinings and remarks in the actual use of the system	Qualitative	User's perception of tasks and system's activities
On-Line Evaluation Tool	Time, Errors, Navigation, Preferred Options	Quantitative	Recording of patterns of interaction
User's Free Exploration	Interaction Momentum	Qualitative	Cognitive insight on preferred pathways
Semi-Structured Interview	Social Impact Cultural Impact Ergonomic Impact	Qualitative	Acceptance Compatibility Usability
SUMI	User's attitude over the artefact in use with a formalised structure	Quantitative	System performance and satisfaction
QUEST	Service overview, equipment, service usage.	Quantitative	Qualities of the service and design features
User Comments Post-Interaction (UCIPI)	Focal points of the interaction as pointed out by users	Qualitative	User's impressions after interaction on features and performance
Usability in Context	Attitude of actors and roles involved in the project	Qualitative	Ecological system evaluation

Table 1 Tools and techniques adopted in the AREA project. (Evaluation tools and targeted information)

- **Usability in context tool:** This qualitative tool was used as a guideline to conduct interviews with a large number of users in the work's context. Based on MITS questionnaire (R2004) this tool provided us with the data which portrayed the progressive changes in the work and interaction patterns of all AREA users throughout the development, experimentation and implementation phases of the system.

Tools and Situation Grid	Prototypes	Evaluation Tool
Direct Observation (Ongoing Interaction) Obtrusive methods	Paper Mock-Ups Working Simulation Operational Simulation	Think Aloud Technique (TAT), Non Participant Observation (NPO), Participant Observation (WT) User Comments (UCI) Free Exploration On-Line Evaluation Tool (*)
User Accounts (Before and after interaction) Unobtrusive methods	None (submission after and/or interaction)	Semi-Structured Interview SUMI QUEST Usability in Context Tool User Comments (UCPI) Questionnaires

Table 2 Application of methods according to the work simulation and timing of the interaction, (*) is an unobtrusive technique active during the interaction.

4 Studies And Results

4.1 Studies

The six studies were completed or planned as follows:

Study 1
The first usability evaluation took place in September in Greece. Fifteen farmers participated in the study, all were novice to the technology and of the AREA services. Paper Mock Ups were used as our initial prototyping during the formative stage of the design. It involved a series of drawings of key screens interfaces. The images represented different ways of presenting the information and structuring the interaction, i.e. information screen, icons message structure and command name. Each user was given a scenario to follow which was composed of four tasks followed by an instruction for each. The user had to accomplish the tasks under two conditions: once following the instruction in the scenario and once without the instruction.

Study 2
The second usability evaluation took place in October in Italy[Bolzoni,94]. Sixteen potential users were involved most having only limited experience with computer technology. Since the system was already partially developed Scripted Simulation

was used. The computer based demonstration highlighted aspects and problems of particular user interface features which were not easily demonstrated using paper-based methods (e.g. navigation, direct manipulation etc.). The user had to accomplish four tasks under two conditions: once following the instructions in the scenario and once without the instruction.

Trials Grid	Place	Time	Module	Technique	Subjects	Subjects #
Study 1	Greece	Sept. 94	Desktop	Paper Mock-Ups	Farmers	15
Study 2	Italy	Oct. 94	Partial service	Script Simulation	ATs	16
Study 3	Italy	Feb. 95	Partial service	Script Simulation	AT, S	36
Study 4	Greece	May 95	Desktop + Partial service	Script Simulation	Farmers	15
Study 5	Italy Greece	June 95	All AREAs	On-line Evaluation	AT, F, S	>200
Study 6	Italy	Sept. 95	All AREAs	Formal Evaluation	AT,F, S	15
Study 6	Greece	Sept. 95	All AREAs	Formal Evaluation	AT, S, E	16

Table 3 Schedule of the Usability trials. AT=Agriculture Technician, F=Farmers, S=Students. E=Experts.

Study 3

The third usability trial took place in February in Italy[Bolzoni.95] and scripted simulation was used again to test services that were not developed for the former trials. Thirty-six subjects took part in the study. Most subjects were familiar with the AREA services; they had either participated in former trials and/or undergone basic computer skills training provided by Technopolis. We followed the same scenario procedure for the testing as before.

Study 4

The fourth usability trial took place in Greece during the month of May and followed the same pattern and methodologies as the one in February in Italy. Fifteen future users with no computer experience and no familiarity with the service took part in the study.

Study 5

A full operational testing of the AREA will be carried out during the month of September in Italy. We will use this form of prototype once the overall design of the services has been completed. We will evaluate both the functionality and the usability of the product in a finished form in a real setting. All subjects will have been familiar with the services as they have participated in the previous trials, have been using the system on a regular basis and have undergone training. Each user is given a specific task to perform which covers all the services offered by AREA.

Study 6
In parallel to the usability studies we will conduct a six month evaluation of real usage of the service in all access points in Greece and in Italy. Each user of the service will fill out an on-line evaluation form after each usage of the system.

4.2 Results

By adapting the proposed usability approach (tools and methods) to the AREA system, we continuously interacted with the developers and relayed relevant information from the user trials. The following are short descriptions of a number of problematic pockets identified during the usability trials and the effect of the different usability methods in overcoming them. The quantitative assessment of instruments such as SUMI and QUEST gave important feed-back on the overall performance of the system in terms of user satisfaction. Ratings resulting from the testing are considered to be satisfactory as they pass the tolerance levels suggested by HUSAT. The users' positive reactions recorded by SUMI and QUEST point to pleasure in interacting with the new system but does not necessarily point to any other benefits such as information delivery or service effectiveness. Measuring performance does not offer any information on what is really wrong with the system; it does not provide us with the information on any specific flaws in the system and its interface. These aspects surface in the results of the qualitative analysis derived from the use of the TAT, NPO, WT, UCPI tools. These tools provided us with an insight to the users' difficulties while interacting with the system. The information collected was detailed enough to be used to modify and enhance the prototypes.

Changes to the appearance of the interface
On a micro level of interaction, throughout the usability studies designers were informed of the users' ability to understand different service interfaces. Throughout the development phase, guidelines for interface designs were created and were submitted to designers to guarantee an ultimate design of the interfaces. Ergonomic problems were discovered using the PMU, WT and Semi -Structured Interview tools. For instance, using these tools we revealed problems with the size, shape, position and colours of icons. Modifications have been applied so that an underlining logic will guide the user throughout the mental "mappings" of the system. This has avoided a sense of loss of control while interacting with different services as pointed out by users and observed by the usability team.

Changes to the cognitive properties of the interface
Formal methods and qualitative evaluations of the desktop and other AREA services offered valid material to the designer concerning important system error and vital suggestions for corrections which could affect the success of the prototypes in testing. The conceptualisation of the services emerged as a primary issue. Naming objects is defining to the users the world and the objects they are dealing with. The appropriateness of these definitions has a strong impact on how users perceive the services and abstract the objects they are interacting with. This aspect emerged and has been shown to be relevant in the choices of icons and metaphors for the AREA services. Icons are very important in the development of the interface as they partially free the system from being text dependent. Icons' representation, i.e. decoding the meaning of the figural aspects of the icons' image, were performed and relevant changes to the interfaces were applied. The utilisation of a specific questionnaire on icon representations overcame the problems of choice and memorability of icons that the users experienced, which need to be adapted to the cultural environment. Simple and familiar images with functional properties have been successfully implemented in the substitution of the previous cognitively "opaque" icons.

Moreover, we discovered through the usability evaluation that minimal syntactic aspects of the interface had an effect on the semantic understanding and use of the interface. Exit, Quit, Go Back features were used inconsistently driving users to confusion, with the risk of rendering unstable interaction if not losing data or elapsing the work session. This information surfaced during the testing of the working simulation prototypes. The utilisation of WT, Direct observation, UCI, and UCIPI tools assisted us in this process. The relevant information was relayed to the designers who applied the changes to the interfaces.

Changes to the Social and Cultural Properties of the interface
Social and cultural aspects of the interface are more visible than one might expect. These factors emerged as vital aspects for both the functionality and acceptance of the AREA services interfaces. Bringing AREA services' technology into a rural environment has a sizeable outcome in terms of interface design. It is evident that the choice of direct manipulation eliminates all the problems involved in the pragmatics of communication with computers (you don't have to input commands to operate the system, neither to wander through menus, but to choose and navigate between available visible options). Options were rendered very visible and "inviting to action" [Norman] by dedicating extra size and privileged position in the interface. Using the Usability in Context tool, appropriate metaphors linked to the working, cultural, social and everyday environment have been analysed and implemented in order to improve system accessibility. These needs fit the background of the users who have limited experience with computers and links the AREA services to existing artefacts of rural everyday life.

Nevertheless, the social and cultural factors not only influence the appearance of the system but also have a profound effect on its functionalities. An example can be taken from the development "Expert Agenda" service. Experts use the " agenda manager" to plan their activities. This computerised agenda is used to manage schedules and appointments with particular attention given to the teleconsultation appointments. Although the service was developed by designers and agricultural experts, the results from the working simulation prototype testing was that everyday situations were not considered in its design and the system could not cope with them. Using Semi structured interviews, WT, UCI, and QUEST we found that features such as changing of appointments were hardly negotiable as it required a complex series of operations which compelled the expert to lose the data provided by the farmers for the consultation meeting. Thus, the service proved not capable of coping with situations emerging from real usage. Information from the semi structured interviews demonstrated that the reconstruction of the service using the paper based agenda (filofax) as a model would have proven to meet user's needs and be more effective. Once again, recommendations were given to the developers who applied them while designing the operational simulation prototype.

Usability informed designers of cultural and social inconsistencies regarding everyday work and social practices. But the impact of the technology can reveal other aspects that usability can bring out. The results of the usability testing of the videoconferencing facilities (teleteaching and teleconsultation) using tools such as NPO, UCI, and UCIPI pointed out ergonomic difficulties in practical terms regarding the control of devices such as the camera, keyboard, phone handset, pen, paper, books and photos while conducting the sessions. Redesigning the teleconsultation/teleteaching work environment helped to solve those constraints. We further noticed that the technology did not improve the communication pattern between the participants or the quality of information exchange but it did reduce distance, improve accessibility, facilitate relationships, transactions and negotiations among all the actors. These aspects were taken into consideration by the designers in order to produce a videoconferencing environment which closely matches current rural practices.

A further example of the role of usability emerges from another teleteaching usability session. A set of experiments using a working simulation prototype was conducted in order to verify the usability of a system for didactic purposes. An accurate examination of the conditions and circumstances revealed that the poor interactive response and low usage of the artefact was not due to ergonomic or other flaws in the system but to the intimidating situation of the trials and to the normal conditions in which the educational school system in Italy has constrained participation of pupils in the classroom. A restricted approach on usability that would have ignored the social and cultural practices of the environment would have resulted in the prescription of useless, costly revisions of the system, with the risk of not detecting the real causes of the problem through an inappropriate methodology based only on a black-box trial and error approach.

5 Conclusion

In conclusion, the usability approach described in this paper provided all AREA's developers with information on the three different properties of the interface. Flexible and complementary methodologies applied at the different phases of the usability trials have proven to be effective in reducing the risks of system's crash, understanding and acceptance by the potential users. Benefits for the development of the system are not only sizeable in system performance but also in costs in terms of revisions and development of the application. Moreover, even if usability cannot assure in itself the full success of the system and cannot predict the outcome in terms of full scale deployment, it can certainly offer, from a short term perspective information and guidelines which can prevent immediate flaws and risks of refusal and re-orientate design in the avoidance of paradoxically unexpected and unwanted results of system use and implementation.

6 Acknowledgement

We would like to acknowledge the help of Tim Hewson, Rosita Mainieri and Lydia Goldberg in carrying out the evaluations for the AREA project. We would also like to thank all the users of the AREA services trials for their participation.

7 References

[Bevan,94]   Bevan N. and M. Macleod. "Usability Measurement in Context." *Behaviours and Information Technology.* 113: 1, 2, 1994: 132-145.

[Bolzoni,94]   Bolzoni M., AREA Project Usability Trials Bari October 1994, Preliminary Report, University of Padova, November, 1994.

[Bolzoni,95]   Bolzoni, M., AREA Project Usability Trials Bari February 1994, Preliminary Report, University of Padova, April 1994.

[Goldberg]   Goldberg M., R. Marom, I. Contardo, F. Corona, and L. Goldberg. "Project Results Evaluation Methodologies.", R2102/THE/000/DS/009, December, 1994.

[LUSI,90]   Hewson, T., "Report describing first collaborative venture between the LUSI project and the AREA project, HUSAT, 1994.

[ISSUES 94]   ISSUE R1065. "Usability Evaluation, Guidelines for IBC Service designers, Volume IV, Race 1994.

[SUMI]   Kirakowski, J. and M. Crobett. "SUMI: The software Usability Measurement Inventory." *British Journal of Educational Technology* 24: 3, 1993: 210-212.

[LUSI 92]   LUSI R2092 "Methodological Considerations of Usability Assessment. December, 1992.

[MUSIC 5429]   "Context Guideline Handbook Version 2.1.1., HUSAT NPL ESPRIT 5429, 1993. Neugebauer, C. and N. Spielmann. "The ergonomics lab: a practical approach." *Behaviours and Information Technology.* 113: 1, 2, 1994: 132-145.

[Norman]   Norman, D. *The design of everyday things.* New York: Basic Books, 1988.

Counting the Costs and Benefits of Metaphor

Chris Condon & Stephan Keuneke

Brunel University Centre for Research in Information Environments
Bremer Institut für Betriebstechnik und angewandte Arbeitswissenschaft
Chris.Condon@brunel.ac.uk   keu@biba.uni-bremen.de

Abstract. It has been demonstrated that the use of suitable metaphors in the user service interface can have a dramatic effect on the way in which the user perceives the services, depending on the category of metaphor chosen. An earlier paper by the authors postulated that interactional metaphors might lead to greater long term usage of services, as they direct the user to thinking about what the services are for, rather than thinking about the services themselves. To test this hypothesis, a model of usage was built, showing the likely impact of using different types of metaphor. The model showed that usage levels over time varied according to metaphor category. These usage patterns were then imposed on existing techno-economic models from specific industry sectors. Interactional metaphors led to the highest long term usage in most industry sectors, although the construction industry showed higher usage with spatial metaphors. In all sectors, inappropriate choice of metaphor would be sufficient to destroy the economic advanatages of advanced communications services.

1   Introduction

Experiments within the MITS project have demonstrated that the use of suitable metaphors in the user service interface can have a dramatic effect on the way in which the user perceives the services and whether they are likely to be used. Two factors were identified as critical to this. The first factor concerns the mapping between the vehicle (the real world artefact) and the service and is discussed by Anderson et al. [AND94] The second factor concerns the categorisation of metaphors and the manner in which the choice of metaphor category affects what the user is aware of.[CON94]

It is this second factor which is considered in this paper which seeks to assess its impact on the actual usage of new services. The changes in the user's perception are qualitative, rather than quantitative but, to understand their impact on usage, they must be translated into quantitative measurements. An earlier paper by the authors had postulated that interactional metaphors might be more successful, as they direct the user to thinking about what the services are for, rather than thinking about the services themselves.[CON94] To test this hypothesis, a model of usage was built, using CRIMP (CRoss IMPact analysis tool),[KRA94] showing the likely impacts of the different types of metaphor.

2 The CRIMP Model

2.1 Cross Impact Analysis

The CRIMP tool models the impacts of trends on other trends and on themselves. Once a model is built, the software steps through the changes in trends in each time frame. Depending on the confidence associated with the predicted impacts, a randomising factor is added to the model and the run-through is repeated a hundred or more times. Actual values are not used by the model as each factor is normalised into a range from $-\infty$ to $+\infty$ with the initial value set to 0. This makes the tool valuable for modelling subjective values, as in the model described below. The underlying principles and mathematics of CRIMP are described elsewhere.[GOR68], [KAN72], [HEL77], [DUI95]

2.1 Factors

Four factors were identified as having a potential impact on usage trends. The first of these summarises the various elements which lead to *resistance* to new technologies. It would be fatal for a company to assume that a new service is used as much and in the way it is supposed to from the day it is installed. Users have to be trained and old working habits have to be overcome, sometime even by those who do not directly work with the new service. There is evidence that even systems that are well accepted after installation show a decay of use in the long run.[HUT93] This factor had already been identified by URSA as a potential inhibitor:

"Human factor problems in the form of psychological resistance are often associated with process re-engineering or company re-organisation as it may be perceived by part of the management and the labour force as a threat to their position or to the control they exert. The old jobs consisted of specialists who did one task. The new case handlers perform a variety of tasks. Therefore people working on case handing process teams will find their work far different from the repetitious performance of one task to which they were accustomed."[SIN94]

At an earlier stage of the MITS project a number of experiments were carried out which identified critical factors in the choice and design of interface metaphors. Of these, the BIBA demonstrator showed a *qualitative* difference between the impact of metaphors on the use of advanced telecommunications services. A system of categorisation of metaphors based on three axes had been proposed:

> Spatial
> Activity-based
> Interactional

Particular metaphors usually embody varying aspects of each of these concerns and can be positioned in a three dimensional space with respect to these three axes. The classification, originally based on an extensive revision of the interface metaphor types identified by Hutchins,[HUT89] is further described in Anderson et al.[AND94] It should be noted that the classification deals with the underlying metaphor, not the medium in which it is presented. For example, spatial metaphors can be presented in verbal form, as in some text-based adventure games, "You are in a room. There are doors on your left and on your right. Stairs lead down." Experiments demonstrated that metaphors typifying these three axes influence the user in more than just *how* the user sees the functionality available; they also influence *what* the user sees.[CON94]

Activity-based interface metaphors can focus on differing levels of generality. For example, collaborative systems can be designed in terms of metaphors for specific tasks, such as project management, or can be generalised, such as 'agents'.[LAU90] Activity-based metaphors turn the user's attention to the functionality of the services offered and, by making the user more aware of the functionality at their disposal, tend to support the general *usability* of the system.

The spatial aspect of the metaphor is often considered as a means of providing a location for tools and methods of communicating and working. However, these spaces can also be more or less explicitly defined. So, it is possible to utilise stereotypical aspects of particular places in the design (e.g. libraries) or, more commonly, general properties of spaces (e.g. rooms).[CON90].[CON92] Spatial metaphors emphasise the interface itself, turning the user's attention towards the presentation of the services. By providing interfaces which attract the user, spatial metaphors encourage initial *likeability* and, it is suggested, are particularly good for new or novice users.

Concerning the interactional aspect, interfaces can support particular forms of communication (e.g. conventional e-mail), or provide less explicit spaces or opportunities for interaction through artefacts such as forms.[HAM91a] Interactional metaphors tend to turn the user's attention towards the activities which the services support and therefore the *relevance* to the users tasks, providing longer term motivation. Although this might appear to point to interactional metaphors as most suitable in the long term, arguments have been presented that a spatial metaphor such as the room is more suitable for real-time interaction in which the interaction itself is clear, whereas an interactional metaphor such as the form is better in supporting non real-time cooperation in which the interaction is less obvious.[HAM91b]

This gives us four trends for the model which impact on a fifth factor, *Usage*. The factors do not have an even effect on usage over time and are therefore filter through time series (see below). The interactions between the trends and time series are shown in the diagram below.

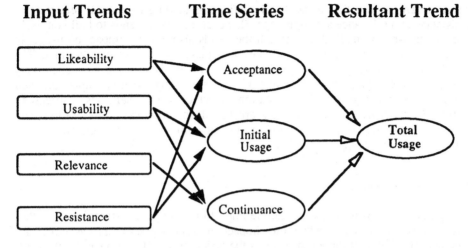

Fig. 1 Cross Impacts in the MITS model

This results in the following matrix defined for the CRIMP model:

	Likeability	Usability	Relevance	Resistance	Usage
Likeability				-0.1	S
Usability					S
Relevance				-0.5	S
Resistance				0.2	-2
Usage					1
PoorL	0				
PoorU		0			
PoorR			0		

The five trends defined above are shown across the top of the matrix as trends which are impacted *on*. These impacts come *from* the eight factors shown in the side column. These consist of the same five trends together with three actions: PoorL (poor likeability), PoorU (poor usability) and PoorR (poor relevance). These actions denote poor implementation of the interface in each of the three categories, but in the *a priori* model their impact is set to zero. Generally, the following is given as a guidance to the scale of cross impacts:[DUI94]

0.1	->	Small Impact
0.5	->	Medium Impact
1.0	->	Large Impact
2.0	->	Very Large Impact

Thus, it can be seen that likeability is defined as leading to a small reduction in resistance, and relevance a medium one. However, resistance tends to build on itself, with a small-medium sized impact. All the trends also directly impact on usage. In the cases of likeability, usability and relevance, these impacts change over time. The 'S's indicate time series which are described below. Resistance has a constant, very large, negative impact on usage, while usage has a strong, positive impact on itself (as more people use telecommunications services, their usefulness increases). Although the scale of these impacts are based on 'common sense' rather than empirical evidence, run-throughs of the model with varying impacts showed no significant difference in final results.

2.2   Time Series

In the cases of resistance and usage, the impact on usage is constant. For the other three trends, the manner in which they impact on usage was set as a time series with an emphasis on one of three time series:- the *acceptance* of the services by the user; the *initial usage* of the services; the longer-term *continuance* of use.

Service Acceptance

It is common to see some enthusiasts taking to new services immediately, with others taking a more cautious attitude. Although acceptance will be affected by all three trends, the dominant factor at the interface at this stage is likeability of the interface. A sufficiently likeable system will attract short term interest, even from users to whom it has no relevance and even if its usability is poor, c.f. the attraction of early VR (Virtual Reality) demonstrations. This can be expressed as a very strong initial impact in favour of using the system which gradually fades away:

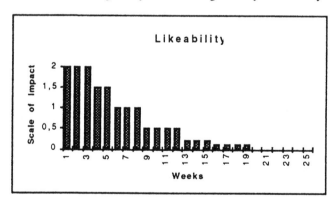

Fig. 2 Service Acceptance Time Series

All time series are shown over a period of 25 weeks, after which it is assumed that usage will settle down at a relatively constant level. The scale of impacts is the same as that given in the main CRIMP matrix, i.e. from 0.1 (weak) to 2.0 (very strong).

Initial Usage

Poor usability may lead to a lack of usage even though the initial likeability of the interface attracted the user's attention and the system is perceived as having high relevance to the user's tasks. Even with poor usability, some users will master the services. Thus, the impact of poor usability will not be immediate but will mainly be in the form of a certain percentage of new users giving up the system within a relatively short time, as shown in the graph below:

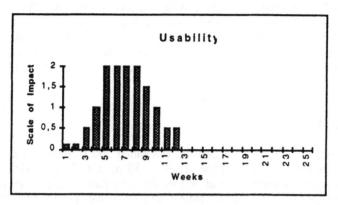

Fig. 3 Initial Usage Time Series

Service Continuance

As with likeability, the relevance of the services will affect the initial willingness of users to take up the service but it will also affect the users' long-term motivation. People who have started to use systems which have been designed to meet their known needs and have been implemented with high usability may still drift away from their use as poor motivation leads to a decreasing awareness of the relevance of the system to their work. This will affect existing users more strongly than those who have yet to start using the services and both will therefore encourage a fall-off of usage among existing users. It should be noted that it is not the actual relevance of the system which matters: it is the relevance perceived by the user.

The impact of the perceived relevance of the system will initially be very strong. As factors such as usability involve the user more in issues of 'how' rather than 'why' to use the system, the effect will fade away. Having mastered the usage of the system, the effects of motivation then come more clearly into play and will continue into the longer term:

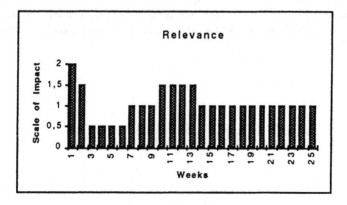

Fig. 4 Service Continuance Time Series

In all cases, the exact formulæ for the graphs cannot be known. However, this does not affect the validity of the model as long as the *relative* time frames are understood and there is some idea of the *relative* impact of the different factors: as explained above, CRIMP works with qualitative data, not just quantitative. For example, the shape of the 'Service Acceptance' curve above will remain the same whether a service is taken up by 10% or 90% of potential users within a given time frame: all that changes will be the gradation of the time axis. It may be noted that relevance shows a greater overall impact: this was taken into account when defining the actions described below.

2.3 Trends

The four trends of likeability, usability, relevance and resistance were all given constant values throughout the time series of 50 on a scale of 0-100. This is an arbitrary figure representing 'typical' values for these factors (the actual figure for a constant trend does not affect the CRIMP model, only changes to the figure will have an impact). For the overall usage it is expected that, typically, it will take time for users to learn to use the services but that usage will steadily climb, with 50% of potential users active within six weeks. After this, usage will continue to grow to a maximum of 85% – a figure of 100% never being achievable, due to factors such as equipment out of service, users on leave, etc. Running the model with this assumption showed a close correlation with the *a priori* figures, indicating a robust, consistent model.

2.4 Results

The constant values for the three interface factors assumed 'typical' interfaces, i.e. just good enough to allow the expected take-up of services. These could, in turn, be affected by poor interface design in any one of three areas (the PoorL, PoorU and PoorR factors). Factors which have a single effect on a model are known in the CRIMP methodology as *actions*.

The model was run, in turn, with a negative impact for each of the actions on the relevant trend, e.g. with a cross impact of -10 of *PoorL* on *likeability*. To compensate for the greater impact of the *relevance* time series, the cross impact for this was -5, i.e. a 10% reduction on the 'standard' interface, whereas the other trends were treated to a 20% reduction. By running each in turn, the following usage profiles were obtained:

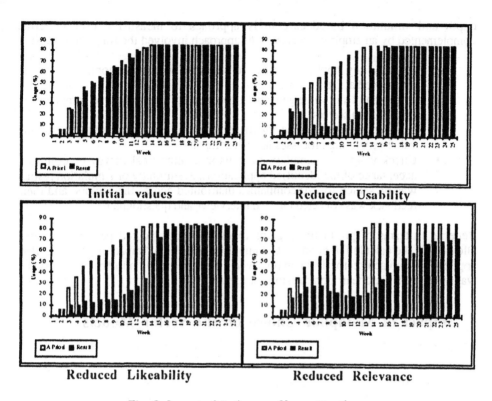

Fig. 5 Impact of Actions on Usage over time

The output shows that, in all cases, the impact on overall usage is much greater than the 10-20% changes made to the interface factors. It also demonstrates that reductions in usability and likeability have immediate effects which are overcome over time, whereas a reduction in perceived relevance has a smaller but longer term impact.

3 Industry Sector Models

To fully understand these changes in usage, it was then necessary to examine their impact within specific industry sectors. RACE project URSA had carried out a number of such studies which were used to provide the *a priori* values. These resulted from a systematic and in-depth study of the innovative usage of advanced communications in many economic sectors. The two objectives of this study were:

- to identify the key applications on which innovative demand for advanced communications is likely to be based in the European economy
- to describe and quantify the benefits generated by these applications

It is difficult to measure the impact of advanced communications on company productivity and other aspects of competitive advantage, as advanced services are still in the development phase and as usage conditions in pilot experiments cannot often be compared with real commercial usage conditions. URSA therefore adopted an in-depth and simulation based case-study approach to measuring user benefits, complemented by an empirical survey. This approach involved the following steps:

- Identify key applications for sectors into which economic activity is aggregated at the EU level.
- Reconstruct key value generating processes of representative companies in the different sectors.
- Simulate impacts of the identified applications on the outcome of the value generating processes in each company.
- Check validity of simulation results in an empirical survey on company acceptance of the identified applications. A summary of simulation results was presented to 120 companies distributed across the sectors, and their feedback was collected through interviews and questionnaires.

Benefits in terms of productivity gains together with expected penetration rates allowed for the calculation of an ECU equivalent of the total impact of the identified applications on each sector. Data on turnover, employment and number of companies was taken from European Commission surveys.[EUR93]

4   Impact of Metaphor Choice on the Sector Models

4.1   Method

The MITS usage patterns shown above were then put into selected models from the URSA project, chosen to accord with the pilots within the MITS project. Although the MITS model appears to show a dominant effect for the relevance factor, this is not necessarily the case when this data is combined with other effects. For example, likeability has a much greater impact in the very short term which could, in some cases, be more important than the longer term effect of relevance.

Two types of action were therefore defined:
- **Delay**: the dramatic impact of poor *likeability* and usability in the early stages can be summed up as a delay in take-up.
- **Reduction**: although not as dramatic in its initial impact, poor *relevance* leads to a small longer-term reduction in usage.

4.2   Electrical Engineering

The chosen application examined in the electrical engineering model was that of EDI, in this case referring to the exchange of information with customers and suppliers, including videoconferencing as well as more conventional EDI. This closely corresponds with the services being examined in the PTT Telecom and Nokia pilots of MITS. URSA's CRIMP model shows the introduction of these services as an action having a direct impact on production lead time. This was therefore taken as an indicator. The results of the URSA model (unmodified) showed a reduction in lead time of 1.5 days as advanced services were implemented. Reductions in likeability reduced this only in the first year, but poor relevance reduced the long term improvement by one third (from 1.5 days saved to 1 day).

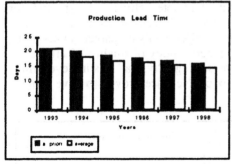
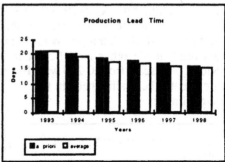

Fig. 6   URSA's results   Impact of poor relevance

4.3   Construction

For construction, the chosen application was more dramatic and more obviously linked with metaphor: the virtual meeting room. This application is one of the central pilots within the BRICC project, using similar service interfaces to those in the BIBA/Vero pilot of MITS. URSA related this application to its potential impact on the national market share of a large construction company, in particular the ability of such services to allow the company to penetrate into parts of the sector in which it currently has little presence. In this case, poor likeability had a greater effect.

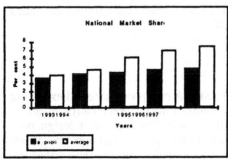
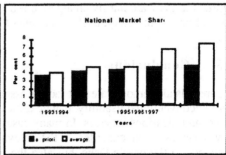

Fig. 7   URSA's results   Impact of poor likeability

In this case, the impact of the new services is predicted as dramatic and steadily increasing, both in absolute and proportional terms. By the final year of the model, the potential market share for the company becomes 7.45% rather than 4.8%. The action continues to have a significant impact for some time after its introduction. Although the potential market share in the final year appears to be little reduced, it is likely that the reductions in the previous years would jeopardise this.

4.4   Transport Equipment

For transport equipment, the application chosen (as in the MITS pilot by LUTCHI and the Automobile Association) was that of remote delivery of expertise for maintenance and interference rectification. As an additional service which could be provided to customers, this was not related by URSA directly to market indices of the types used in the other models, but to a general index of quality of after-sales services, on a scale from 0-200 with 100 representing a 'typical' quality level which can currently be maintained. In this case, poor relevance proved to have a much stronger impact:

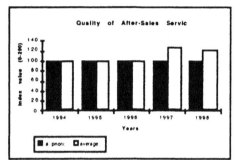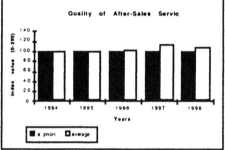

Fig. 8   URSA's results   Impact of poor relevance

As a more advanced service which will not be available for some time, its impact does not come into effect until 1997. The impact of this service is therefore not as strong as those of the applications examined in the other sectors. In this case, the figures for 1997 are hardly changed, but by 1998 the impact of the service has faded away almost completely, back to a value of 107.

4.5   Conclusions

The models all demonstrate that factors associated with metaphor usage and extending beyond conventional usability can be critical in supporting the potential for advanced telecommunications services. In many cases, the most critical factor that designers of new services must consider is the relevance of the service, not just the actual relevance to the user's tasks, but also the manner in which the interface makes the user aware of the relevance. In some cases, likeability can also be a critical factor, as in the example from construction.

These categories are not exclusive, nor should the designer concentrate on a single factor to the exclusion of others. For example, the 'virtual meeting room' used in the construction sector example could be designed to emphasise the 'room', the *spatial* aspect, or to emphasise the 'meeting', the *interactional* aspect, and ideally should project both aspects for their separate usefulness. In this case, the *spatial* aspect could be used to present the intial impression of the system with a high the likeability factor, whereas *interactional* aspects could be supported more strongly as the users begin to use the services, emphasising their relevance.

5 References

[AND94]   Anderson, B; Smyth M; Knott, R.P; Bergan, J; Bergan, M. & Alty, J.L. "Minimising Conceptual Baggage: Making Choices about Metaphor", *proc* HCI'94, Cambridge University Press, 1994

[CON90]   Condon, Chris "Networked Cooperative Work: Usability Issues of MILAN", *proc* Telematics '90, BIBA, Bremen, Germany, 3-5 Dec 1990

[CON92]   Condon, Chris, "Multimedia & CSCW in Manufacturing: Working with MILAN" *proc* Conferencia ESPRIT: CIME & Competividad Empressarial, Bilbao, 1992

[CON94]   Condon, Chris & Keuneke, Stephan "Metaphors and Layers of Signification: The Consequences for Advanced User Service Interfaces" in "Towards a Pan-European Telecommunications Service Infrastructure - IS&N '94", ed H-J Kugler, A Mullery, N Niebert, Springer-Verlag, 1994, pp 75-88

[DUI94]   Duin, Heiko & Kiessling, Silvia "CRIMP for Windows 3.1 User Manual", BIBA, 1994

[DUI95]   Duin, H. "Object-Oriented Scenario Management for Simulation Models" To be presented on: IMACS European Simulation Meeting on Simulation Tools and Applications, 28-30 August, 1995, Gyor, Hungary

[EUR93]   Eurostat, "Enterprises in Europe", Second Report, Commission of the European Communities

[GOR68]   Gordon, T. J. , Hayward, H. "Initial Experiments with the Cross Impact Matrix Method of Forecasting", Futures, December 1968, S. 100-116

[HAM91a]   Hämmäinen, Heikki "Form-based Approach to Distributed Co-operative Work" PhD Thesis, Department of Computer Science, Helsinki University of Technology, 1991

[HAM91b]   Hämmäinen, Heikki & Condon, Chris "Form and Room: Metaphors for Groupware", *proc* COCS '91, ACM Conference on Organizational Computer Systems, Atlanta, Georgia, 5-8 Nov 1991

[HEL77]   Helmer, O. "Problems in Futures Research: Delphi and Causal Cross-Impact Analysis", Futures, February 1977, S. 17-31

[HUT89]   Hutchins, E. "Metaphors for Interface Design" in Taylor, M.M, Neél, F. & Bouwhuis, D.G. (eds) "The Structure of Multimodal Dialogues, Elsevier: North Holand, 1989, pp11-28

[HUT93]   Hutchinson, Chris & Rosenberg, Duska "Cooperation and Conflict in Knowledge-Intensive Computer Supported Cooperative Work" in "CSCW: Cooperation or Conflict?" Springer-Verlag, London 1993

[KAN72]   Kane, J. "A Primer for a New Cross-Impact Language - KSIM", Technological Forecasting and Social Changes 4, 1972, S. 129-142

[KRA94]   Krauth, J. , Hamacher, B. , Duin, H. "A Set of Tools for Rough Enterprise
           Modelling", *Proc* CISS (First Joint Conference of Simulation Societies),
           1994

[LAU90]   Laurel, Brenda "Interface Agents: Metaphors with Character" in B. Laurel
           (ed.) The Art of Human-Computer Interface Design Addison-Wesley 1990

[SIN94]   Sinnigen, Marita; Cools, Jan-Pieter & Sampera, Jaume "URSA Deliverable
           5" RACE Project R2091, 1994

Technologies for Management Systems

Session 3a Introduction

Chairman - George Pavlou

University College London, UK
Tel : +44 171 380 7215
Fax : +44 171 387 1397
G.Pavlou@cs.ucl.ac.uk

In recent years, the combination of relatively cheap computing power and the increasing transmission capability of telecommunications infrastructures have made possible new advanced services. However, the complexity and sophistication of these services and of the underlying supporting infrastructure have posed new management requirements. The Telecommunication Management network (TMN) has been developed as the framework to support the introduction of sophisticated 'open' management systems. The TMN supports network and service planning and deployment, end-to-end performance management, fault diagnosis and recovery, static and dynamic (re-) configuration, accounting and security services.

TMN proposes an architecture to support the standardised exchange of information between management applications and the managed network and service elements. This is achieved through standard reference points and interfaces which are transaction-based, involving both agreed protocols and relevant 'exported' information models. The TMN applications 'see' each other through object-oriented abstractions of relevant resources across reference points; they are also likely to be structured internally in an object-oriented fashion. Despite that, in the standard ITU-T practice, only 'on-the-wire' interactions are standardised. As such, the TMN is object-oriented only from an information viewpoint across reference points.

Open Distributed Processing (ODP) standardisation is reaching a level of maturity. A number of RACE projects and other important consortia in the management arena have been looking into the applicability of ODP methodologies and concepts to TMN. In fact, the ODP Information, Computation and Engineering viewpoints have a definite relationship to the TMN Information, Functional and Physical architectures respectively. ODP however addresses also the internal structure of distributed applications from and object-oriented perspective. The third paper in this session looks at the engineering viewpoint for TMN systems and in particular, at the relationship between the computation and enterprise viewpoints. This is particularly important as the ODP viewpoints are simply different ways to see a distributed system and a consistent view across all of them is necessary.

A very important parallel standardisation activity relates to Intelligent Networks (IN), which provides a framework for the repaid introduction and operation of new telecommunication services. The relationship between IN and TMN is subtle, the former being concerned with the operation of services (control plane) while the latter with the management of the network and services (management plane). While the TMN is object-oriented and distributed, the current IN is functional and centralised. The TINA Consortium is looking at a long-term architecture for the rapid interaction, operation and management of future sophisticated services. In fact, the TINA architecture recognises the need for a distributed, object-oriented IN infrastructure and tries to bring together IN, TMN and ODP concepts.

The first paper in this session describes and applicability of the TINA architecture in its current stage to ATM Connection Management. The latter is a new management functional area introduced by TINA to complement the OSI/TMN ones and is necessary for future services with sophisticated session and connection requirements (e.g. multiparty, multimedia conferencing etc.). Though it is referred to as management, it is mostly concerned with setting up and operating such services, so in our view it relates more to the operation/control rather than the management plant. Of course, the target of the TINA architecture is to integrate the two through uniform underlying mechanisms, but in doing so, it has to take into account current sophisticated 'management plane' functionality

While the previous two papers look at future dimensions of management systems and technologies (the applicability of ODP and TINA respectively), the last paper in this session looks at an important current dimension: the simulation of ATM network technology for the support of current TMN systems. Network and service simulation is extremely important as it allows to address issues of scaling, flexibility in operation and configuration e.g. artificial generation of faults, as well as providing support for comprehensive measurements, all of which are not possible in real environments. A particularly novel aspect of the described simulation is that while network and service control/operation aspects are simulated, management plane functionality may be exercised by a real TMN system. The latter may operate over the simulator in a totally transparent fashion i.e. exactly the same TMN system may also operate in a real environment. The provided flexibility for experimentation is obviously enormous.

ASTERIX: The TINA-C Architecture Applied to ATM Connection Management

Luis A. de la Fuente[1]

Telefónica Investigación y Desarrollo (TELEFONICA I+D)
Emilio Vargas, 6 - 28043 Madrid - Spain
Phone: +34 1 337 4382; Fax: +34 1 337 4222
alberto@tid.es

Abstract. This paper presents the activities undertaken in TELEFONICA I+D to implement the connection management functionality defined in the TINA-C Architecture for the management of an ATM network. These activities are part of the ASTERIX project, whose final objective is to deploy advanced telecommunication services in the short-medium term. In order to achieve this objective, the TINA-C Architecture concepts and ideas are used.

Keywords: ASTERIX, RECIBA, TINA-C, connection management, element management layer-connection performer, network management layer-connection performer, distributed processing environment

1 Introduction

ASTERIX is a project of Telefónica Investigación y Desarrollo (the R&D company of the Spanish public network operator) for Telefónica of Spain whose final objective is to deploy new advanced services (i.e., multimedia, mobile and/or personal services) in the short-medium term. It is concerned with the design and implementation of such services and the associated management activities, and with the service creation environment that will be used in the construction of these services. ASTERIX stands for Telecommunications Services Advanced Architecture for Information Networks.

Currently, when offering, developing and/or managing these sophisticated services, some inefficiencies arise: low level of reuse of components, low interaction level among services, more than one service model, dependencies from the underlying supporting environment, dependencies from the services supplier/developer, high deployment and management costs, difficulties when adding changes or new services, etc. In order to cope with these inefficiencies, the architecture defined by TINA (*Telecommu-*

---

1. The author wishes to thank the rest of the members of the ASTERIX connection management group for the interesting discussions held. These people are: A. Barbero, J. Cañadilla, M. Rangel and F. Ruano.

nication Information Networking Architecture) Consortium (TINA-C) will be used as reference in ASTERIX. This architecture provides solutions to solve such inefficiencies, specially in a multi-vendor, multi-supplier environment, covering both services-related and network-related aspects, aspects that ASTERIX is covering as well.

This paper focuses on the network management aspects and, more specifically, on how the TINA-C Architecture has been applied in ASTERIX for the connection management of a pre-existing ATM network called RECIBA. Section 2 presents an overview of the TINA-C Architecture today and highlights the main differences with last year Architecture [1]. Section 3 describes the network-related management functionality defined in TINA, focusing on the connection management aspects. Section 4 covers the advantages that can be obtained through the application of the TINA-C connection management functionality in ASTERIX, while Section 5 discusses the practical aspects of its application to the RECIBA network. Finally, Section 6 provides an overview of the future work to be done in this area.

2   An Overview of the TINA-C Architecture

TINA-C is a worldwide consortium formed by network operators, telecommunication equipment suppliers and computer suppliers, whose purpose is the definition of an *open* architecture to support the *rapid* and *flexible* introduction of new telecommunication services in the emerging broadband, multimedia and information society era, the ability to manage these services and the network infrastructure in an *integrated* way, and the *independent* evolution of this architecture from the switching and transport infrastructure [4]. The TINA-C Architecture allows the construction and deployment of applications independently from specific underlying networks technologies and defines a distributed processing environment (or DPE) that allows application *interoperability* in a transparent way, enabling different software components -that can be contained in several heterogeneous computing nodes- interact across different network domains in a distribution transparent way [9].

In TINA-C, service is understood in a broad sense including the traditional concepts of telecommunication service -any service provided by a network operator, a service provider, etc., to customers, end users or subscribers- and management service -any service needed for control, operation, administration and maintenance of telecommunication services and of the network and computer infrastructure used to provide these telecommunication services-. The management services in the TINA context refer to operations on network resources and also on telecommunication services. Moreover, the TINA-C Architecture also defines a common framework for the specification, design and development of both telecommunication and management services, allowing a smooth integration and convergence of both types of services [1][2].

The concepts and principles of the TINA-C Architecture, that can be used to design and implement any telecommunications software application, are classified in four technical areas (in contrast with the three defined previously and described in [1]).

These technical areas, by extension, are also called architectures: Service, Network, Computing and Management Architecture (Figure 1).

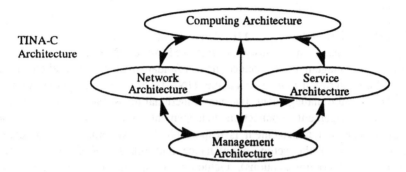

TINA-C
Architecture

Figure 1. The TINA-C Architecture

The *Computing Architecture* provides a set of modelling concepts for the information, computational and engineering viewpoints defined in the ODP standard [6]. It also defines a Distributed Processing Environment (DPE) that provides the support for the distributed execution of software components in a transparent way. The *Service Architecture* defines a set of concepts and principles for specifying, analysing, reusing, designing, and operating service-related telecommunication software components. The *Network Architecture* provides a set of generic concepts that describe transmission networks in a technology independent way. It includes a high level view of network connections that can be used by the services to satisfy their connectivity needs, and generic descriptions of network elements technology independent that can be specialised to particular technologies and characteristics. The *Management Architecture* provides a set of generic management principles and concepts to be applied to the Service, Network and Computing Architectures to obtain the desired management functionality. In other words, each one of these Architectures is responsible for the management of the resources, elements and/or components that are under their scope, being therefore outside the scope of the Management Architecture the definition of the concrete management activities on them.

In this paper the interest is on the management applied on the Network Architecture and, specifically, on the connection management functionality, and a brief description is presented in the next section.

3   Management in the Network Architecture

This section describes the main characteristics of the management in the Network Architecture, focusing on the connection management functionality, discussing the relevant functional layers and areas, the information model and the connection management functionality. More information can be found in [2] and [3].

The *TMN functional layers* [5] relevant in Network Architecture management are the Network Management Layer (NML) and the Network Element Management Layer (EML), since both networks and (network) elements are the resources being considered in the Network Architecture.

The *functional areas* defined in OSI Management (namely fault, configuration, accounting, performance and security) are followed in the TINA-C Architecture. But considering that management of connections is a fundamental activity and the special relevance of configuration management, the TINA-C Architecture defines, only for Network Architecture management, two new functional areas that replace and refine configuration management: connection management and resource configuration (Figure 2). Connection management is responsible for providing the functionality required to deal with the setup, maintenance and release of connections (i.e., connection-oriented communications), including the specification of a connection model, the signalling and routing methods, the management of the resources needed for the connections, and the methods for handling resource failures and overloads. Resource configuration includes the rest of activities within the "classical" configuration management functional area.

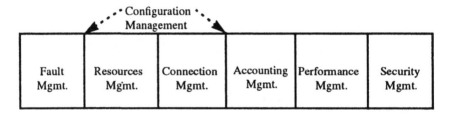

Figure 2. Functional Areas for the Management of the Network Architecture

Concerning the *information aspects*, the Network Resource Information Model (NRIM) is the specific model defined for the Network Architecture. It contains the object classes needed for the representation of network resources and is independent of the technology of them (e.g., SDH, SONET, or ATM). The NRIM can provide support to different types of services (e.g., VPN, PSTN, multimedia, ...), and it is concerned with how individual network elements are related, topologically interconnected, and configured to provide and maintain end-to-end connectivity. The NRIM is mainly based on the ITU-T information model described in Rec. M.3100 [10] (extended with new object classes to include aspects not covered by M.3100, that is mainly oriented to network element management) and in the Rec. G.803 [7] for the concepts of layering (a layer network is a set of compatible inputs and outputs that can be interconnected and can be characterized by the information that is transported) and partitioning (a layer network may be decomposed into subnetworks connected by topological links, and each subnetwork can be further decomposed into smaller subnetworks). Although this recommendation is focused on SDH, these concepts are extended to cope with other network technologies. The NRIM is presented in fragments -like in the Rec.

M.3100-, and contains, among others, the common managed object classes relevant for the connection management, resource configuration and fault management functional areas, and also the classes needed to express the structure of a network and the connectivity on it. The NRIM does not define a model for network elements, only relationships to existing standards in this area.

About the *connection management (CM) functionality* [3][12], it spans only the Element and Network Management Layers. Clients of this functionality can be both telecommunication services (providing them the necessary connectivity between terminals and/or processing nodes) and management services (providing them the connectivity needed to access specific network elements and also the connectivity needed to support the desired management policies, like re-routing policies in case of failures, etc.). Its activities can be classified in connection manipulation (creation, modification, and destruction of network connections including control of network resources and locating connection end points) and connection resource management (identification of resources used to implement connections and management of the information needed to select resources and routes through the network). The connection management functionality can be modelled as a set of computational objects. An example can be found in Figure 3.

The computational objects identified in the CM functionality can be classified in two groups: those that are independent of the underlying networks technologies and structures, and those that deal with the particular characteristics of each layer network and, in consequence, are specialized for each one of them (for an ATM layer network like in RECIBA, for instance). The Communication Session Manager (CSM) and Connection Coordinator (CC) belong to the first group, and the Layer Network Coordinator (LNC) and Connection Performers (CP) belong to the second one.

The CSM transforms a request from the user of the CM functionality (a service application or, in general, a service component) into a request for a network connection. This request is provided in terms of a logical connection graph [3]. The CSM offers the service for setting up, maintaining and releasing logical connections - logical because the user of this functionality identifies namely computational object interfaces instead of addressable points in the network, and because the request from the user is distribution, network structure and network technology independent. This request is delivered to the CC as a physical connection graph [3] that identifies addressable network termination points. The CC deals with the network complexity (at layer network level) and is not associated with any particular layer network. The request identifies the network access points to be connected and the desired bandwidth and quality of service, but it is independent of information concerning the underlying transmission and switching technology and the structure of the underlying networks. Analysing this request, the CC identifies which layer network to use, and request the proper LNC to set-up a trail connection [3] in its layer network. There is one LNC per layer network in a management domain and, if needed, it can interact with LNCs in other domains through a federation mechanism.

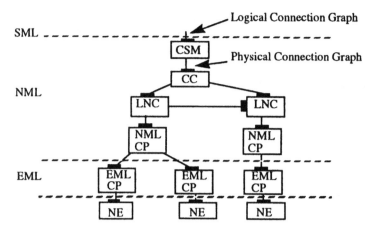

Figure 3. Connection Management Computational Model (Example)

The LNC relies on the CPs and, in particular, in the NML-CPs, that are able to provide subnetwork connections for the subnetwork in their scope, interconnecting termination points. Every subnetwork is managed by one NML-CP, and the structure of NML-CPs reflects the partitioning [7] of the layer network. The last level belongs to the EML-CPs, which directly control network elements (NEs) managing the connections on them through the NEs agents.

4 Advantages of TINA-C Connection Management in ASTERIX

As first step towards the achievement of the ASTERIX goals, it was decided to implement the TINA-C connection management functionality due to the fact that, in the TINA-C Architecture, the services (components) need to use this functionality to express their connectivity requirements, as mentioned before. Although the TINA-C Architecture specifications are still evolving (the time frame of the TINA project is 1993-1997), the connection management functionality is considered stable enough to be implemented.

Another reason was that some configuration management functionality was already available in the pre-existing ATM network infrastructure and it was considered interesting to improve it to be TINA compliant. This network is called RECIBA, that stands for Broadband Integrated Communications Experimental Network, and is an already running network on top of which multiparty and/or multimedia services -videoconference, broadband videotex, videotelephony, teleteaching, ...- and non multimedia services -typically data transfer services, like high speed data transmission or LAN interconnection- are provided. The TINA-C connection management is flexible enough to support the connectivity types required by the previous services (point-to-point for videotelephony and point-to-multipoint for teleteaching, for instance).

The main characteristics of the previously available configuration management functionality in RECIBA (that covers both the network and the network element management layers) are the following: creation, modification and destruction of virtual paths; establishment and modification of cross-connections in the NEs; modification of virtual paths bandwidth; representation of the physical configuration of the network; and establishment and maintenance of network resources configuration states.

The implementation of the TINA-C CM functionality over RECIBA is seen as a way to improve the previously available functionality for the following reasons:

- In addition to management services and/or human operators in a management station (that are the current users of the management of connections), applications in general can also be clients of the CM functionality.

- As mentioned before, the TINA-C CM functionality is flexible enough to support the connectivity needs of both the multimedia/multiparty services and the data transfer services. Moreover, it allows connectivity requests of any type for any network topology.

- Telecommunication services developed "à la TINA" -that is another of the goals of ASTERIX- can make use of these management capabilities, knowing nothing about the underlying network technology.

- The TINA-C CM functionality allows the management of virtual circuits (not only virtual paths), enabling a more powerful management with a higher resolution level.

- The current RECIBA management functionality is technology dependant, that is, directly attached to the RECIBA network elements and, thus, it is not valid for the management of other types of ATM networks. As the TINA-C CM is technology independent (i.e., it supports different network technologies and isolates their specific characteristics in a way in which the clients of this functionality do not need to know the underlying technology nor the physical location of the resources), its implementation can be used to manage not only RECIBA, but also other networks in a distribution and technology independent way.

The following subsections present the CM functionality for RECIBA that is considered in the first phase of ASTERIX (ending June 1996), and the computational, information and engineering related aspects.

5 Implementation Aspects

5.1 Connection Management Functionality in ASTERIX

It is considered that, in the RECIBA ATM network, the virtual paths (VPs) can be seen as semipermanent connections among switches that deal with virtual circuits (VCs), Therefore, two different layer networks have been identified: an ATM-VC layer network and an ATM-VP layer network. This is depicted in Figure 4:

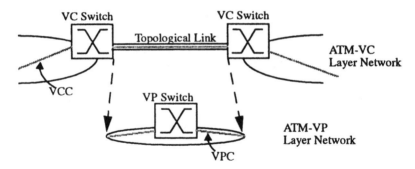

Figure 4. Layer Networks in ASTERIX

The ATM-VP layer network acts as a server layer network for the ATM-VC layer network. In this layer network, the VP connections are established. The VP connections can be considered as semipermanent connections. The ATM-VC layer network acts a client layer network for the ATM-VP layer network. Connections are established in the ATM-VC layer network and they can be considered as switched connections. Although the management of the VP layer network is also considered, the focus will be on the management of the VC layer network.

The general characteristics of the CM functionality for RECIBA are the following: management of point-to-point bidirectional connections; management of point-to-multipoint unidirectional connections (it will be possible to add or remove branches during its life time); bandwidth dynamic management; and management of semipermanent connections (i.e., virtual paths, usually managed by the management plane) and of switched connections (i.e., virtual circuits, usually managed by the signalling plane).

5.2   Computational Aspects

In the ASTERIX project, the emphasis has been put on the CM computational objects that have dependencies on the underlying ATM network (i.e., LNC, NML-CP and EML-CP). In this way, a complete connection management can be easily offered providing, in addition to these objects, the generic CSM and CC computational objects. Also, inside the scope of the ASTERIX project and in order to achieve a complete TINA environment, a resource adapter for the RECIBA network element has been developed to obtain a "TINA-compliant" network element, i.e., a network element offering a computational interface using the DPE as infrastructure. Figure 5 shows these computational objects, and the following is the description of them:

- LNC: As mentioned before, the LNC controls the connections in a single layer network. Since all the network elements to be managed are considered to belong to the same management domain, no federation mechanism among

LNCs will be needed (remember that there is one LNC per management domain). The LNC receives requests for trails in its layer network and provides interconnection of termination points relaying connections requests to the appropriate NML-CPs.

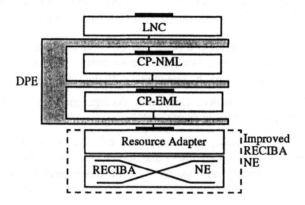

Figure 5. Computational Objects in ASTERIX

- NML-CP: Each NML-CP controls one subnetwork, providing interconnection of termination points of the subnetwork (i.e., subnetwork connections). A NML-CP can have other NML-CPs as subordinates if its subnetwork is decomposed in other subnetworks, and can interact with peer NML-CPs. As mentioned before, the NML-CPs are specific for the layer network they are managing and they know the technology of the layer network (i.e., ATM-VC layer network). For RECIBA, the NML-CPs perform the following:

  - Establishment of connections in the ATM-VC layer. This can be done concurrently (the CPs of the high levels direct and control the establishment of the connections) or sequentially (the connection is controlled and directed by the CPs of the low levels). This should be fixed at configuration time and can not be dynamically changed.

  - Dynamic routing in a subnetwork using a simplified version of the Dijkstra algorithm that looks for the short route between two nodes using the less busy link in that route.

  - Dynamic reconfiguration. Some configuration aspects that can be changed along time, like adding/removal of topological links, are also considered.

- EML-CP: There is one EML-CP per NE. This object adapts the requests received from its NML-CP to the functionality and model that the NE presents. This object accesses to the management interface that the NE offers and contains detailed information about the NE (i.e., channels allocated, channels available, bandwidth available, etc.).

- Resource adapter: This computational object offers the management interface of the NE to the CM computational objects (and, in the future, to computational objects implementing other management functionality). Therefore, the access to the NE is done through this adapter, that can be considered as its computational representation (for management purposes). There is one resource adapter per RECIBA network element.

Concerning the configuration aspects, the functionality identified for the Connection Management Configurator (CMC) and the Resource Configurator (RC) computational objects will be implemented by one object with the following characteristics:

- CMC functionality: Consistent configuration of the CPs based on the information coming from the RC.

- RC functionality: Maintenance of the network configuration (i.e., relationships among client and server layer networks, CPs assigned to each layer network, addition/removal of topological links in the client layer network if trails in the server layer network are added/removed, etc.).

5.3 Information Aspects

The information model used is based on the NRIM. The following fragments [2] of the NRIM have been used: *network* (it shows the structure of the network to be managed in terms of layer networks, subnetworks and topological links); *connectivity* (it models the different types of connections -i.e., point-to-point and point-to-multipoint- that can be established across the network in terms of connections, subnetwork connections and edges); and *termination point* (TP) (it models the end-points of the connections in terms of link TPs, network connection TPs and network TP pools).

Figure 6 shows the inheritance tree used, and Figure 7 depicts the information model used in OMT notation [11].

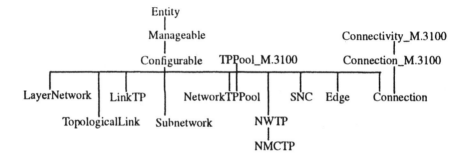

Figure 6. Inheritance Tree in ASTERIX

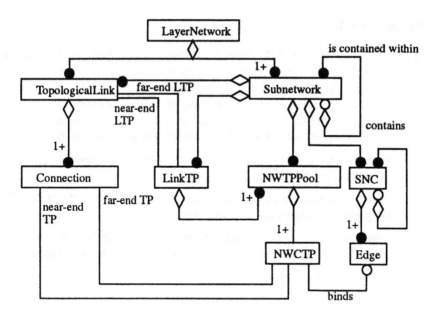

Figure 7. ASTERIX Information Model in OMT Notation

5.4 Engineering Aspects

The DPE used is the ORBIX product from IONA Technologies that is compliant with OMG's CORBA [8]. The computational objects are implemented as C++ objects, and the IDL-to-C++ ORBIX translator is used. ORBIX offers an object locator server to whom it's necessary to provide the identification of the node in which the object to be accessed is. For this reason, it has been considered necessary to incorporate a trader [1] to this DPE. As the trader provides a match making function of interfaces for objects providing a service and objects that want to use it, and facilitates the dynamic binding between these objects, no previous knowledge is needed about where the object server is located to access it. The trader implementation performed by ANSAware is incorporated to ORBIX.

The notifications generated by the objects (events and/or alarms) are currently logged in a file, although the need for a notification server [1] is recognized. This server acts as a broker between emitters and receivers of notifications in a way in which that neither of them need to interact explicitly and individually in order to receive or send the notifications.

Explicit distribution and resource allocation policies units (i.e., capsules [1]) have not been implemented yet. Instead, an administration tool has been developed to move objects from one node to another, add new objects to a node, add new nodes, etc.

The kernel transport network is a LAN Ethernet with TCP/IP as transport protocols.

6 Future Work

Based on the results of ASTERIX Phase I, and once the computational objects that implement the connection management functionality are available, new networks from different suppliers can be managed with this implementation if the corresponding EML-CPs are developed. Regardless the concrete underlying network, the same connection management service can be offered to the client telecommunication services applications. Other activity lines that will be considered in the future are the following:

- Interconnection among different management domains through federation mechanisms at LNC level.

- Integration of the connection management functionality with the fault and configuration management functionality.

- Extension of the resource adapter interface for fault management, configuration management, etc., and not only connection management.

- Quality of service aspects influence.

- Incorporation of a notification server to the DPE.

- Implementation of engineering deployment units (clusters and capsules).

All these activities will be performed in ASTERIX Phase II (June 96 - June 98).

References

1. de la Fuente, L.A., et al.: Application of the TINA-C Architecture to Management Services. IS&N '94, September 1994, Aachen, Germany.

2. Berndt, H., de la Fuente, L.A., Graubmann, P.: Service and Management Architecture in TINA-C. TINA '95, February 1995, Melbourne, Australia.

3. Bloem, J., et al.: TINA-C Connection Management Components. TINA '95.

4. Chapman, M., et al.: An Overview of the Telecommunications Information Networking Architecture. TINA '95.

5. ITU-T Recommendation M.3010, Principles for a TMN. 1992.

6. ISO/IEC 10746-3, Open Distributed Processing Reference Model - Part 3: Architecture. 1995 (draft).

7. ITU-T Rec. G.803, Architectures of Transport Network Based on the SDH. 1992.

8. OMG, Common Object Request Broker Architecture and Specification, 1993.

9. Graubmann, P., et al.: TINA-C DPE Architecture and Tools. TINA '95.

10. ITU-T Rec. M.3100, Generic Network Information Model. 1992.

11. Rumbaugh, J., et al., Object Oriented Modelling and Design. Prentice-Hall. 1991.

12. de la Fuente, L. A., et al.: Application of the TINA-C Management Architecture. ISINM '95, May 1995, Santa Barbara, California, USA.

ATM Network Simulation Support for TMN Systems

Matthew Bocci, Eric Scharf - Queen Mary & Westfield College, UK
Panos Georgatsos - ALPHA SAI, Greece
Michael Hansen - GN Nettest, Denmark
Jakob Thomsen - DELTA Software Engineering, Denmark
Jim Swift - University of Durham, UK

Abstract. RACE Project R2059 ICM is developing cell rate ATM network simulators for use in the development and testing of a generic Telecommunications Management Network (TMN) test bed. The simulator is used as a substitute for a physical ATM network and enables the TMN to be tested in a controlled environment before interfacing with the real network. The simulator is able to provide a number of major enhancements over current ATM networks enabling more extensive and thorough evaluation of the TMN. This paper describes the functionality of the simulator and the underlying modelling principles. It emphasises the interaction of the simulator with the TMN system and the facilities of the simulation models to aid this interaction. The paper shows how the simulator is being used to model a Virtual Private Network for the final ICM case study, in particular concentrating on the use of generic reusable components (User, Call Handler and Switch models) for simulating additional entities such as the Customer Premises Network, the Interworking Unit, and the Gateway between two ATM networks.

Keywords. Simulation, ATM, Cell Rate, Burst Level, VPN, TMN

1 Introduction

1.1 Importance of TMN Research

The dramatic developments in the last two decades in telecommunications have opened the door to many new user services. Powerful network management tools are needed to support the provision of these new services and to maintain the increasingly complex networks. The ICM project was initiated to produce a prototype of an Integrated Communications Management System tailored to the management of IBC networks and services, making use of Advanced Information Processing (AIP) technologies, and conforming to the emerging TMN standards. Selected TMN functions have been implemented and validated in a laboratory test bed which includes an ATM simulator as well as interfaces to real ATM networks such as the RACE Exploit ATM Test Bed (ETB).

1.2 Role of Network Simulation in TMN Studies

Traditionally the main application of simulation has been in ATM research and the design of networks and equipment. More recently, simulation has found application in the development of TMN systems such as those of CEC RACE projects NEMESYS [NEME92], MIME [MIME93] and ICM [ICM92]. Simulation has a number of distinct advantages over the use of real networks. Large ATM networks are not yet available for experimentation and development work, and in any case, access to a revenue-earning network is generally restricted. Additionally, such networks would be difficult and costly to instrument in a manner that allows them to host realistic experiments. Unlike with a simulator, access to the parameters for configuring a real network and its traffic would be limited. Experiments run on a simulator can also be frozen in time and are potentially repeatable. For evaluating new TMN systems, simulation reduces the time and cost, especially when the network set-up phase is taken into consideration.

2 Network Simulation for TMN Studies

2.1 General Requirements for Supporting TMN Experiments

In the ICM project, the simulator is regarded as both a substitute for, and a complement to, a real network for exercising the ICM test bed in an ATM environment. The ICM simulator thus supports both the development of the ICM TMN test bed, and of the applications for exercising the test bed. The simulator provides the following features which are required to support TMN experiments.

Scaling
The simulator can simulate large high-speed ATM networks. Present day demonstration ATM networks are usually small scale and laboratory based.

Flexibility
Compared to a real commercial network, the simulator can offer flexibility in the following respects:

- *Functionality*: A variety of network functions, technologies and traffic types can be supported by the simulator.
- *Measurement*: The simulator provides access to a variety of network data and parameters which may be difficult or impossible to obtain in the real system.
- *Scenarios*: The simulator can simulate a variety of traffic scenarios as well as eventualities such as node and link failures and buffer overflows.

Portability
The simulator is a software product written in ANSI C and designed to be ported easily between different computing platforms. This means that the TMN platform can be exercised at many different experimentation sites with a minimum of specialist equipment.

TMN Interface
The simulator provides a standard TMN interface.

Simulation Speed
Ideally, the simulator should run at speeds comparable to those of the real network in order to enable results to be gathered rapidly and also because the simulator should appear as a real network to the TMN. Hence the ICM simulator employs cell rate (burst level) modelling techniques in preference to cell level modelling. In cell rate modelling each event represents a change in the rate of flow of ATM cells [PITTS90], speed-up over cell level modelling being achieved by the consequent reduction in the number events which have to be processed. Speed-up can be achieved

while maintaining an accuracy sufficient for the purposes of exercising the TMN test bed. Additionally, the Simulator/TMN system has a mechanism to ensure a common perception of time.

2.2 Requirements Specific to the ICM Project

In ICM the Performance Management, Configuration Management, and Monitoring functions of the TMN are being exercised through three main case studies which the simulator must be able to support:

1. Network Monitoring
2. Virtual Path Connection and Routeing Management (VPCM)
3. Virtual Private Network Management (VPNM)

The Monitoring case study was a prerequisite for the VPCM case study, which was in turn needed for the VPN case study. In the Monitoring case study, the main network statistics of concern were cell loss (generated by observing buffer overflow), cell delay and bandwidth utilisation on a link and in a VPC. Alarms generated by buffer overflow and equipment malfunction were also of interest.

Virtual Path Connection and Routeing Management requires the following functions in the simulator:

1. Connection admission control algorithms
2. Usage parameter control algorithms such as the 'leaky bucket'
3. Routeing algorithms
4. A call control mechanism to co-ordinate all three of the above
5. A management interface for receiving the management actions regarding VPCs (creation, deletion, bandwidth modification) and routes (creation, deletion)

In addition the TMN operator must be able to change the bandwidth and routeing of virtual channels and paths.

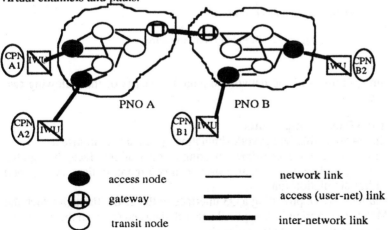

Figure 1: Network Environment for the VPN Case Study

The third ICM case study involves the management of Virtual Private Networks. The environment to be simulated is shown in figure 1 and consists of a number of interconnected ATM networks. Network traffic is generated from a number of network users accessing the network at the network access nodes; network users may belong to CPN (Customer Premises Network) sites of specific organisations. Network traffic is carried between source-destination access nodes and/or between CPN sites of the same organisation, through the establishment of switched connections and/or through leased lines. This case study is described in more detail in [MOUR95].

3. Architecture of the ICM Simulator

3.1 Overview

To fulfil the requirements of the ICM case studies, a network simulation tool has been developed that allows users to view the system as a instantiation of a real network through a set of well defined interfaces.

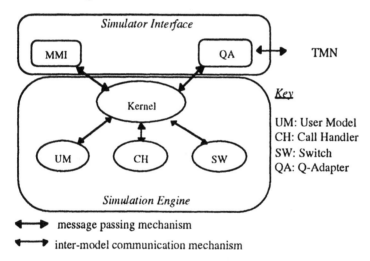

Figure 2: Simulator Architecture.

The architecture (figure 2) of the network simulator consists of the following two sets of components.

Simulator Interface Components
These enable interaction with the network simulation system and comprise a:
- *Man-Machine interface (MMI)*, offering a graphical interface, through the menu-driven facilities of which users can configure, initiate, control and monitor simulation runs.
- *Q-adaptor (QA)*, providing a Q3 interface to the TMN, through which the TMN can (a) retrieve the required network data and (b) send the necessary management actions to the network.

Simulation Engine

This receives network configuration information from the MMI and the QA interface components, runs the simulation and produces the required network data for the TMN and the MMI. The Simulation Engine (SE) contains a real-time, distributed kernel which controls the simulation and handles the flow of data in the simulator system. The SE contains all the models required to simulate the operation of the network. The SE comprises the following:

- The *Kernel*, responsible (a) for handling the flow of messages exchanged between the different components in the *Simulation Engine* and (b) for the overall time synchronisation of the simulation.
- The *User Model (UM)*, simulating (a) the generation of network traffic comprising call requests and the generation of the cell streams associated with particular services, and (b) the Usage Parameter Control (UPC) mechanism.
- The *Call Handler (CH)*, simulating the Call Control (CC) part of an ATM switch or interworking unit (IWU); this includes the Connection Admission Control (CAC) and Routeing functions.
- The *Switch (SW)*, simulating the cell switching and transmission functions of an ATM switch.

The above components include measurement functions for calculating the required network performance data. The CH and SW components make up a transit network node or an ATM-to-ATM gateway node. The UM, CH and SW models make up a network access node or a CPN including its interworking unit. The number and configuration of these components can be defined at the start of a simulation run to suit the particular simulation task.

Time advances in the simulator on an event driven basis. Components within the simulation engine communicate using kernel packets that are delivered to each component in time order by the kernel. These packets can be used to represent, for example, cell rate changes or signalling messages, and their processing in a model can result in the generation of further packets. Simulated time is propagated outside the simulation engine through the simulator interface. This allows, for example, the TMN to obtain knowledge of the simulator's perception of time. Time synchronisation across the whole simulation is maintained using a synchronous time stepping scheme [BOCCI94] that enables the simulator to be ported easily between sequential and parallel computing platforms.

Communication between the simulation engine components and the simulator interface components is done by exchanging messages of an agreed format; the payload of these messages is not interpreted by the kernel. In order to guarantee the correct chronological order in packet processing, communication between the components inside the simulation engine or between the simulation engine and simulator interface components is allowed only via the kernel.

The following sections briefly describe the main functionality of the architectural components of the whole simulator.

3.2 Kernel

The aim of the simulator kernel is to provide the environment for the execution of the component models that represent the network under study, and the necessary communications to the MMI and TMN systems, which are outside the simulator. In real networks, all the equipment and software executes in a distributed fashion, so one of the major functions of the kernel is to provide the mapping of the distributed execution of the network hardware onto the one or more processors used by the simulator. For network simulation, the fundamental requirement for any equipment model is to be able to process and exchange information with adjacent nodes, the processing being triggered by a scheduled event or the arrival of information from an adjacent node.

The simulation of complex equipment and protocols is a non-trivial task, requiring significant development effort. It is therefore important that such models should be as independent as possible of the computer hardware or operating systems on which they execute. The kernel therefore is responsible for providing a complete environment for these models. All operating system and hardware dependencies are restricted to the kernel, allowing the investment in model development to be retained if the simulator is ported to a different computing platform. Example platforms considered for the simulator at the outset were a single processor UNIX workstation and a Transputer array, for which the hardware and operating system are completely different. The current version of the simulator is hosted on a single processor UNIX workstation.
To meet these requirements, the kernel functions as an event-driven operating system. The kernel has knowledge of the network topology and the location of the various elements in that network. The model writer has only to consider the operation of a single instantiation of a model, the kernel dealing with multiple instantiations thereof.

3.3 Node Models

3.3.1 User Model

The user model allows the simulator user to configure a number of traffic sources generating the required network traffic. The simulated traffic sources generate calls (corresponding to a user population) using pre-defined service call generation patterns. Successful service calls generate appropriate cell streams according to the network bearer service they are using. Cell level traffic is modelled as a series of rate changes, or *bursts*. The User Model can operate in one of two modes, either as a model of specific users of a network, or as a generator of the network traffic produced by a CPN (see below) interworking with an ATM WAN.

The user model protocol stack consists of the following three layers as well as interface to the kernel.

User Application Layer The user model supports a wide range of services, a service being defined as a collection of between 1 and 4 connections. The notion of a *user group* has been introduced to specify a user population that uses the same service type. This concept makes it easy to specify a large number of users in a simple way.

The connection types are specified by a burst model type together with parameters such as cell rate, burst duration and silence time.

User Call Control Layer This maps a user application to the access signalling layer, which can only deal with single connection at a time, and keeps a record of this mapping.

Access Signalling Layer This is an implementation of the user side of the User Network Interface (UNI) signalling according to the ITU-TS specification Q.2931 as described by the RACE project R2081 TRIBUNE.

The user model, when used in CPN mode, generates the traffic produced by an ATM CPN with a number of users. The assumption of ATM means that the same code can be used for both user model and CPN. The services carried by the CPN are assumed to be the same as those already defined for previous ICM case studies. In the VPN case study, the main area of concern is the traffic on the VPCs linking the CPNs, as opposed to the traffic within the CPN itself. Therefore only the inter-CPN traffic from each terminal as it appears at the CPN/WAN access is modelled. Provided that there is sufficient bandwidth on the VPCs, calls are set up through simple agreement between CPNs. However, if the VPC bandwidth needs to be increased (or decreased) in order to allow more traffic between the CPNs, then this is done through negotiation with the TMN. In order to cater for the addressing of specific CPNs an addressing scheme has been implemented which makes use of a unique addressing triplet (Network number, node number and UNI (CPN) number).

3.3.2 Switch Model
The SW model simulates the switching and transmission functionality of an ATM switch, IWU or gateway node. Single-stage non-blocking switching architectures are modelled; these include (a) a single buffer per outgoing link and (b) a buffer shared between all outgoing links.

The switch is modelled at the burst level using a fluid-flow technique [MEIS91] [MEIS92]. In this model, the rates of the flow into and out of the buffer(s) are the combined rates of all cell streams arriving at, or departing from, the switch. In cell rate simulation the traffic rates remain constant between successive cell rate change events (bursts), and hence the model allows the calculation of link/buffer/cell level statistics as well as the prediction of certain events/alarms (e.g. buffer full events). The model operates on burst arrival events (coming from other SWs), connection establishment/release (coming from the CH) and link queue-empty/busy events, and buffer full/not full events which are local events scheduled by the model itself.

Cell switching is accomplished by a table look-up procedure. VP-cross-connection and VC-switching tables are maintained per input port, associating input VPI or <VPI,VCI> pair values in the cell header with the corresponding output port and output VPI or <VPI, VCI> pair values. The VP-cross-connection tables are updated by the TMN at VPC creation/deletion events and the VC-switching tables are updated by the CH model at connection acceptance/release events.

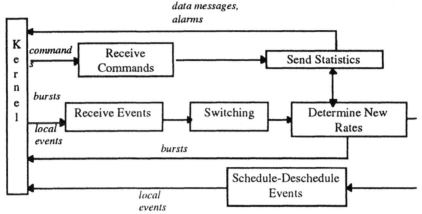

Figure 3: Switch Model Architecture

The SW model (figure 3) includes the following functional blocks:

Receive Events: This function receives and identifies incoming traffic events and updates performance data associated with input links, and VP links are updated accordingly. Simultaneous bursts (cell rate changes) are processed together.

Receive Commands: This function receives messages from interface components (MMI/QA).

Switching: This switches the incoming bursts to the appropriate output link.
Determine New Rates: This function determines the status of link queues and buffers for all outgoing links and buffers for which an event has occurred. New cell rate changes are generated on output links accordingly and the performance data is updated.

Schedule-Deschedule Events: For all output links and buffers for which an event has occurred, any previously scheduled local events (link-queue empty, buffer full) are cancelled and the new times when these events are to occur are calculated.

Send Statistics: This function retrieves required performance data from the appropriate parts of the switch model and sends them to the interface components (MMI, QA). Statistics are sent to the MMI at the time of their occurence in the form of 'alarms' corresponding to the events link queue empty/busy and buffer full/not full. Sample statistics concerning links and buffers may also be sent to the MMI at the end of the simulation run. Statistics are sent on request to the QA and include data associated with links, buffers and VP links.

3.3.3 Call Handler

The call handler model (CH) is responsible for the setup of connections through the network based on connection setup requests via a UNI with a user model. Signalling proceeds to other CHs in nodes along a connection's route via Network-Network Interfaces (NNIs) until the destination node is reached where the user model is signalled via a UNI.

The architecture of the CH is shown in Figure 4. The most significant layer is the call control layer which selects routes, performs CAC, administers VPC bandwidth, and communicates with the MMI and QA. A variety of CAC algorithms can be used, depending on the selection made via the MMI. Call control sends notifications for call setup, rejection and release as well as scheduled notifications of statistics of VPC usage. The notifications sent to the MMI and the QA can be selected by using the MMI or the TMN via the QA interface.

The access signalling layer lies below the call control layer. The signalling implemented in the simulator is based on a subset of the access signalling protocol specified in ITU-TS Q2931 as described both by RACE project R2081 TRIBUNE and in recommendations from the ATM Forum. This signalling closely resembles that used in real networks. In the simulator this protocol is used at both the UNI and NNI. When a call is set up or cleared, the CH model sends messages to the SW model via the SW model interface enabling the SW model to update its VC-switching tables.

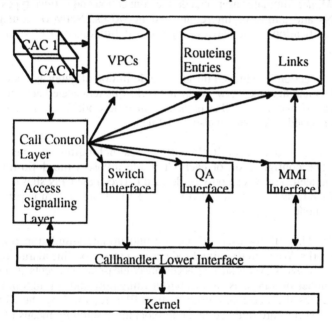

Figure 4: Call Handler Architecture

The physical topology of the network to be simulated is distributed between the various simulator models and is defined and initialised through the simulator MMI. The logical topology, comprising routes, route selection entries and VPCs, can either be defined by the MMI or the TMN.

The operation of the call handler model can be supervised by the TMN through the Q Adapter, making it possible to alter and tune the logical topology of the network. As well as filtering notifications, the call handler QA interface accepts messages to add, delete and change VPCs, routes, route selection tables and CAC parameters.

The simulator is able to simulate multiple-domain networks and let different TMN systems manage each domain. When simulating CPNs or more than one network, the call handler model can be configured to operate as a limited IWU or a gateway respectively.

3.4 Simulator MMI

The MMI component is the front-end of the simulation system. Through its menu-driven and mouse-driven facilities, users can configure, initiate, control and monitor simulation runs. The MMI operates in four main functional modes:

Administration Mode: In this mode, simulator users are offered facilities for defining their working environment and file-handling facilities for handling their defined connection types, service types, networks and experiments. Simulator user administration facilities are also offered.

Configuration Mode: Simulator users can define connection and service types and can configure CPNs and single or interconnected networks. Network configuration includes the specification of the network topology, definitions of VPCs, routes, and network traffic.

Network interconnection is achieved through the definition of gateway nodes at each network boundary and the links between the networks. Interconnected networks can be treated as separate networks as well as parts of a super network. The traffic and routes between interconnected networks can also be defined.

Simulator users can define multiple CPN sites belonging to one or more organisations. The configuration of each CPN site involves the definition of the traffic to be generated. Traffic can only be defined between the CPN sites of the same organisation. Each CPN site is associated with an IWU which in turn is connected to a network access node.

Running Mode: This allows users to prepare and initiate simulation runs of the configured networks. While in Running Mode, users can see on-line alarms indicating the occurrence of network events such as acceptance/release/reject events at connection level at CPNs, nodes or VPCs, as well as buffer full/empty and cell loss indications. The MMI produces the necessary initialisation files required by the Simulation Engine. Simulation runs are initiated in a non-blocking fashion by means of a client-server networking architecture in which the MMI acts as a client. These runs may be distributed on any computer connected with the machine hosting the MMI, which also allows users to activate the *QA* interface.

Result Mode: At the end of a simulation run, the MMI may enter this mode, allowing users to view statistics for network entities such as nodes, buffers, links, and VPCs. The type of statistics to be displayed are defined before each simulation run.

The MMI also offers facilities common to the Configuration, Running and Result modes for providing different views of the network (e.g. physical, logical, super-network and CPN connectivity) and for querying configuration information.

4 Interfacing the Simulator with the TMN System

4.1 Overview

In order to exercise the TMN testbed adequately, the simulator must appear to the TMN as a real network. The ability to change transparently between the real network and simulator, as far as the TMN system is concerned, is the cornerstone of the approach taken by the ICM project.

It was decided early on to base the ICM TMN system on the ISO CMIP protocol. From this decision, a detailed study and evaluation led the project to choose OSIMIS as the TMN platform. OSIMIS provides a very high level, object-based API to CMIP/CMIS facilities, removing the need for the TMN application writer to concern himself with the details of the CMIP protocol. In addition, OSIMIS provides an extensive range of tools to facilitate software development, such as GDMO and ASN 1 syntax compilers, MIB browsers and information source and sink tools. Since OSIMIS is UNIX-based, all the TMN applications used by ICM run on UNIX platforms.

Ideally for communication between the TMN system and individual network elements (NEs) such as switches, the devices within the NEs would directly interpret the CMIP protocol. However, CMIP requires computing resources more powerful than those usually found in NEs, and alternative protocols such as SNMP are still widely used by equipment manufacturers. Where alternative protocols are used, the CMIP-based TMN accesses the NEs via a Q adapter function (QAF) which at one side provides the CMIP interface for the NE, and at the other, communicates directly with the NE in whatever proprietary protocol (M interface) the latter uses.

4.2 Design of the QAF

As a simulator can consume considerable computing resources, the option to run the simulator on non-UNIX machines, such as a Transputer Array was required. This means that the protocol used for the M interface should be usable across different processor architectures without high encoding and decoding costs. To minimise such hardware dependencies the simulator uses a simple protocol based on ASCII strings. The protocol can then use a transmission method appropriate to the hardware on which the simulator is running. For a UNIX-based simulator, the simulator kernel communicates with the QAF via sockets. However, if the simulator were Transputer based, INMOS links could be used. ASCII strings are capable of providing adequate performance since messages in the management plane are on a time scale of seconds rather than the milliseconds associated with the control plane.

The major difference between using the TMN with a real network and the simulator is that of time. TMN systems normally use the clock on the host processor to schedule events and calculate rates of change of network parameters. However, a simulator will not normally run in real time. Depending on the size of network being simulated, the nature of the traffic, the granularity of the simulation and the power of the host computing platform, the simulator may run faster or slower than real time. It is therefore necessary to ensure that the TMN runs against simulated rather than real time. Since one of the aims of the simulator was to test the TMN software, ICM has extended the facilities of OSIMIS so that TMN applications run transparently in real or simulated time, depending on whether or not time is broadcast from the QAF. Thus a TMN process binary can be run with the simulator or the real network without any changes to the program code.

The M interface to the simulator has been designed to support any number of QAFs, allowing complex TMN configurations to be used. Since it is not necessary to simulate in detail the flow of the TMN messages in the network, a pseudo TMN network has been implemented within the simulator kernel to deliver and transmit the messages to and from the individual node models.

The managed objects (MOs) within the TMN side of the QAF must be identical to the objects presented by the simulated NE. If the MOs are described for the NEs in GDMO terms, then OSIMIS allows this definition to be compiled to create the TMN side of the simulator QAF for that NE. As OSIMIS compiles MOs into C++ objects, the main hand-crafted parts of the QAF consist of mapping changes in object attributes into the M-interface protocol and vice versa, transmission and reception of the messages between the QAF and the simulator and the delivery of the messages to the appropriate MOs.

5 Further Developments

Ongoing work on the simulator concentrates on performance evaluation, both in terms of simulation speed and in terms of the scalability of the problem domain. Further work envisaged in ICM aims to maximise the simulator's performance on a specific computing platform by the enhancing the modelling components and optimising both the time synchronisation mechanism and event list management.

6 Conclusions

Despite the increasing number of real ATM networks, network simulation continues to play an important role in TMN research and development. Simulators act as both a substitute for, and an extension to the scale and facilities obtainable with, real networks and enable management systems to be tested in a controlled environment at low cost, avoiding the need for experimentation on revenue earning networks. This paper has described the architecture and basic features of an ATM network simulator that can be used in a TMN test bed, and shows how the simulator interworks with the TMN. By making use of generic and reusable modelling components, the TMN can be exercised in a wide variety of network scenarios which include not just the basic ATM network, but also the attached CPNs. The use of the cell rate modelling technique in the simulator maximises the simulation speed whilst still allowing cell level traffic statistics to be obtained. The inherent portability in the simulator design gives potential for further speed-up by enabling the simulator to be ported to a multiprocessor environment.

7 Acknowledgements

The authors gratefully acknowledge the support and funding of CEC RACE project R2059 Integrated Communications Management.

8 References

[BOCCI94] M Bocci, J M Pitts, E M Scharf, "Performance of Time Stepping Mechanism for Parallel Cell Rate Simulation of ATM Networks", 11th IEE UK Teletraffic Symposium, April 1994

[ICM92] CEC RACE R2059 ICM Deliverable 3; "Selection of Networks, Network Interfaces and Definition of Testbed Laboratory"; R2059/VTT/TEL/DS/R/004/b1; December 1992

[MEIS91] B.Meister, N.Karatzas, P.Georgatsos "Modelling of ATM networks", 5th RACE TMN Conference, London, 1991.

[MEIS92] B.Meister, P.Georgatsos, N.Karatzas "Switch models for TMN applications", 6th RACE TMN Conference, Madeira, 1992.

[MIME93] CEC RACE R1084 MIME Deliverable 26, "Final Experiment Reports - Recommendation on Definition of a Common TMN Testbed", R1084/AL/ISD/DS/C/026/B1, Feb 1993

[MOUR95] K.E.Mourelatou, D. Griffin, P. Georgatsos, G. Mykoniatis, "ATM VPN services: Provisioning, Operational and Management Aspects", IFIP TC6, 3rd Workshop on Performance Modelling and Evaluation of ATM Networks, West Yorkshire, UK, July 2-6, 1995

[NEME92] CEC RACE R1005 NEMESYS Deliverable 11; "Conclusions of Project NEMESYS"; 05/KT/AS/DS/B/028/b1; December 1992

[PITTS90] J M Pitts, Z Sun, "Burst Level Teletraffic Modelling and Simulation of Broadband Multiservice Networks", 7th IEE UK Teletraffic Symposium, April 1990

Engineering a TMN in an Open Distributed Processing Environment

Raymond Larsson - Telia Data Ab, Sweden
Don Cochrane - Cray Communications, Uk

Abstract. The Engineering Viewpoint of Open Distributed Processing (ODP) is concerned with the provision of an environment in which the computational description of a system can be interpreted. It provides distribution mechanisms to support the transparencies selected in the Computational model. This paper presents the results of the PRISM project work on the Engineering Viewpoint. The paper provides guidelines for how to proceed from a Computational model to an Engineering model taking into account Enterprise requirements. It also provides correspondences between Engineering Viewpoint concepts and TMN concepts, and a set of guidelines for how to proceed from an Engineering model to a TMN target architecture.

Keywords: TMN; ODP; Engineering Viewpoint; BEO; Cluster; Capsule; Node; Channel

1 Introduction

During the four years of RACE II, the PRISM project has been applying the concepts of ODP (Open Distributed Processing) to the management of services. The work commenced at the Enterprise Viewpoint and has proceeded through the Information and Computational viewpoints. This work has been presented in earlier conferences. The SMRC (Service Management Reference Configuration) developed by the project is completed by the addition of this work on the Engineering Viewpoint. This provides a generic management framework for the TMN-compatible management of services.

The work of the last year has focused on the Engineering Viewpoint. As with the other viewpoints, the application of the ODP concepts to a subject as complex as Service Management has not been straightforward; the subject is easily the largest test of ODP attempted. The ODP concepts have had to be individually considered and reinterpreted if they were to be used for management at all. This paper presents the results of this work in the PRISM project, including how the Engineering Viewpoint objects can be derived from the Computational Viewpoint and what and how the Enterprise Viewpoint impacts Engineering.

2 Aim and Scope

The aim of the PRISM work and this paper is to clarify the Engineering Viewpoint of the SMRC as much as possible. The goals of the paper are to:

- define the Engineering Viewpoint (in a Service Management context).
- explain the necessary concepts and the relationships between these concepts.

- state the relationship of the Engineering Viewpoint to TMN (Telecommunications Management Network).
- give guidelines for the transition from the Computational Viewpoint concepts into Engineering.

The scope of the PRISM Engineering Viewpoint work is shown in Figure 1.

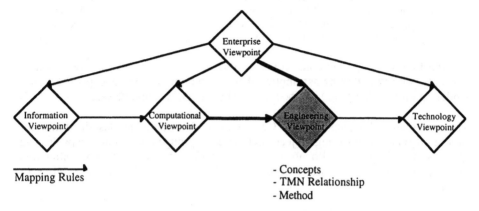

Fig. 1. Scope of the PRISM Engineering Viewpoint Work

Figure 1 shows the five ODP Viewpoints and the relationships between them. Note that this is the PRISM approach, which is a bit more restrictive than ODP, since ODP doesn't constrain the RM-ODP by explicitly specifying any relationship between viewpoints. The link between Enterprise and Engineering is seen as progress towards ensuring that the Engineering Viewpoint fully reflects the requirements expressed in the Enterprise Viewpoint and is particular to this work. Details of the Technology Viewpoint itself are out of scope of the project.

The work has taken into account both the standards work [2,3,4] and that of TINA-C [5] to present a standards-compliant set of concepts that are relevant and meaningful to Service Management.

The paper describes a number of concepts in terms applicable to Service Management and how these concepts corresponds to the TMN concepts. The work has been verified by case study application to VPN (Virtual Private Network) Accounting. The concepts of Basic Engineering Objects (BEO), clusters, capsules, nodes and channels are described and presented.

While the ODP standards are becoming complete, earlier work has found that they are weak on the practical task of relating objects and information in one viewpoint to those in another: the ODP Viewpoints are not intended to be hierarchical but to be orthogonal ways of looking at a distributed system. The question of migrating impacts from one viewpoint to another and assuring consistency while so doing is a major goal of this work and is fully described.

3 The PRISM Engineering Viewpoint Approach

PRISM has defined the view of the Engineering Viewpoint as follows:

> "The PRISM Engineering Viewpoint defines a framework for specifying an abstract view of the underlying physical mechanisms for communication, distribution and processing of management applications that are spread over a heterogeneous distributed system. The framework defines a set of concepts and mapping rules to the other viewpoints." [1]

For the Engineering Viewpoint concepts PRISM has focused on the deployment concepts, with the working assumption that the communication concepts (i.e. stub-, binder-, protocol-object etc.) will be provided by the selected middleware.

The computational objects from the Computational Viewpoint are described in terms of a set of Basic Engineering Objects and these are grouped into nodes, capsules and clusters for describing allocation and distribution transparent processing and deployment. Distribution transparent object interaction is described in terms of channels.

The concepts are taken from [2,3,4,5]. We have based our work mostly on the ODP definitions since [3] and [4] have now been accepted as International Standards by both ISO and ITU it is better to use the standard definition.

3.1 Deployment Concepts

The following list of concepts to be used for deployment have been chosen:

- basic engineering object (BEO) [4].
- cluster [4].
- capsule [4].
- node [5].

In the case of the *node* concept the TINA definition has been chosen instead of the ODP definition because it expresses the same thing as the ODP definition, but in a more concrete and understandable form. The TINA definition is also more focused on the network.

3.2 Communication Concepts

Not all of of these concepts have been fully explored and are left for further study beyond RACE. It will be assumed that the communication concepts not addressed are hidden and transparent through the middleware used. The following communication concepts are however needed in order to establish communication between BEOs:

- channel, PRISM adaption.
- stub object [4].
- binder object [4].
- protocol object [4].
- interceptor [4].
- engineering object interface [4].

A *channel* provides a *binding between interfaces* to engineering objects through which interaction can occur in a distribution transparent manner using a client-server architecture. The rationale for having our own definition of a channel is that the channel concept as it is defined by ODP and TINA is more complex than is necessary for Service Management. The definition above is therefore an adaptation of the ODP definition excluding these references to unnecessary concepts.

3.3 Conformance Concepts

The interworking reference point concept is used to allow for a mapping to a TMN interface.

- interworking reference point (IRP) [4].

3.4 Concepts Relationship

This section will provide the context for the concepts. This is done by showing the relationship between the concept and providing an abstract example.

The relationship between the concepts is shown in Figure 2. This uses the Rumbaugh OMT graphical syntax [6].

As can be seen from Figure 2 a node consist of a number of capsules, a capsule consist of a number of clusters, a cluster consist of a number of basic engineering objects which may (if they are in different clusters) or may not (if they are in the same cluster) communicate through a channel binding together two engineering object interfaces. Communication in a channel is done through the use of stub, binder, protocol objects and interceptors which are communication and transparency providers. For conformance testing purposes there is an interworking reference point between any two interworking providers located in different nodes.

Figure 3 shows an abstract example of how the concepts relate to each other. All interaction is done between BEOs, and all interaction that is done between BEOs in different clusters are done through a channel. It doesn't matter if the clusters are within the same capsule, within the same node or in different nodes. How interaction between BEOs within the same cluster is performed is not prescribed.

Figure 3 also shows the internals of a channel. Note that the interceptor is only present at a domain boundary and is used to translate between protocol objects supporting different protocols, i.e. CMIP and SNMP.

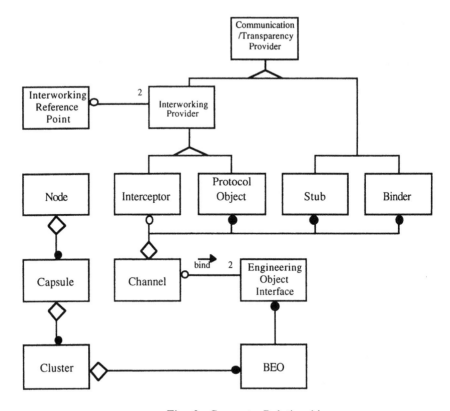

Fig. 2. Concepts Relationship

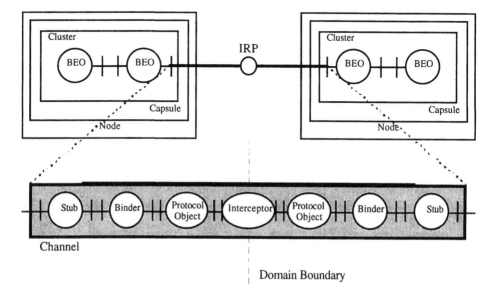

Fig. 3. Abstract Example

3.5 Transition to the Engineering Viewpoint

This section contains comparisons between the concepts of the Enterprise and Computational viewpoints with the concepts of the Engineering Viewpoint. Correspondence rules for how the Computational model relates to an Engineering model are provided. No real correspondences between the Enterprise concepts and the Engineering concepts have been found, however the Enterprise requirements will put constraints on how the system is distributed in the Engineering Viewpoint. These aspects are covered in the next section.

Enterprise Viewpoint Impact on the Engineering Viewpoint

The Enterprise Viewpoint of the SMRC focuses on the specification of Service Management Services from the perspectives of user and Enterprise requirements. Certain aspects of this specification are not (completely) embodied in the Computational Viewpoint. They relate to distribution and are properly considered in the Engineering Viewpoint. Of course, although most of the transparencies defined for the Engineering Viewpoint will be needed when engineering Service Management Services, we are interested in those that are especially relevant to Service Management, whether these need refinement, and whether any new transparencies need to be defined.

Two areas of transparency are particularly important for Service Management these are the use of network resources and the policing of Service Management contracts. A customer who negotiates a particular QoS should expect to have that QoS provided invisibly - he will not usually be especially interested in how it is provided. The provider supplying that QoS will want the service policed invisibly - he will not be particularly interested in how it is policed. The former is a specialisation of replication transparency; the latter is related to federation and the concept of a (distributed) interceptor.

A number of requirements stated in the Enterprise Viewpoint will have impact on how the system being developed should be distributed in the Engineering Viewpoint. Below follows a list of candidates that should be taken into consideration when specifying an engineering model (some of them might already have been addressed in other viewpoints):

- *QoS requirements* (e.g. the service have to be up and running 24 hours/day).
- *security requirements* (e.g. account data have to be stored in a secure node).
- *technology constraints* (e.g. we have to use existing hardware or we can only buy HP hardware).
- *geographical constraints* (e.g. actors are located in different geographical locations and certain functionality which is maintained by an actor have to be in a node at the actors geographical location).

This all comes down to different types of management *domains*.

Computational Viewpoint Impact on the Engineering Viewpoint

In the Computational Viewpoint, an ODP system is specified as a set of interacting objects providing application specific functions. The details of the infrastructure which supports these interactions are not visible in the Computational Viewpoint.

They are described in the Engineering Viewpoint. Therefore the Engineering Viewpoint can be seen as a refinement of the Computational Viewpoint in which distributed transparent interactions and deployment requirements are taken into account.

The following correspondences between the Computational Viewpoint and the Engineering Viewpoint have been found:

- a computational object would correspond to a BEO (which can be replicated).
- a computational interface would correspond to a BEO interface.
- a computational binding object would correspond to a channel.
- a computational building block could correspond to a cluster.

Note that it is not clear if a computational building block should be mapped over to a cluster or a capsule (or something else). The many properties of a computational building block make it hard to find a correct mapping. This have also been identified by TINA-C. The mapping to a cluster seems to be the closest match there is, but this mapping needs more study.

4. From an Engineering Specification to a TMN Target Architecture

This chapter addresses how the Engineering Viewpoint relates to TMN, and how an engineering specification can be realised on a TMN target architecture, creating a TMN system which can run in an ODP environment.

4.1 The TMN Physical Architecture

The TMN Physical Architecture is described in [7] so this description won't be repeated here. The following TMN concepts are of interest for mapping purposes between the Engineering Viewpoint and the TMN physical architecture:

- TMN Building Block (physical component)
- TMN Standard Interface
- Communication Path

These concepts are however not enough for implementing an engineering specification. To achieve this the relationship between TMN concepts from the functional and information architectures to those of the TMN physical architecture also have to be understood. The next section gives an interpretation of what is stated in [7] and provides correspondences between engineering concepts and TMN concepts.

4.2 Correspondences Between Engineering Concepts and TMN Concepts

To start with the concepts of the TMN physical architecture are examined. A *TMN Building Block* could correspond to a *node*. The *channel* would correspond to the TMN *communication path* realised by the specific TMN functional components MCF (Message Communication Function), DCF (Data Communication Function) and ICF (Information Conversion Function) (exactly how these functional components could be mapped onto ODP communication concepts are for further study). *TMN standard interfaces* would correspond to a grouping of *interworking reference points*.

An overview of how PRISM and TMN concepts relate to each other is shown in Figure 4.

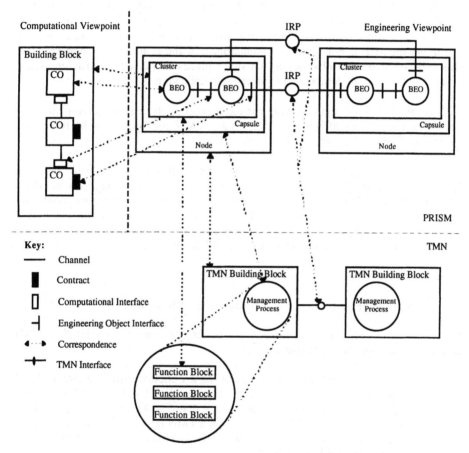

Fig. 4. PRISM and TMN Concepts Relationship

The engineering model does not just describe the communicating devices and the interfaces between them, but also addresses the internal structure of a node. To map the engineering concepts BEO, cluster and capsule to a TMN architecture it is needed to examine how these relate to concepts of the TMN functional and information architectures and how TMN concepts themselves are related.

In [7] TMN function blocks are allocated to TMN building blocks. TMN function blocks are composed of functional components (even though these are presently not subject to standardisation). TMN function blocks interact over reference points and it should be noted that not all reference points are realised as TMN interfaces, only those external to a TMN building block. It should also be noted that it is stated in [7] that the information that actually needs to be conveyed may be only a subset of the information possible at a reference point and that the scope of the information at an interface is determined by SMK (Shared Management Knowledge). The MAF (Management Application Function) functional component is the component implementing the TMN management services and this component will act in the role of Manager or Agent. This is the link to the information architecture which talks about management processes which for a specific management association will take on either the manager or agent role. There is however no explicit statement in [7] that clarifies the relationship between a TMN function block and a management process. It is however stated that the relationship within a system between the Manager, Agent and the MIB is not subject to standardisation and is an implementation issue. This type of statement also occurs on other places in [7] which provides some freedom for how to implement an engineering specification.

With the above in mind the following correspondences between engineering concepts (and some computational concepts as well) and TMN concepts from the functional and information architectures are proposed.

ODP has no real correspondence to the TMN functional components, however in the PRISM approach BEOs are grouped into clusters which could be mapped onto TMN function blocks. This would indicate that functional components could be encapsulated in BEOs. A *capsule* would correspond to a *management process* (i.e. a manager or an agent). The SMK could also correspond to a selection of computational interfaces from COs offering multiple interfaces.

These correspondences are also shown in Figure 4.

4.3 TMN Constraints on the Engineering Viewpoint

This section identifies the constraints on the Engineering Viewpoint that TMN implies. These constraints are listed below:

- TMN does not allow distribution of a Function Block between TMN Building Blocks (distribution of a Function Block within a TMN Building Block is not prescribed). This means that distribution is restricted to be within the same node for a cluster on a TMN target architecture. Note that this is only restricted to clusters involved in management activities which are visible at a TMN interface.
- Protocol Objects within a TMN channel are restricted to support CMIP. Communication over a TMN X-interface could imply that a CMIP protocol object communicates with e.g. an SNMP protocol object using an interceptor to do the protocol conversion.

More are probably to be found, but these are left for further study.

4.4 TMN Engineering Example

This section provides a somewhat more concrete example showing the correspondences of Engineering and TMN concepts. In Figure 5 a line between what can be considered to be functionality provided by some selected Middleware is drawn.

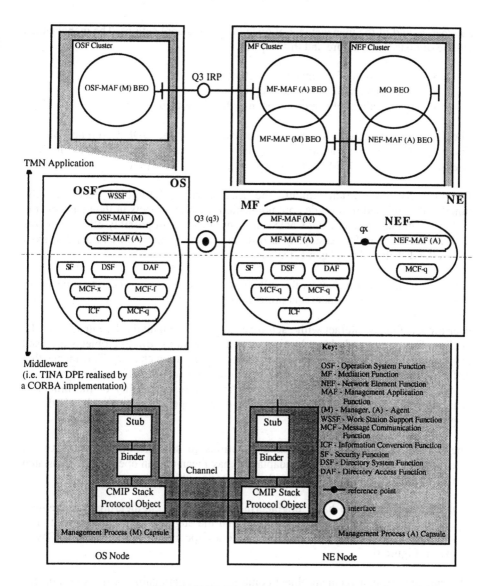

Fig. 5. TMN Engineering Example

The relationship between the Engineering Viewpoint and OSI management concepts (Manager, Agent and MO) has to some extent already been addressed in [2]. Figure 5 also summarises this relationship with some additions from PRISM. An interceptor containing ICF functionality could be placed between the protocol objects if needed.

5. The Engineering Viewpoint and ODMA

Open Distributed Management Architecture (ODMA) is part of the international standardisation initiative ISO/SC21/WG4. ODMA provides an architecture for the specification and development both of systems management as an distributed application and of the management of open distributed applications. The management will be of distributed nature and this implies:

- distribution of the managing activity
- management of distributed applications
- management of resources that may be distributed

The ODMA architecture is based on the concepts of the ODP Reference Model. Therefore, ODMA is structured according to ODP viewpoints. The following is a brief overview of how ODMA relates to the Engineering Viewpoint.

For OSI Systems Management, the Engineering Viewpoint describes the functionality needed to provide communication between objects in the manager role and in the managed role using CMISE in conjunction with possibly other protocols such as Transaction Processing. It also describes the application entities that exist within the distributed management system.

For OSI management purposes, the following three types of engineering objects are of importance: the stub, binder and protocol objects.

Peer binder objects interact with each other in order to maintain the integrity of the binding. Every protocol object has its own set of bindings. Binding objects are used to specify a desired quality of service of the communication channel. This means that different binding objects are needed when different quality of service levels are supported within a communication node. Another responsibility of the binder is distributed to specific addressed engineering objects operations that are handled by a single protocol.

The stub object provides adaptation functions to support interaction between interfaces in different nodes. This adaptation may mean that an operation is translated to a coding that is understandable for the engineering objects.

ODMA identifies three basic engineering objects: the engineering management object, the management association object and the engineering support object. The relationship between these objects and the ones specified by ODP and PRISM is shown in Figure 6.

The engineering management object combines the functionality of the stub into the engineering representation of a computational object. The management association object combines the functionality of the protocol and binder object into one engineering object that corresponds to an existing management association. Finally, the engineering support objects represent additional systems management functions like, for instance, the notification server.

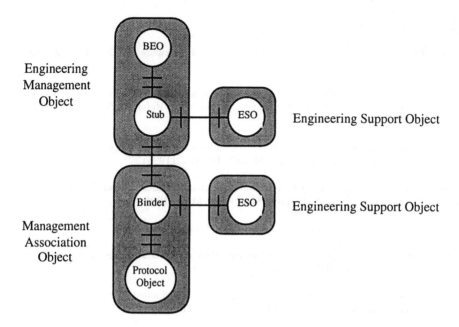

Engineering
Management
Object

Engineering Support Object

Engineering Support Object

Management
Association
Object

Fig. 6. Identification of ODMA Basic Objects

5   Conclusions

There are correspondences between Engineering concepts and TMN concepts which make it possible to realise a TMN system in an Open Distributed Processing environment, as well as an Engineering specification on a TMN target architecture. Some correspondences need further study, especially regarding TMN Function Blocks.

The transition from a Computational model to an Engineering model is non-trivial. The correspondences between a computational building block and engineering concepts are unclear and need further study.

Since the PRISM project is due to finish in 1995, these are open issues to be addressed by workers in ACTS projects.

7   Acknowledgements

We would first like to thank all members of the PRISM project for their good work, and especially Ståle Wolland for his constructive criticism of this paper. We would also like to thank Prof. Gerald Maguire for his comments, as well as those within Telia and Cray who have helped with comments.

8 References

[1] PRISM Deliverable 11: Service Management: Framework for Co-operation, PRISM (1995)

[2] CCITT Draft Recommendation X.901, "Basic Reference Model of Open Distributed Processing - Part 1: Overview and guide to use of the Reference Model", ISO/IEC (1995)

[3] CCITT Recommendation X.902, "Basic Reference Model of Open Distributed Processing - Part 2: Descriptive Model", ISO/IEC (1995)

[4] CCITT Recommendation X.903, "Basic Reference Model of Open Distributed Processing - Part 3: Prescriptive Model", ISO/IEC (1995)

[5] "Engineering Modelling Concepts", TINA-C (1994)

[6] "Object-Oriented Modeling and Design", Rumbaugh, Blaha, Premerlani, Eddy, Loresen, Prentice Hall (1991)

[7] CCITT Draft Revised Recommendation M.3010, "Principles for a Telecommunications Management Network", ITU (1995)

Telecommunications Management Systems Design Issues

Session 3b Introduction

Chairman - Nikos Karatzas

Alpha Systems S.A., Xanthou 3 str, Tavros,
177 78, Athens, Greece
Tel: +30.1.48.26.014-16
Fax: +30.1.48.26.017
e-mail: nikos@alpha.com.gr

The development of network and service management systems has emerged from the on-going and increasing need to provide intelligence to the existing and future broadband networks. Intelligence is required mostly for realising the main objective of future telecommunications: integration of services and guarantee of their quality. Nowadays, we witness the deployment of a number of management systems over realistic network implementations. There is a trend to increase the intelligence of the management systems beyond their static and administrative capabilities of configuration and monitoring.

The design of effective management systems is a complex task, requiring careful design effort. Apart from the task of conceiving and designing the related management algorithms, other critical issues need to be equally taken into account, such as the design of appropriate MIBs and the design of an efficient physical management architecture. The design of appropriate information models should target the enabling of interoperability between the different management components, and supporting encapsulation of management functionality and segregation of management concerns. The task of designing efficient physical management architectures requires the allocation of the management components into physical blocks and the design of a management network to host the identified physical blocks. Taking into account the anticipated interactions of the existing management components, the management system should be designed to best meet various performance requirements such as minimising management overhead, avoiding duplication of information, resilience in fault conditions and fast interactions with the network elements.

The design of management systems must keep up with the requirements emerging from the open communications market. This means designing for open systems providing interoperable interfaces and adaptation facilities, thereby providing management system components produced by various vendors with the ability to inter-work with other management components produced by other vendors and over heterogeneous platforms and networks. Proprietary management systems implementations limit their exploitation potential and similarly network equipment offering inadequate management interfaces will not satisfy market needs in the near future.

A methodology providing the necessary guidance for designing open and efficient management systems is clearly of immense importance. Furthermore, a conceptual framework providing the essential ingredients for building management components and their interfaces is also required. Recognising this need, a number of standardisation bodied (ITU-T, ETSI etc.) and international initiatives (Internet, NMF, ATMF, TINA-C etc.) have started working and producing results in management system design architectural frameworks and methodologies.

Within the above framework, the theme of this session is: *Telecommunications Management Systems Design Issues*. A significant amount of work, and in certain cases pioneering work, regarding service and network management has been undertaken by European projects within RACE I and RACE II. The papers in this section come from leading projects in the area of service and network management. Each addresses certain aspects of the overall management system design process. The experiences gained through practical implementation is a valuable dimension of the work presented in this session.

The first paper, entitled "Experiences in Multi-domain Management Service Development" is based on work carried out in the RACE II PREPARE project. It presents experiences in the development of management services that span several administrative domains and which are therefore representative of the complexities of the open service market. The work in this paper involves the process of developing TMN based management systems in support of multimedia teleservices operating over broadband networks.

The second paper, entitled "The Relationship between Information and Computational Objects in VPN Charging Management", is based on work from the RACE II PRISM project. It addresses the issue of modelling telecommunications service management applications in a distributed environment. The paper discusses the modelling design approach followed; it was based on ODP, TINA-C and TMN approaches. The solution suggested is explored in the context of a VPN related case study.

Last, the third paper, entitled "Managing the TMN", focuses on the architecture of a metamanagement system for managing a management system. Following a TMN approach, the paper deals with the issues of management system configuration, monitoring, software maintenance and fault tolerance, presenting a suitable architecture for tackling these issues.

Experiences in Multi-domain Management Service Development

David Lewis and Thanassis Tiropanis
Department of Computer Science, University College London
dlewis@cs.ucl.ac.uk, ttiropan@cs.ucl.ac.uk

Lennart H. Bjerring, L.M. Ericsson A/S, lmdlhb@lmd.ericsson.se
Jane Hall, GMD-FOKUS, hall@fokus.gmd.de

Abstract. The developers of management systems and the management services that operate over them will be faced with increasing complexity as services are developed for the open service market. This paper presents experiences in the development of management services that span several administrative domains and which are therefore representative of the complexities of the open service market. The work described involved the development of TMN based management systems that provided management services in support of multimedia teleservices operating over broadband networks.

1 Introduction

As the structure of the telecommunications industry moves towards that of an open services market the role of telecommunication service developers will undergo a marked change. Developers will no longer be building services and management services for a single network but must utilise open interfaces to develop services over heterogeneous networks and platforms. Significantly, components that deliver these services will not just be under the administrative control of the service providers, but of the customer and the providers of other constituent services also. This requires, therefore, that the developers of services and, as addressed in this paper, of the accompanying management services must be aware of inter-domain issues.

This paper presents some of the experiences of the PREPARE project in developing inter-domain management services for deployment and demonstration on broadband testbed networks used for multimedia teleservices. This work was performed in two overlapping phases, the first started in January 1992 and resulted in a successful demonstration over a testbed based in Denmark in December 1994, the second started in January 1994 and will result in a demonstration in November 1995, operating on a testbed spanning three countries via the European PNO ATM Pilot [BBI94]. This paper will first outline some of the organisational and other non-functional requirements that characterised this work. It will then detail the experiences of the first phase, their influence on the second phase and the techniques that are therefore being used in this latter phase. Finally an assessment of our experiences and their relevance to future work will also be given.

2 Context

The aim of the project was to gain experience in inter-domain management through t implementation of systems managing multimedia teleservices and the broadband testb networks over which they operate. The project's consortium of collaborating partn contained representatives of many of the players in an open service market, includi public network operators, public network equipment vendors, customer premis equipment vendors, multimedia teleservice developers and management platform vendo The consortium structure, however, introduced an artificial flavour to the process developing the management services by effecting the way the design and implementati work was distributed between different organisations. The pattern of a provic analysing and developing a service for commercial customers was not followed, t instead, as in any collaborative research project, the work had to accommodate t aspirations held by and restrictions imposed on each of the partners involved.

In general the development of a methodology for system design, whether the system an inter-domain management system or something else, can be seen as an application the overall principle of "separation of concerns", i.e. the partitioning of a compl problem space into a number of smaller, more manageable problem spaces. A usel methodology for a project like PREPARE is therefore deeply dependent on the nature the system to be designed, and at least as dependent on the background knowledge a experience of the people who are to use the methodology. What we do hope to provi here is therefore an illumination of the complexity of the overall task of designing int domain management systems by identifying the aspects of the task which in o experience are the most relevant and important ones. The necessary backgrou knowledge of people is in-depth knowledge about TMN [TMN] and OSI-SM [X70 architectural principles and communications mechanisms.

A further influence on this development work was the relative immaturity of standar dealing with inter-domain management. The ITU-T's TMN development methodolo [M3020] did not at the start of the project fully address the issues involved in developi service level inter-domain management systems but rather concentrated on t development of network and network element level information models. The ODP w also considered [ODP]. It was seen initially as being too broad to apply to the speci inter-domain interface development work, but has been referred to more in the later, mo complicated phases of the project, though not in a rigorous manner. The use of NN Ensembles [Ensemble] was also considered in the second phase. It was regarded providing an excellent way of expressing the requirements and designs and has bee adopted for their documentation. The lack from the start of PREPARE of existii Ensemble definitions relating to inter-domain management has however precluded development path closely integrated with the OMNIPoint strategy, since we have, fro the start of the first phase, had to produce our own base specifications (e.g., informatic models) which then were reused and further enhanced in the second phase.

3 Experiences from the first testbed

3.1 Requirements and limitations

The main focus of the first development phase was on the end-to-end management resources of diverse networks owned and administered by different organisation domains. This was reflected in the first phase testbed which was made up of ATM a Token-Ring LANs, a DQDB MAN and an ATM WAN [BBI93]. The PREPARE team that time had experience in network resources management and deployment of management systems. The biggest challenge was to demonstrate end-to-end management of resources over heterogeneous equipment and management software. This required a effective, well-defined methodology and, at the same time, an awareness of the practical limitations.

One area where practical limitations combined with individual partner requirements in shaping the approach taken was that of management platforms. TMN recommendations and OSI management were chosen as the basic technology and different platforms were available to different partners, namely OSIMIS, HP OpenView, IBM's Netview/6000 and later TMN/6000 and Ericsson's TMOS. However the current state of OSI management platform technology does not provide for a standard high level API. The industry standard XMP/XOM was found to be too low level for the volume of development required and in some cases was augmented with a higher level interface. A high level API had been developed for manager applications which provided managed object location transparency by providing a single interface to X.500 and X.700 objects [CFS H430], but this was not ported to all platforms in time to support the project's development cycle as a common API. This lack of a common programming API therefore led to a situation where individual TMN operations systems (OS) were developed by different partners on different platforms, precluding the development of common service software modules that could operate on several platforms. This in turn led, in the initial stages of the project at least, to a concentration of effort on the careful definition of OS functionality as governed by its behaviour at inter-domain interfaces (X-interfaces in M.3010). As will be explained below, as experience of inter-domain interface definition grew, the emphasis in the second phase changed towards specifying the broader management functionality of co-operating systems also.

3.2   Approach

The approach to be taken for the design and implementation of the PREPARE inter-domain management system was not clear at the beginning of the project. The standardised methodologies discussed above did not seem to provide an approach that integrates the service specification, design and implementation phases of inter-domain management services.

The first stage of the initial plan was the definition of management scenarios to be demonstrated, the definition of management services and the definition of initial information models for inter-domain management. After the completion of this stage, the next step was the implementation of the intra-domain systems for the sub-networks and then the implementation of the inter-domain components.

For the first stage, four groups were formed:

- *Scenarios group.* Its aim was to produce a set of management scenarios to reflect a representative subset of management functionality to be demonstrated, set in a realistic enterprise context.

- *Architecture group.* It focused on specifying an implementable TMN based framework for interfacing the components in each domain.

- *Management Services group.* Its work was to define a set of services to operate between the OSs of the different domains.

- *Information Modelling group.* It worked on the definition of information models required for the different OSs involved in inter-domain management according to the GDMO recommendations [GDMO].

For the second stage the output these groups produced had to be refined to a form that could be implementable. The scenarios were refined to describe the flow of information identified in the initial information model between the management OSs identified in the TMN architecture. These *information flows* then provided the basis for refinement of the information models for each OS.

When the information model became stable, the team produced Test Design Specifications (TDS) according to the IEEE standard 829-1983. The TDSs described all the tests that involved components from two or more partners. The test cases derived from the scenarios produced by the scenarios group after the refinement through the information flows and the information modelling. The final level of detail involved defining the CMIP operations between the OSs and all the parameters of the information to be exchanged. The fact that all this detail was specified before the implementation proceeded too far was of great importance, because it gave the opportunity to address and resolve many technical issues that had not been considered during the early stages of the design.

The main stages of the methodology followed by PREPARE in its first stage can therefore be summarised as:

- *Definition of scenarios* to clarify the aims of the whole development and the relationships between the different actors on the enterprise level.

- *Information flows and Information Models* to outline the interactions that take place between the OSs to provide end-to-end management as well as the information structuring of the components in each OSs for inter-domain management.

- *Test Design Specifications* which describe in detail the lower level management interactions between OSs as well as the information to be exchanged between them during the execution of the scenarios.

4 Experiences from the second testbed

The second phase of the project was intended to build upon the first phase both in terms of experiences gained in the inter-domain management service design process and in the extension of the physical testbed and the specific services that operated over it [ISN94].

4.1 Requirements and Limitations

The second phase differed from the first in that the enterprise situation was more complicated, involving more service providers and relationships between service providers and the range of scenarios being addressed was to be more ambitious. There was also a change in emphasis from simply producing service level OSs in each to domain, to adding WSFs with rich functionality for service and network administrators in each domains. It was however similar to the first phase in that the physical management architecture had a TMN-based structure and that the effect of implementing on multiple management platforms was still in evidence.

The actual enterprise situation involved separate multimedia conferencing teleservice and multimedia mail global store teleservice providers providing their services to users on remote customer premises networks (CPN). The teleservice providers used the services of a separate virtual private network (VPN) provider to manage end-to-end network resources over several public network domains and the CPN domains to support the teleservices' communication needs.

4.2   Approach

Based largely on the experiences of the first phase, it was felt that a more cohesive working approach was required. All the analysis, specification, design and implementation was therefore performed in one homogenous group which would split into subgroups at various stages to address clearly defined functional areas rather than splitting into groups addressing the different stages themselves. The approach taken again had to combine both top down and bottom up aspects, though in this case more attention was paid to some of the top down design aspects due to the increased complexity of the enterprise situation. The work can be broken down into the following areas:

- *Enterprise Modelling and Scenario Description:* this laid out the organisational context in which the work of this phase would be and defined, in the form of scenarios, the scope of the management functionality to be addressed within this context.

- *Management Architecture Design:* this provided the physical architecture of the TMN systems that would provide the framework for the more detailed design work.

- *Role Specifications:* These provided a way of describing in more detail the requirements of the involved organisations through the definition of responsibilities for individuals identified in the enterprise model and a way of mapping these requirements to lower level management function requirements.

- *Information Modelling and Information Flows:* This involved the identification of information required to perform the required management functions, their definition as information models representing each domain and the definition of inter-domain operations performed on this information model in support of the required management functions.

- *Operations System Design:* This involved the functional design of the various TMN operation systems. These were identified in the management architecture in accordance with the requirements defined by the role specifications as specified by the inter-domain interfaces, which in turn are defined by the information models and information flows.

- *Workstation Function Design:* This area addressed the particular requirements of the user interface for the individuals operating the management services. It was therefore closely related to both the Role Specifications and the Management Function Design.

It should be made clear that these various areas were not addressed sequentially but were to an extent interleaved with some being revisited after the initial work on others had provided clearer insight into the requirement upon them. Each of these areas are described in more detail in the following sections.

Enterprise Modelling and Scenarios

This area identifies the organisational stake-holders, their characteristics (e.g. core business area) and focus on the objectives for their involvement in order to identify their main high-level requirements on the system and each other. Based on these organisational structures are considered and lead to the identification of roles. We concentrated on inter-domain aspects, i.e. inter-organisational relationships where agents of the stake-holder organisations interact on behalf of their organisations. The overall domain structure was also considered here, and motivates our use of the term inter-domain management.

Role Specifications

In an inter-domain service management system it is very important to distinguish between the roles in the various domains and to establish the relationships between these roles as it is here that inter-domain management functionality is required that must be supported by the inter-domain service management infrastructure. Role specifications were therefore adopted as a means to ensure that the management functionality required by the role holders in the scenarios was adequately described and could be supported by the inter-domain management system infrastructure that was being implemented.

The specific roles required for the demonstrator were first identified via the scenario descriptions. A common role specification template was adopted as a means to structure in a similar manner for all roles what the actual role entails, and by refinement to be able to map from the role specification down to the operations on the resources, and finally as a result of information modelling to the managed objects.

This template was developed based on the work of the ESPRIT project ORDIT, which investigated the organisational requirements for information technology systems by examining roles and responsibilities within an organisation [ORDIT]. In PREPARE the ORDIT concepts were adopted for the specific needs of the role specification work and the particular aims of PREPARE regarding inter-domain service management for the second testbed. Therefore the role specification work was concerned with specifying the responsibilities of the various roles and the relationships between them, according to their depiction in the scenarios. On the basis of the responsibilities, obligations were described, and based on these, the activities associated with each obligation could be described. The role specification template therefore included for each role holder the responsibilities of the role holder and to whom, the obligations that need to be discharged by the role holder in order to meet the responsibilities of the role, the activities which need to be carried out to enable the role holder to fulfil the obligations deriving from the responsibilities and the resources and access rights required to enable the role holder to carry out the activity.

Management Architecture Design

The management architecture referred to here is the physical architecture of the total inter-domain management system. The architecture is designed, in common with that of the first phase, based on the following principles: each organisational stakeholder has its own associated TMN; organisations which own physical network technology has within their TMN an operations system (OS) which implements management functions (OSFs) specific to the particular network being managed by that TMN, a Network Operations System (N_OS); each TMN has a Service Operations System (S_OS) implementing service management functions associated with that particular domain, and which takes part in the overall distributed end-to-end service management function execution. This means that N_OSs inter-operate with the S_OSs in their own TMN (via Q3-type interfaces) and that S_OSs inter-operate with other S_OSs in other TMNs (via X-type interfaces). Management end-users (e.g. CPN administrators, VPN administrators) access the pertinent management function by TMN Workstations interfacing the pertinent operations system. The figure 1 presents an overview of the resulting physical management architecture.

Attention was paid in this area to explicitly addressing the non-functional requirements imposed by the scope of partners interests, the platforms available to partners and to reducing the overall complexity of the information modelling and information flow definition tasks by minimising the number of inter-domain interfaces involved. Significantly by addressing these issues at the architectural stage of the design process it was found that it was subsequently easier to split the work between relatively independent groups addressing functionality in different areas of the physical architecture.

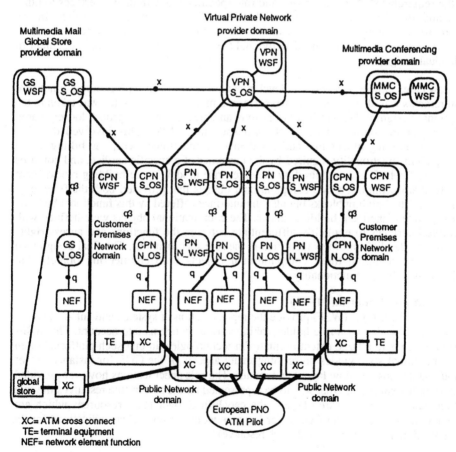

Figure 1: Second Phase TMN Architecture

Information Models and Information Flows

Initial information models based on the requirements imposed by the enterprise model and the scenario descriptions were made more concrete by the role specification process. Following the pattern well established in the first phase, this information model was mapped onto the physical architecture. Information flows were then generated, detailing how the management activities outlined in the scenarios are accomplished by operations on managed objects. An information flow describes a management function implementation in terms of CMIS message flows between two management applications, thus it describes which MOs are being managed, which operations are sent to them, and which attributes and parameters are transferred between the two management applications. Information models and flows need to be designed in an iterative fashion, since information flows will identify missing information which needs to be included in the information model specification and verified through updated information flows.

Operations System Design

The functionality of the OSs identified in the TMN architecture was governed generally by the requirements of the scenarios and role specification with their interfaces to other OSs and the functional interactions between different interfaces defined in the information models and the information flows. Once this level had been achieved however the functional design of individual OSs was left largely to the judgement of individual designers.

In a few cases, however, where the development of several OSs with common functional requirements was shared between two partners a more formal approach was taken to the functional design of the OSs. This work took an object oriented approach loosely based on some of the output of the TINA Consortium [TINA]. This involved defining functional building blocks that addressed specific functional areas, e.g. billing or the management functionality required by the customer of a specific service, and could be used in different OSs as required by the enterprise model. These building blocks were then further decomposed into computational objects that provided both the functional structure of the building blocks but also the interfaces offered by this functional building block to other functional building blocks. The computational objects were defined with multiple interfaces to explicitly differentiate between the functions and access rights required by the different roles identified in the role specifications. The final design then consisted of mapping these computational objects onto engineering objects that implemented the OS functionality.

Workstation Function Design

Workstation Functions (WSF) provide the representation of systems and sub-systems as relevant and needed by a role holder, taking various concerns into account. The WSFs design depends to a large extent on platform technologies, in that such platforms often have individual style guides prescribing many aspects of the GUI, for instance use of colours and maps, window layout and menus. The role specifications however provided important indications of what was to be represented on the screen (the resources the role holder manages or is aware of), and the capabilities over these resources which are available to the role holder (which for instance would provide clear indications for the contents of menus associated with each resource).

5 Assessment and Further Work

The PREPARE project did not have as one of its objectives the development of an explicit methodology for developing inter-domain management systems and the work presented here should therefore not be assessed on this basis. However this work has provided, we believe, some useful insights into and experiences of the processes involved in such development. Some of these insights are implicit in the evolution from the first to the second phase described above, but others are assessed in more detail below, together with areas of further work the project consortium aims to follow in this area.

The ORDIT based role specification aspect of the second phase allows a top-down method that is intended to ensure that all the functionality needed to be implemented in the scenarios is covered. As a top-down approach, the role specification process allows the services and their users to be perceived in a service environment rather than in a network environment. The responsibilities, obligations, and activities that are defined exist regardless of the specific underlying technology, and by focusing on the service a more long-term, service-oriented view is determined, i.e., the emphasis is on what the service offers rather than how it is implemented. Especially in an open service market it is important to consider services from the user and customer viewpoint and the role specification approach supports this by examining the requirements that users have on the management system in order to carry out their tasks. Therefore role specifications provide a suitable methodological approach to capture this very important customer and user-oriented aspect of service management. Role specifications can be seen as a means to define and analyse the main business objectives, interests and motivations for various stake-holders' involvement. This enables design of solutions which fit into the application context of the customers and users. Therefore, domain/stake-holder analysis, followed by work role analysis was found very useful.

The area of functional architecture and specification has not been addressed fully in the project. Issues concerning the definition of functional components for inter-domain service management must be dealt with in more detail if functional components of different services are to be integrated to manage value added services and if functional component reuse is to be encouraged. This requires the specification of a functional architecture to complement the physical TMN architecture and information models.

The PREPARE consortium plans to continue working together in the ACTS programme to investigate the application of inter-domain management to new services and will in the process continue to examine and refine the methodologies used to develop working systems. In the time frame of the ACTS programme work from standards bodies and industrial consortium will provide more relevant input into PREPARE (some of it influenced by PREPARE initially). The NMF is investigating issues of service management [NMF] and some of PREPARE's results in this area expressed in Ensemble format are being prepared as input to this group.

The TINA Consortium is also promising some relevant results that could be applied to multi-domain service design. Its highly object oriented approach to defining service components containing integrated management services as utilised for some OS functional design, may prove useful in the future. Currently however TINA's reliance on a Distributed Processing Environment providing object location transparency, tends to obscure some issues of inter-domain information modelling related to ownership of resources, protection of data from failure in other domains and associated security. There is nothing, however, preventing these issues being addressed as specific services in the TINA-C architecture, and this would be a likely area of research for the PREPARE consortium.

6 Conclusions

The PREPARE project is now in its fourth year of designing and implementing TMN based management systems operating inter-domain management services. This period has seen an increase in the experience of techniques required to perform this task successfully together with a greater understanding of the specific problems invoked by the inter-domain nature of this area. Techniques of iteratively refined information and scenario-based information flows have proven adequate for designing simpler systems, but more complex enterprise situations and wider ranges of management activities need more attention to top down techniques. In particular, where multiple stake-holders are involved in a service, the use of role specifications provides a suitable method for mapping the overall management service functionality to individual role activities and OS functions. The application of TMN principles still largely rotates around the definition of interfaces between components in the physical architecture however. The efficient use of management services in an open service market will require more emphasis on the definition of functional architectures and components.

7 Acknowledgement

This work was partially funded by the commission of the European union under the RACE II program, project number 2004. The view presented in this paper are not necessarily those of the PREPARE consortium.

8 References

[TMN]   ITU-T Recommendation M.3010 (1992), *Principles for a TMN*.

[X700]   ITU-T Recommendation X.700, *OSI Systems Management*.

[ISINM]   D.Lewis, W. Donnelly, S. O'Connell, L.H. Bjerring, "Experiences in
           Multi-domain Management System Development", Proceedings of the
           International Symposium on Integrated Network Management, Santa
           Barbara, 1995

[ORDIT]   R. Strens, J. Dobson, "Responsibility Modelling as a Technique for
           Organisational Requirements Definition", Intelligent Systems
           Engineering, 3 (1), 1994, pp.20-26

[ISN94]   L.H.Bjerring, J.M.Schneider, "End-to-end Service Management with
           Multiple Providers", Proceedings of the IS&N Conference, Aachen, 1994

[M3020]   ITU-T Recommendation M.3020, "TMN Interface Specification
           Methodology"

[TINA]   H. Berndt, L. de la Fuente, P. Graubmann, "Service and Management
            Architecture in TINA-C", Proceedings of TINA Conference, Melbourne,
            1995

[NMF]   Network Management Forum, "A Service Management Business Process
            Model" , 1995

[BBI94]   D. Lewis, P. Kirstein, "A Broadband Testbed for the Investigation of
            Multimedia Services and Teleservice Management", Proceedings of the 3rd
            International Conference on Broadband Islands, Hamburg, 1994

[BBI93]   D. Lewis, P. Kirstein, "Multimedia Applications and Services in the
            PREPARE Testbed", Proceeding of the 2nd International Conference on
            Broadband Islands, Athens, 1993

[ODP]   ITU_T, Recommendation X.900 Series, "Basic Reference Model for Open
            Distributed Processing", ISO 10476

[Ensemble] Network management Forum, "The Ensemble Concepts and Format",
            Forum 025, Issue 1.0, August 1992

[GDMO]   ITU-T Recommendation X.722, "Guidelines for Definition of Managed
            Objects"

[CFS H430] RACE CFS H430, "The Inter-Domain Management Information Service
            (IDMIS)". Issue D, 1994

The Relationship Between IOs and COs in VPN Charging Management

Joan May, Cray Communications (joan@cray-communications.co.uk)
Aurelio Maia.Companhia Portuguesa Radio Marconi (afmaia@cprm.pt)

Abstract: This paper addresses the issue of modelling telecommunications service management applications in a distributed environment. The modelling approach is based on ODP [1] and TINA-C and is harmonised using the PRISM Service Management Reference Configuration (SMRC) [2] to the TMN approach. ODP structures a specification of a distributed system according to five viewpoints. Each viewpoint represents a different abstraction of the system. As TMN Services are designed using these ODP Viewpoints a dilemma occurs as the development process moves from the Enterprise Viewpoint to develop the models of the Computation and Information Viewpoints. On one hand the Computation Viewpoint looks at the set of activities and their actions which will provide the functionality of the TMN service. The actions derived from this are placed in Computational Objects. On the other hand the Information Viewpoint looks at the relationships, attributes and life cycles of Information Objects. To produce consistent Engineering Objects which fully reflect the designs of both the computation and the information viewpoint some method is needed to map between Information and Computation Objects. This paper explores the solution suggested by the PRISM project and describes how this has been tested out by work in its VPN case study.

Keywords: Service Management,ODP, ODP Information Viewpoint, ODP Computational Viewpoint, Information Object(IO), Computational Object (CO), TMN.

1 Introduction

VPN Charging Management is being studied as part of a case study [3] to validate the PRISM SMRC. The PRISM SMRC uses ODP to partition the analysis and design activities necessary to identify management components and management functions. For VPN Charging Management examples of Management Functions include:

- Correlation of charging data
- Calculation costs
- Call charging

To implement the functionality to support such management functions, the analyst needs to understand the operation of the function in terms of the separate actions that will take place in distributed systems through the techniques of the computation viewpoint. In addition, using the techniques of the information viewpoint the analyst needs to determine the attributes and relationships of a static object model, understanding the dynamic impact of such a management function within the lifecycle of these objects. Finally, the results need to be put into MIM [4] and GDMO [5] formats so that they are usable in a TMN environment.

The method used in PRISM to try to overcome these problems can be described in a number of steps which take the analyst from the decomposition of Management Service Components through to the start of the Engineering Viewpoint.

- In the Enterprise Viewpoint identify Management Functions(MF) using the PRISM decomposition method [See 2]. Management Elements are selected by examining the plain text of the Enterprise description. Normally, a method of selecting appropriate nouns is used.
- In the Computation Viewpoint draw Computational Diagrams for each MF using the identified Management Elements as prototype COs.
- In the Information Viewpoint draw OMT[8] object diagrams based on scenarios of management functions and identified management elements which can be used as prototype information objects.

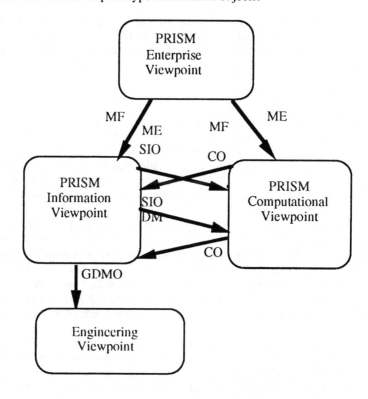

Legend

MF- Management Function   ME - Management Element
SIO - Static Information Object   CO - Computation Object
DM - Dynamic Model

Fig. 1. Mapping Information Model to the Computational Model

- Compare the models produced in the Computation and Information Viewpoint. New objects identified should be included in the models of the other viewpoint.

- Perform Dynamic modelling to check the consistency of the object models produced. New objects, relationships, attributes and operations may be identified at this stage.
- Redraw Computation and Information Model based on the derived changes.
- Convert these models into MIM/GDMO descriptions.

2 Information Model for VPN Charging Management

The VPN case study took three steps to model the Information Viewpoint for Charging Management:

- A static object model using the concepts and notations of OMT.
- A dynamic model using OMT state transition diagrams.
- The resulting models are then converted in to GDMO descriptions.

2.1 Static Object Model

VPN charging will involve several business participants such as customers, network providers and value added service providers (VASPSs). The VPN charging object model has to cater for all those involved. However, all participants need not be involved directly in a charging relationship. For instance, a customer may pay a VASP for a service, whilst a VASP will subsequently pay the network provider. We therefore decided to develop a general charging model to reflect this.

The central relationship is between a supplier and a consumer who agree a contract. The contract is an attribute of the agreement relationship between the supplier and consumer. In the more specific case a supplier may be a network provider(BSP), a VASP or a customer who maintains accounting details of users. The consumer may be a VASP, a customer or a group of users. The contract once agreed will specify the service to be provided. The service has the relationship with a supplier, in that the supplier supplies the service, and with the consumer who pays for the service.

The service will be described in a service specification from which will be derived a set of profiles describing sub-sets of the service which will be provided for particular types of individual users. The account will describe how a service consumer, whether they be a VASP, customer or user, has exploited the service. The service consumer is shown in the diagram as being either a consumer or client. A client is any actor who benefits from a contract without being a direct party to it. So if the consumer is a VASP the client could then be a customer, a user or other VASPs.

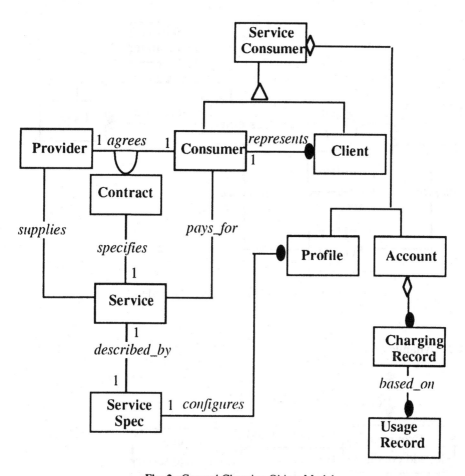

Fig. 2. General Charging Object Model

The generic model may then be exploited to produce a specific model. The case descibed below is that of the VASP/Customer, in which a customer pays a VASP for a VPN service. A customer will represent one or more users.

The relationship of a customer to a user may exist in several forms:
- The customer is the user.
- The customer represents a number of small businesses or residential users.
- The customer represents a small company whose employees are users.
- The customer represents a medium to large organisation.

In each of these forms except possibly the first the customer will want to have sufficient information from the VASP to be able to distribute the costs.

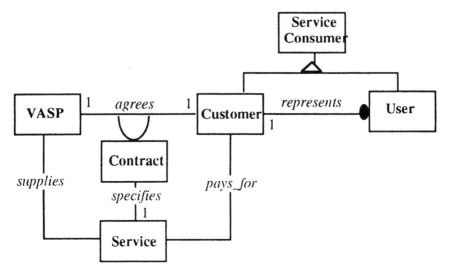

Fig. 3. VASP Specific Charging Object Model

2.2 Dynamic Model

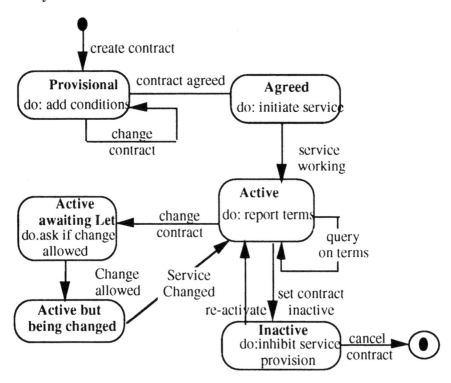

Fig 4. Contract Object State Model

Dynamic modelling is then performed to explore the completeness and life cycles of the proposed objects. The Contract State Model has been chosen as an example of the dynamic modelling done in the case study. OMT state transition diagrams were used as a modelling tool.

The Object Contract may react to one of the events, Create Contract, Agree Contract, Make Contract Active, Make Contract Inactive and Cancel Contact.

- Provisional - The parties to the contract are in the process of negotiating the contract, no agreeement can be presumed on either side but the current object contains details of the components of the agreement outlined so far.
- Agreed - The contract has been agreed but has not yet been put into effect.
- Active - The contract is now in operation , there is an active service with which it is associated.
- Active but Waiting for permission to allow change - A change has been requested to the contract but has not yet been agreed.
- Active but in the process of being changed. - A change is being made to the service.
- Inactive - The service has been suspended, normally prior to one or more parties to the contract cancelling the contract.

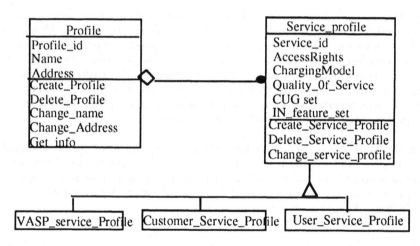

Fig.5. Profile Object Class Description

The models shown above have been restricted to depictions of the objects and their relationships so that the general design is not confused by detail. However, a detailed analysis of the attributes and operations required can also made. The figure 5 gives more detail of the profile object.

The diagram describes the Profile object using OMT. Access Rights, charging model, quality of service, CUGS and IN feature set are represented as relationships to other objects which are only referenced here. The top third indicates the name of the object, the middle third the attributes and the bottom third the relevant operations. From this information and the models GDMO definitions were then constructed.

3 Computational Model for VPN Charging Management

In order to relate specifications of TMN Management Functions (MFs) of the Enterprise Viewpoint to specifications of the Computational Viewpoint the relationship between MFs and computational concepts has been investigated. It is suggested that an MF corresponds to an operation on a CO and that execution of an MF triggers (and terminates) the execution of an activity in the Computational Viewpoint. Hence specification of activities can define the relationship between (user) requirements for functionality and the COs and their interactions which support the functionality.

The analysis will be restricted to the VASP management domain and will concentrate on the following Management Functions:

MF1 Correlation of charging data
MF2 Calculation costs
MF3 Call charging

To provide external users with the functionality of these management functions associated Management Activities can be considered to exist. A Management Activity links together the distributed actions of several objects. Together these actions perform all the necessary processing to satisfy the user request. The links which connect actions together are messages in the form of interactions (request/response pairs) or notifications. An interaction or notification will trigger an action whose operation is offered through an interface on a CO. Actions are either observable actions (operations) or CO internal actions.

A modelling method for the Computational Viewpoint has been used including several steps and iterations. The relationship to the enterprise viewpoint, i.e. identification of MFs, are the starting point of the modelling method. Then initial Operation Signature corresponding to MFs, Management Activities and CO Types and Interface Types are specified.

3.1 Computational Objects

In order to fulfil all the management functionalities required by the selected MFs it is necessary to identify the COs that will support those functionalities.

The following COs were identified:

MF1 Correlation of charging data
- Resource Usage Collector (NL)
- Usage Records (SL)
- Data Correlator (SL)
- Customer Records (SL)

MF2 Calculation costs
- Costs Calculation Manager (SL)
- Tariff Manager (SL)
- Contract Manager (SL)
- Charging Records (SL)
- Customer Records (SL)

MF3 Call charging
- Charging Manager (SL)
- Charging Records (SL)
- Customer Account (SL)
- Contract Manager (SL)

The purpose of identifying COs is to specify the Computational Interfaces, that is the set of Operations provided to a user (man or machine) hiding the details of communication and distribution. Such Computational Interfaces are the interaction points between COs and may produce TMN Reference Points after a binding process has been applied. An interface at a CO acts as both an access port to the services offered by that object and also as a controlling mechanism. A client object need only be aware of the interfaces of an object that it wants to use. However, other interfaces may well exist which offer different services to other client objects.

For management interactions Computational Interfaces have been proposed as operational interfaces, which are interfaces in which interactions are structured in terms of operations invocations and responses.The following is a textual description for each COs.

Resource Usage Collector: This CO will be able to identify the utilisation of the available resources. It will classify the usage of any resource by type, by volume (e.g. amount of time, amount of bytes, etc.) by customer.

Usage Records: This CO will be the responsible entity for recording the appropriate format and place all the relevant data provided by the Resource Usage Collector.

Data Correlator: This CO will be able to correlate all the resource usage information recorded by the Usage Records CO. It will classify that information by type of records and by customer. The result of the correlation will be used to update customer records.

Customer Records: This CO will retain all the resource usage information for each customer. That information will be one of the main inputs for the calculation of call costs.

Costs Calculation Manager: This CO will be able to perform several functions to calculate costs for the utilisation of the VPN facilities of a given VPN call.

Tariff Manager: This CO will be able to select the appropriate tariff for the utilisation of the VPN facilities. The available tariffs are predefined by the tariff policy.

Contract Manager: This CO will be responsible for maintaining up to date details concerning the VASP-Customer contract. It will be able to provide such information whenever required by other COs.

Charging Records: This CO will be able to record all call costs produced by the Costs Calculation Manager. The recorded information will be the base for production of account data for each customer.

Charging Manager: This CO is responsible for all details relevant to financial matters. It will be responsible for providing customer account data relevant to producing customer bills. It will be able to answer queries as to the credit worthiness of a customer and to charge for the uses of a particular management service

Customer Account: This CO will be responsible for updating and producing a customised bill using the account data provided by the Charging Manager.

3.2 Computational Diagrams

The graphical notation for abstract computational specifications consists of the following three diagrams:

- Computational activity Diagram;
- Computational Object Type Diagram;
- Building Block Type Diagram.

Computational Activity Diagram
The purpose of this diagram is to provide the semantics of the Management Activity, describing all the operations and actions involved. In order to validate the Activity diagrams and testing out that the Activity designed will provide the correct behaviour event-trace diagrams are also drawn. Only one representative diagram for each type of diagram will be shown.

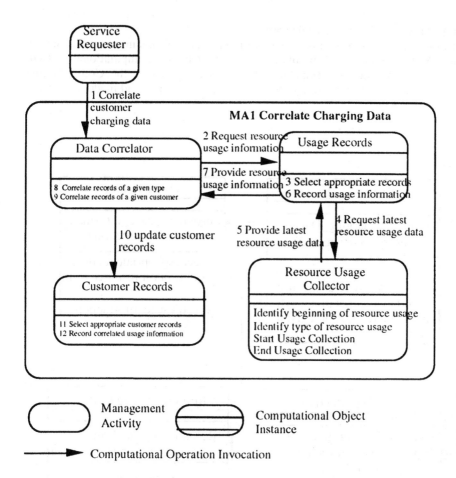

Fig. 6 Computational Activity Diagram related to MF1

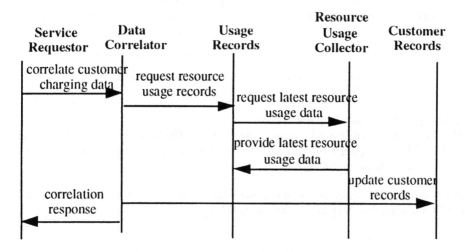

Fig. 7 Example of Event-Trace Diagram related to MF1

Computational Object Type Diagram

The Computational Object Type Diagram groups and collects information captured in the Activity Diagram. This diagram is similar to the Computational Activity Diagram but instead of a sequence of operations, it shows the Computational Interfaces of each COs. Therefore the operations are grouped and made available at specific interfaces.

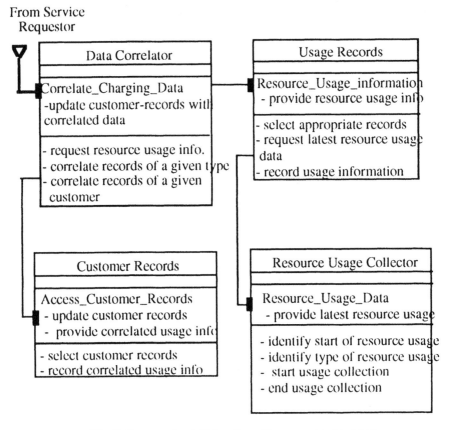

Fig. 8. Computational Object Type Diagram related to MA1

Building Block Type Diagram

The Building Block Type Diagram describes the Building Block types that are packages for grouping Computational Objects types. Externally visible interface types are defined in contract types.

Fig. 9 shows the referred diagram contain Building Blocks that are the packages for the COs involved in the provisioning of the MFs.

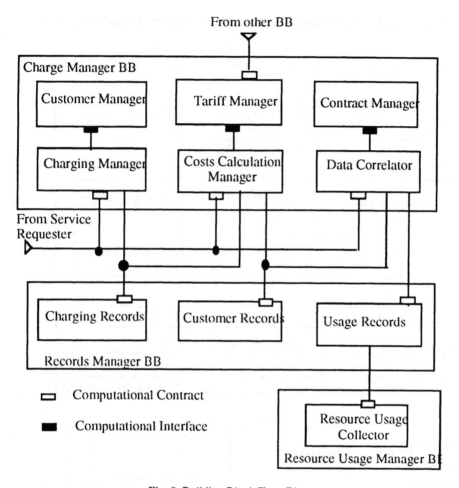

Fig. 9. Building Block Type Diagram

It is important to notice that each of the identified Building Blocks could contain other COs that support other functionalities that are related with other MFs.

4 Mapping between Information and Computational Objects

The following diagram shows a mapping between Computational and Information objects.

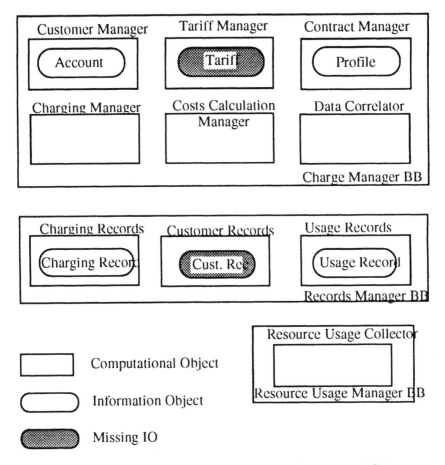

Fig. 10. Mapping of Information Objects onto Computational Objects

From the diagram it can be seen that some information objects are missing in particular Tariff and Customer Record that need to be added to the information model. For the remaining objects (Charging Manager, Costs Calculation Manager, Data Correlator and Resource Usage Collector) there is no direct mapping to any Information objects because they are "pure" COs that make some calculations or provide some mechanisms for data collection.

According to the previous mapping, a new information model can be drawn. The next Figure shows that new model.

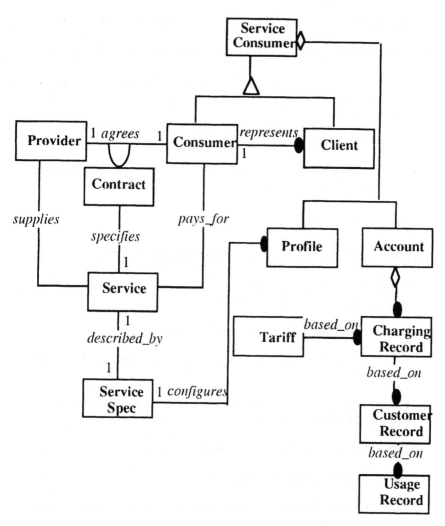

Fig. 11. Amendment of General Object Model

Mapping the Information Viewpoint to the Computational Viewpoint identifies new attributes and relationships and interactions. This is however, too detailed to expand on here.

5 Conclusion

In the VPN case study we have produced two orthogonal models for the information viewpoint and computational viewpoint for a Charging Management Management Service. In mapping these two models together we have been able to improve on each model and so produce a better base on which a engineering/technology implementation can be developed. It is possible that using such a mapping the following benefits may be achieved:

- verification - unneccessary objects and relationships can be identified;

- completeness - unspecified objects and relationships can be identified;

- logical correctness - logical faults in the sequnce of operations can be spotted;

- scheduling control - can identify whether events are handled correctly irrespective of the state of the object;

- simplification - unnecessary complexity can be revealed such as using the services of intermediary objects when not required.

6   References

[1]   ITU-T X.901 ISO/IEC 10646 -1 ODP Reference Model Part 1.

[2]   PRISM Deliverable 8, "Reports on selected Areas of the Service Management Reference Configuration" [1994]

[3]   PRISM Deliverable 7 Volume 2 "Second VPN Case Study".

[4]   ITU-T X.720 Structure of Management Information: Management Information Model

[5]   ITU-T X.722 Structure of Management Information Guidelines for the Definition of Managed Objects

[6]   TINA-C Information Modelling Concepts

[7]   TINA-C Computational Modelling Concepts

[8]   Object-Oriented Modeling and Design, RumBaugh, Blaha, Premerlani, Eddy, Lorensen Prentice Hall ISBN 0-13-630054-5

Managing the TMN

Stelios Sartzetakis, Costas Stathopoulos, Vana Kalogeraki, David Griffin
Institute of Computer Science, Foundation for Research and Technology - Hellas
PO Box 1385, Heraklion, GR 711-10, Crete, Greece.
tel.: +30 (81) 39 17 27, fax.: +30 (81) 39 16 09.
e-mail: stelios@ics.forth.gr

Abstract. Metamanagement provides a means to manage the management processes, systems and software comprising a TMN system. In this paper we present the basic requirements of metamanagement, the design and implementation of these TMN metamanagement extensions. The OSI Directory Service is used to store information about the TMN resources, CMIS/CMIP is the basic management service/protocol, and FTAM is used for distribution of software and configuration files. Metamanagement agents communicate with other specialised management processes, providing the TMN metamanagement service, according to the OSI manager/agent model. We show how this model is used to aid the configuration of prototype TMNs and how this can be automated to evolve to a self-managed, self-configured TMN.

1. Introduction

Whereas the scope of the Telecommunications Management Network (TMN) [M. 3010] is to manage networks and services, the scope of metamanagement is to manage the TMN itself. Although the management of the TMN is identified in the standards as one of the basic TMN management services, it is marked as "for further study" [M. 3200]. The purpose of this paper is to describe an architecture that provides metamanagement functionality, initially intended for use by a TMN operator, and when extended, by TMN processes.

It is quite often argued that the TMN is very complex and its implementations based on OSI management technology, despite being very powerful, are difficult to implement because of the complexity of the underlying service/protocol (CMIS/P) and the power and expressiveness of the associated information model. The RACE II ICM project has managed to design and implement large TMN systems [Gri95] and has successfully demonstrated the use of these systems both in ATM testbed experiments and in larger simulated ATM networks. The OSIMIS TMN platform [Pav93b] developed earlier in the project is now enriched with *metamanagement* agents and specialised management processes for manipulating management applications and for accessing them in a distributed fashion. We introduce the term *metamanagement* to refer to the *function of managing the management processes, systems and software comprising the TMN*. This includes the start-up and shutdown of management processes, recovery procedures in case of failures, as well as automatic code and configuration files distribution and updates, so that an easily configured and fault tolerant TMN is provided.

The TMN is a distributed management environment [M. 3010]: a variety of distributed processes interact over a Data Communication Network (DCN) in order to manage the services and resources of an underlying managed physical network. Management information is maintained on a large number of distributed agents. These agents interact with a variety of management applications (managers) over the DCN by using the OSI manager/agent model which forms the basis for both intra-TMN (Qx/Q3 interfaces) and inter-TMN (X interface) communication. The collection of agents and managers (or, with one name, management processes) implement the Operations Systems (OSs), Mediation Devices (MDs), Q Adaptors (QAs), Workstations (WSs) and Network Elements (NEs) of the TMN. The main objective of the TMN is to offer a variety of management services that will provide for effective service and network management.

Within this paper we will assume that the various TMN management processes interact according to the enhanced OSI manager/agent model proposed in [Sta95]. This means that each management process is capable of not only updating the Directory with information about its state, functionality, location etc., but also interrogating the same information about other process via Directory queries. This provides the critical information about which processes exist in a TMN, the management services they provide and a location transparent way for associating with them. This kind of information, also called Shared Management Knowledge, forms the basis for our metamanagement system which should be able to maintain a global view for the various TMN building blocks and the way they interact. The metamanagement system described in this paper is implemented on top of OSIMIS [Pav93a] which supports the mechanisms described in [Sta95].

The TMN building blocks (OS, MDs, QAs, NEs and WSs) are installed on a network of computers which communicate over the DCN. In this paper, we call these computers "*TMN hosts*". These may be independent workstations, like in our testbeds, or special management single board computers inside the network elements themselves. The only requirement for a *TMN host* is to run a multitasking operating system which uses OSI to communicate with other hosts in the TMN. Within certain limitations any TMN building block may be instantiated and run on any TMN host. These limitations are related to:

performance issues, e.g. QAs and some other building blocks may need to be physically close to NEs or other building blocks in order to minimise TMN traffic.

software availability, e.g. the latest version of each binary image executable and the necessary configuration files need to be available to the host.

capabilities of the host, e.g. RAM size, processing capability, current processing load.

Being another form of management, metamanagement covers all of the five management functional areas: fault, configuration, accounting, performance and security.

Configuration management of the TMN deals with the instantiation and termination of TMN building blocks on TMN hosts.

Fault management of the TMN is concerned with handling failures of either TMN hosts or TMN building blocks or link failures in the DCN.

Accounting management of the TMN keeps track of the overall TMN activity by maintaining and processing a variety of log data files.

Performance management of the TMN ensures that the allocation of TMN building blocks to TMN hosts satisfies certain TMN performance criteria.

Security management of the TMN certifies the authorised exchange of management information between the management processes and ensures user authentication and authoritative only access to the management processes.

The above areas are closely related. For example, configuration must take into account performance issues and also the current state of failures, subsequent failures in the TMN will result in reconfiguration, faults in the TMN will cause performance degradation and so on.

The system presented in this paper is a first step to this direction, and for the moment aids TMN operators in the tasks of configuration, fault, and performance management of the TMN.

There are two important issues in a distributed TMN environment:

The provision of an initialisation - configuration system to start TMN's management processes

To ensure availability of the management processes, coping with failure cases.

In essence, we are dealing with relocatable TMN application processes (that is, processes that are not constantly bound to some specific TMN host). Having such processes in the TMN is a major advantage, because for example reconfiguration in case of failures is possible, but makes metamanagement a more challenging task.

The fundamental requirement of metamanagement is to provide the mechanisms for the efficient operation of the TMN. The overall functionality of the metamanagement service can be decomposed to the functional management areas described above. The user of the service is not only the TMN operator who has to configure, and occasionally reconfigure, the TMN but also the various management processes that have to report whenever there is a problem in communicating with other processes.

In the first implementation we dealt with the case of the TMN operator who wants to initially configure and start the TMN. We offer a tool to the TMN operator that helps him configure the TMN. After the configuration, the metamanagement system monitors the operation of the TMN and informs the operator of any exceptional conditions. It is the responsibility of the operator to take appropriate action. This initial 'manual' mode of operation provides the basic capabilities of metamanagement, and permits an evolutionary path towards its eventual automation by incorporating intelligence into the metamanagement components themselves.

In the next section, we present the metamanagement model for the TMN management processes. Section 3 gives the design details on the implementation that we developed for the first version of this work. We conclude by discussing the open issues in managing the TMN and our plans for future enhancements.

2. The Metamanagement system: model and current operation

Figure 1 depicts the metamanagement system model. Two new processes are introduced: the metamanagement agent (MMA) and the metamanagement operations system (MMOS). It is assumed that every TMN host runs a MMA acting in the agent role containing a MIB presenting managed objects representing the management processes running on that TMN host. Additionally, the MMA contains an FTAM client that is capable of connecting to an appropriate FTAM server in order to retrieve software and configuration files not present locally on the TMN host.

The MMOS acts in the manager role. It implements an Operations System (OS) equipped with an HCI (Human Computer Interface) through which the TMN operator is capable of performing appropriate management operations on the MMAs and therefore the management processes on the TMN hosts. Both the MMA and the MMOS contain a Directory User Agent (DUA) that performs the Directory updates and queries described in [Sta95].

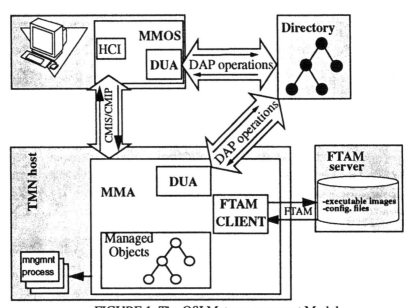

FIGURE 1. The OSI Metamanagement Model

The MMA and MMOS are critical for the operation of the TMN and are therefore implemented as fault tolerant processes as discussed later in this section.

The MMOS obtains information concerning the running MMAs, and therefore the available TMN hosts, through Directory search operations. Additionally it keeps the current load of each TMN host and an estimation of the load of the DCN links interconnecting the MMOS to that particular MMA. The load of each TMN host is retrieved from the corresponding MMA's MIB. The load of the DCN links is currently measured by periodic round trip time estimations for MMOS messages to MMAs. In the future more sophisticated measurements based on performance monitoring of the TMN will be used.

The TMN operator, using the MMOS HCI, determines the appropriate TMN host on which to run a new management process. This decision is based on the information about MMAs displayed on MMOS HCI, on the knowledge that the operator has about the topology of the managed network, the topology of the TMN DCN, and the points where the TMN connects to the underlying managed network elements.

After an appropriate MMA has been identified, the new management process is started via a CMIP "create" operation from the MMOS to the chosen MMA, i.e. the MMOS creates a managed object within the MMA's MIB. The attributes of the new managed object (MO) include the Distinguished Name (DN) of the process that this MO represents and the DN of the software package that the new process should run. The process's software package consists of the necessary executable, configuration and data files and this information is stored in the Directory (described in Section 3.)

The MMA reads from the Directory the entry corresponding to the process's software package and checks whether the necessary files are available locally. If the software package is already available, the MMA executes the process and updates its MIB with the newly created MO. After receiving the appropriate object creation message the MMOS updates the Directory with information about the started process. This information includes the MMA that is responsible for the process, the software package that the process runs and the execution vector used to start the process. Note that the MMOS updates the directory rather than the newly instantiated management processes since these are not usually aware of all the required update information.

On the other hand, if the software package is not available on the TMN host (or it is out-of-date), the MMA retrieves the files from the TMN FTAM server located in one of the TMN hosts (for increased availability, more than one FTAM servers could be available, mirroring the TMN data). After retrieving the necessary files the MMA proceeds as above.

As well as creating new management processes, the TMN operator may need to terminate already running management processes (e.g. reconfiguration of the TMN). A prerequisite for this is that the MMOS should know which processes run and on which TMN hosts. Again, this information can be retrieved from the Directory. To terminate a management process, the MMOS issues a CMIP "delete" operation on the appropriate managed object in the appropriate MMA.

It is critical that the MMOS and the MMAs are implemented in a way that ensures TMN reliability and fault tolerance. Fault management of the TMN refers to the mechanisms that ensure that faults in the TMN (failures in management processes, TMN hosts and the DCN) are detected and appropriate actions are taken by the metamanagement system. One of the important issues here is that the metamanagement system has to manage itself, that is the MMAs and MMOS have to deal with their own failures. In order to achieve this, each MMA communicates with a lightweight process[1] (running in the same TMN host as the MMA) by periodically sending "I am alive" messages. If the MMA does not respond for some time, the lightweight process assumes that it has died, and is responsible for restarting it. There is provision for a log file that keeps checkpoints describing the latest state of the MMA. This log file is read by the MMA on start-up, ensuring that it recovers its latest state to maintain consistency. Conversely, if the lightweight process dies, the MMA will restart it. A similar approach can be followed for the MMOS, although for the time being MMOS faults are simply detected by the TMN operator.

Since the MMOS periodically requests the load status of each TMN host by issuing CMIP requests to every MMA, it has a simple way for detecting either a faulty TMN host or a DCN failure whenever a MMA does not respond (bearing in mind that MMAs are reliable processes as described above). In this case, the TMN operator has to take appropriate action by either reconfiguring the TMN (in case of a faulty TMN host) or fixing any failing DCN links. Note that in order for the MMOS to distinguish among the above two failures special functionality should be implemented (e.g. alternate routes for connecting to MMAs) that help determining whether an MMA is unreachable because of a host crash or a link failure.

An alternative approach to the polling method, is for the MMAs to provide event reports such as 'overloaded TMN host'. In this case other management processes required to communicate with failed management processes on a crashed TMN host would be responsible for informing the MMOS that the required management process is not available. This way the MMOS will be informed of TMN host crashes.

Each MMA periodically checks whether the management processes running on its TMN host are alive (e.g. by issuing a CMIP get request for the top - namely, the system - MO). In case of a process failure, the MMA sends an appropriate notification to the MMOS. The TMN operator (through the HCI of the MMOS) is always able to have a complete view of the TMN. He can query the directory for TMN hosts, the processes that run or are available, where they run and their state. As soon as the meta-management system informs him that a process is dead (or unreachable), he can decide to restart it. Additionally, he may decide to reconfigure the system, so to improve performance, by terminating and restarting a management process in another TMN host. This will usually involve complicated recovery procedures, especially when a TMN process has to recover its previous state. As mentioned above, other management processes also report to the MMOS whenever they fail to associate with a peer management process. For the present, the TMN operator has to track the reason for that failure.

---

1. In this context, lightweight process does not refer to a thread but to a special separate process that implements some simple function without supporting any "heavyweight" OSI application protocol (e.g. CMIP). It is achieved via a local communications method (internal to the host).

3. Design Considerations

3.1 Directory objects for metamanagement

The MMOS is an OSI application process, hence it will be stored in the OSI Directory just like other management processes [Sta95]. The communication with the MMOS is realised through one application entity. The same holds for the MMA. The attribute that distinguishes a metamanagement process is the management service that it implements (namely, "metamanagement"). This is the service that both MMAs and MMOSs provide, the difference being the management functions they performs.

One question here is to which TMN building block the MMAs belong. Although this is still open to discussion, the MMA can be considered a QA since it provides a Q interface to the operating system dependent process invocation mechanisms.

Two extra attributes will be used in the directory object of each management process. The first points to the MMA containing the MO representing the management process:

```
metaManager ATTRIBUTE
WITH ATTRIBUTE-SYNTAX DistinguishedNameSyntax
SINGLE VALUE
```

The above attribute is an optional attribute of the *managementProcess* object class (see [Sta95]). The value of this attribute is chosen during the initial configuration procedure of the TMN and is updated by the MMOS as described in the previous section.

The second attribute points to the software package necessary to run the process:

```
usesSoftwarePackage ATTRIBUTE
WITH ATTRIBUTE-SYNTAX DistinguishedNameSyntax
SINGLE VALUE
```

The software package is represented by a Directory entry containing attributes for the required executable, configuration and data files and the generic argument vector that should be supplied on starting the process. Additional attributes will contain the version of the software package, description of the purpose of the process, operating system requirements, hardware requirements etc. The ASN. 1 definitions for the above entry follows. A new object class is defined as:

```
icmSoftwarePackage OBJECT CLASS
SUBCLASS OF top
MUST CONTAIN{execVectorFormat, binaryImage, configurationFile}
MAY CONTAIN{spVersion, description, OSrequired, hardware}
```

The definitions for the attributes defined in the *icmSoftwarePackage* object class are:

```
execVectorFormat ATTRIBUTE
WITH ATTRIBUTE-SYNTAX caseIgnoreString
SINGLE VALUE
binaryImage ATTRIBUTE
WITH ATTRIBUTE-SYNTAX unixfileSyntax -- a unix file path for first version
MULTI VALUE
configurationFile ATTRIBUTE
```

```
WITH ATTRIBUTE-SYNTAX unixfileSyntax -- a unix file path first version
MULTI VALUE
spVersion ATTRIBUTE
WITH ATTRIBUTE-SYNTAX caseIgnoreString
SINGLE VALUE
OSrequired ATTRIBUTE
WITH ATTRIBUTE-SYNTAX caseIgnoreString
MULTI VALUE
hardware ATTRIBUTE
WITH ATTRIBUTE-SYNTAX caseIgnoreString
MULTI VALUE
```

Additionally, the QUIPU *execVector* attribute [Kil91] will be used in every management process entry keeping a character string with the actual arguments that were used at process invocation.

The file attributes, mentioned above, can also be pointers to entries below a subtree of the DIT that keeps file information. This will not be the case for our first implementation but it will be used in the future. A general description of the directory objects representing files can be found in [Bar94]. Based on the information contained in this subtree an FTAM client will be able to retrieve the files from an appropriate FTAM server. For our metamanagement system this FTAM server will make available all the software packages needed for the TMN management processes. Security and access control are currently under design as we briefly mention in the open issues section.

3.2 Definition of managed objects

The Metamanagement Object Class definitions used in our first implementation are the following:

Managed Object Class definition

```
ManagementProcess            MANAGED OBJECT CLASS
DERIVED FROM                 top;
CHARACTERIZED BY             ManagementProcessPackage;
REGISTERED AS                {icmManagedObjectClass 1};
```

Name Binding definition

```
ManagementProcess-system     NAME BINDING
SUBORDINATE OBJECT CLASS ManagementProcess AND SUBCLASSES;
NAMED BY
SUPERIOR OBJECT CLASS        system AND SUBCLASSES;
WITH ATTRIBUTE               ManagementProcessTitle;
BEHAVIOUR                    MngmtCreateBehaviour;
CREATE                       WITH-AUTOMATIC-INSTANCE-NAMING;
DELETE                       ONLY-IF-NO-CONTAINED-OBJECTS;
REGISTERED AS                {icmNameBinding 1};
```

Package definition

```
ManagementProcessPackage PACKAGE
BEHAVIOUR                    MngmtProcessClass;
ATTRIBUTES
ManagementProcessTitle       GET,
```

ManagementProcessExecVector   GET;
NOTIFICATIONS
objectCreation,
objectDeletion;
REGISTERED AS   {icmPackage 1};

Attribute definitions

ManagementProcessTitle ATTRIBUTE
WITH ATTRIBUTE SYNTAX
UCLAttribute-ASN1Module.SimpleNameType;
MATCHES FOR EQUALITY, SUBSTRINGS, ORDERING;
BEHAVIOUR   MngmtProcessTitleBehaviour;
REGISTERED AS   {icmAttributeID 100};
ManagementProcessExecVector ATTRIBUTE
WITH ATTRIBUTE SYNTAX
UCLAttribute-ASN1Module.GraphicString;
MATCHES FOR EQUALITY, SUBSTRINGS, ORDERING;
BEHAVIOUR   MngmtProcessExecVectorBehaviour;
REGISTERED AS   {icmAttributeID 101};

Behaviour definitions

MngmtCreateBehaviour BEHAVIOUR
DEFINED AS
"The objects could be created either when initializing the agent or during the
program execution as required by the manager. There is no limit on the number
of object instances that can be simultaneously generated. " ;

MngmtProcessClass BEHAVIOUR
DEFINED AS
"This is a class of managed objects that represent running processes. It
includes the attributes of these objects, the actions that are performed by
them and the notifications that are generated upon object creation and deletion." ;

MngmtProcessTitleBehaviour BEHAVIOUR
DEFINED AS
"The ManagementProcessTitle is an attribute type whose distinguished value
can be used as an RDN when naming an instance of the ManagementProcess
object class." ;

MngmtProcessExecVectorBehaviour BEHAVIOUR
DEFINED AS
"This attribute type represents an execution vector. It contains the actual
arguments used when starting the process." ;

The *ManagementProcessExecVector* attribute has the same value as the *execVector*
directory attribute. Although in the above definitions we do not have any actions
defined it is expected that actions like process recovery will be identified in the future.

4. Open issues

As we stated in the introduction, the metamanagement service has yet to be formally defined in the standards. Until then, a continuous evolution of the described system is under way implementing many features that, from our experience, we consider useful.

Although the ultimate responsibility for managing the TMN processes belongs to the TMN operator, some responsibility may be delegated to the MMAs. This means that, if one of the management processes running on a host die abnormally, the host's MMA might be able to restart it itself.

Additionally, whenever a new version of a software package is available in the TMN FTAM server, the related directory entries are updated from the MMOS. It is the TMN operator's responsibility to find which management processes use this data, decide whether existing processes need to be restarted with the new version of the software (or data), and instruct the appropriate MMA to retrieve the new version of the software package and restart the processes.

Another area of further work is in the application of security and access control mechanisms. It is important to restrict the access of managers to the MMAs to ensure that only the authorised MMOS is able to initiate new processes and, probably more importantly, to terminate existing processes. Directory access control lists will ensure that only the metamanagement processes will be able to retrieve the relevant information from the Directory. Authentication information for the FTAM sessions will be also kept in the Directory. At the time of writing security mechanisms are being incorporated into the OSIMIS platform, when they are available, they will be used by the metamanagement system.

Other remaining open issues are related to the inclusion of more intelligence into the MMOS. Currently it is assumed that a human TMN operator will decide which TMN hosts should run the management processes. But, by providing algorithms and heuristics for choosing appropriate TMN hosts according to: the TMN topology, loading of the TMN hosts and the DCN, the specific requirements of individual management processes; the MMOS will be able to make TMN configurations with little or no human intervention.

Once the MMOS is able to make automatic decisions on the placement of new management processes we are presented with the opportunity for the automatic, self-configuration of the TMN. In this way, only a single management process needs to be instantiated "manually", from that point, whenever this process needs to associate with peer management processes, and it is discovered that the required management process is not running, it will contact the MMOS, who will then decide where it should run and instantiate it via the appropriate MMA.

5. Conclusion

Starting from the fact that a management system for the management system itself is needed, we designed the system described in this paper. We implemented it, and it is currently in operation in experimental TMNs consisting of 10s of management processes. We have shown that a management model based on OSI systems management concepts is rich enough to support the requirements of a metamanagement system. This type of management is typical in distributed systems contexts but we have shown that the OSI management and directory models provide a very good solution in an OSI management environment.

6. Acknowledgements

The original idea for the management of TMN components using specialised agents was contributed by George Pavlou. The work was carried out under CEC RACE II project R2059 ICM (Integrated Communications Management). The authors would like to thank all the ICM members for their feedback and support (especially the UCL OSIMIS team).

7. References

[M. 3010] ITU/CCITT Recommendation M. 3010, *Principles for a Telecommunications Management Network*, Geneva, October 1992.

[M. 3200] ITU Recommendation M. 3200, *TMN Management Services: Overview*, Geneve, October 1992.

[D10040] ISO/IEC DIS 10040, *Information Technology - Open Systems Interconnection - Systems Management Overview*, June 1991

[X. 500] ITU/CCITT Recommendation X. 500, *The Directory - Overview of Concepts, Models and Services*, December 1988.

[Pav93a] Pavlou, G., S. Bhatti and G. Knight, *OSIMIS User Manual Version 1. 0 for System Version 3. 0*, 02/93

[Pav93b] Pavlou G., *The OSIMIS TMN Platform: Support for Multiple Technology Integrated Management Systems*, Proceedings of the 1st RACE IS&N Conference, Paris, 11/93

[Gri95] Griffin D., Georgatsos P., *A TMN system for VPC and routing management in ATM networks*, Proceedings of the fourth international symposium on integrated network management, ISINM95, ed. Sethi A., et al. Chapman & Hall 1995. p356

[Sta95] Stathopoulos C., Griffin D., Sartzetakis S.: *Handling the Distribution of Information in the TMN*, Proceedings of the fourth international symposium on integrated network management, ISINM95, ed. Sethi A., et al. Chapman & Hall 1995. p398

[Bar94] Barker Paul, *Implementing FTP archive searching using X. 500*, Internet Draft, (aka OSI-DS 40), version3, October 1994.

[Kil95] Kille, E. S. (1991) *Implementing X. 400 and X. 500: The PP and QUIPU Systems*, Artech House, Boston MA.

Infrastructure Demands of Advanced Services

Session 4 Introduction

Chairman - Norbert Niebert

Ericsson Eurolab Deutschland GmbH
Ericsson Allee 1, D-5120 Herzogenrath 3, Germany
Tel : +49 2407 575 122
Fax : +49 2407 575 150
Norbert.Niebert@eed.ericsson.se

New communication networks and services will push the telecommunication industry to become the most important industrial sector in the information society. The work in the IS&N domain focuses on service creation, service management and service engineering especially for new broadband networks and multimedia services. This work tries to be as generic as possible, leading to universal solutions capable of supporting existing as well as new services which have not yet been invented. The results are documented throughout this book.

At the same time the information society's basic services find already their way to the users. Many of them are not yet integrated and many of them are based on slow connections and internet protocols. It is an important worktask in the IS&N domain to analyse advanced services and abstract their requirements for optimal routing, transport and quality of service support towards the future advanced telecommunication infrastructure.

In this section, three examples of advanced services are given, their various demands identified and conclusions are drawn on the necessary developments to support their future development towards more intelligent services. This bottom-up approach is certainly needed for many more key service examples. This section gives a snapshot on the issues for hypermedia retrieval, personal mobility and service intelligence allocation. The area will expand. If the upcoming ACTS projects have started to create an infrastructure which supports advanced services - initially based on island trials but increasingly interconnected to form a trial *service infrastructure* spanning both wireless and fixed networks.

The first paper in this section investigates the impact of new services on the signalling system in personal communication systems based on personal numbers and Intelligent Network infrastructure. In this scenario, every call will be an Intelligent Network call involving a database interaction as the number only indicates the person and its profile not the location or the final service. Several solutions are compared including the centralised, decentralised and outside network implementation. This performance investigation allows also conclusions for the proper placement of the service intelligence in a future network architecture.

A similar question drives the second contribution which investigates the possibilities for even more dynamic configuration for a comparable mobility service. By taking results initially developed for defence systems with the objective of robustness and complete decentralisation, "self organisation" as a structuring principle for Intelligent Networks is examined. The paper focuses on organisational intelligence which is also claimed by the transparency approach of TINA architecture.

The third paper comes back to advanced service examples and their infrastructure demands. Beside communication services, the information society will heavily rely on information retrieval services. The exponential growth of the "World Wide Web" service is just the first of these services which rely on access to all information types via hypermedia links. Current networks and transport protocols are not designed for the transport requirements these services will have. The paper reveals and categorises these requirements and points to solutions especially for broadband environments. This is an example how QoS driven connection control can be made beneficial also for the application itself.

These three papers provide a snapshot on infrastructure demands by taking three prominent example services: mobility, network intelligence and hypermedia information services. These are seen as important constituents of the upcoming service infrastructure.

Personal Communication System Realizations: Performance and Quality of Service Aspects on SS-No.7

Stephan Kleier and Carmelita Görg

Communication Networks
Aachen University of Technology
E-Mail: skl@dfv.rwth-aachen.de

Abstract. The "Personal Services Communication Space" (PSCS) is a telecommunication service providing personal mobility. The users can initiate and receive calls on the basis of a personal network transparent number across multiple networks, at any terminal, fixed or mobile, irrespective of geographic location. This paper describes a performance study of the RACE project Mobilise[1] for the personalization, mobilization and integration of advanced wide area network communication services. With the expected number of PSCS users, the impact on the signalling network could be dramatic. The paper presents the numbering scenarios for PSCS and the use of a Personal User Identity (PUI). Different distributions of the Intelligent Network entities and the impact on an existing SS-No.7 network and the call setup time have been studied.

1 Introduction

This paper describes a performance study of the RACE project Mobilise for the personalization, mobilization and integration of advanced wide area network communication services. Mobilise defines the Personal Service Communication Space - PSCS - which is an extension and contribution to the UPT approach [9]. PSCS is a framework for a wide range of telecommunication services from narrowband to broadband services. The approach is driven by the user's needs and deals with everyday communication tasks:

"Personal Communication offers to end-users, in their different roles, the ability to communicate and organize communication according to their own preferences, in such areas as: time, space, medium, cost, integrity, security, accessibility, privacy, and quality."

---

[1] WWW: http://www.comnets.rwth-aachen.de/project/mobilise;
Project Partners: Ericsson Eurolab Deutschland GMBH (EED, D) , CAP Gemini Innovation (CAP, F), PTT Research (PTT, NL), Telia Research AB (TVT, S), Aachen University of Technology (CN, D), Ericsson Telecommunicatie B.V. (ETM, NL), empirica Ges. f. Kommunikations- und Technologieforschung mbH (emp, D), Vodafone Ltd. (VOD, GB), CAP Sesa Telecom (CSTL, F), Ascom Tech. Ltd. (ASCOM, CH), and GIE-COFIRA (COF, F).

With the expected number of PSCS users, the impact on the signalling network could be dramatic if the PSCS service is designed without forethought. The paper presents the numbering scenarios for PSCS and the use of a Personal User Identity (PUI) as an indirect address between the PSCS number and the PSCS service profile of the called PSCS user (termed Flexible Service Profile, FSP).

Three models for distribution of PSCS service logic in SS7 are analysed and evaluated with respect to mean values for the setup times for ISDN calls, outgoing calls (PSCS → ISDN), and incoming calls (ISDN → PSCS). Also the increase of the needed processing power of the exchanges and the needed capacity of the signalling links between them is taken into account.

The signalling network contains a real traffic mix of local, regional, long distance and international ISDN calls, and the outcome of the setup attempt is taken into account. The roaming of the users in the network is assumed to be like the call distribution. About 60% are roaming in the local area, 25% in the regional area, 12% nationally, and 3% of the users travel internationally. The simulations were performed on the structure of the telecommunication network of the "Deutsche Telekom AG" and can easily be adapted to other network structures.

2 Numbering

The UPT/PSCS numbering scenarios introduced in ITU-TS Recommendation E.168 [8] are based on the numbering plan for ISDN, E.164 [4], beginning in 1997, which allows a number length of 15 digits. Each scenario requires a different level of network capability, starting with fairly basic call forwarding mechanisms and ending with a fully integrated global intelligent network.

In the *home-related* scheme ([**CC**]-[**NDC**]-[**SN**][2]) the telephone number does not contain any indication of the PSCS service accessed by this number. The service profile, which contains all information related to the subscriber, is stored in the home domain of the subscriber. A caller could not distinguish, if he calls a PSCS number or a PSTN or ISDN number.

In the *national scheme* ([**CC**]-[**PSCS NDC**]-[**SN**]) the *National Destination Code* (NDC) contains a PSCS indicator and a service provider indicator followed by the *Subscriber Number* (SN). A calling party of the same country and the network can recognize the number as a PSCS number.

In the *international global* scheme ([**PSCS CC**]-[**NDC**]-[**SN**]) a *Country Code* (CC) is assigned to PSCS numbers as a PSCS indicator. With this indicator one gets globally recognizable PSCS numbers. The NDC element may contain a CC, in which case there is national number administration. With no CC in the NDC, the administration of numbers is a global matter.

---

[2] **CC:** Country Code; **NDC:** National Destination Code; **SN:** Subscriber Number

2.1 The Use of a PUI

A *Personal User Identity* (PUI) can be used as an indirect address between the *PSCS Number* (PSCSN) dialled by the calling party and the service profile (termed *Flexible Service Profile* (FSP)) of the called PSCS user. The PSCSN and the PUI will, although being different numbers/identities, normally have a one-to-one relationship. The PSCS user need not have knowledge of the PUI related to his PSCSN, nor does the calling party when calling the PSCS user. The PSCSN-to-PUI relationship is stored in a database, and it must be looked up before the FSP can be accessed.

The advantages of the PUI are that network evolution causing the PUI to be changed does not influence the PSCSN (only the PSCSN-to-PUI mapping is changed) and that PSCS subscribers can retain their PSCSN when changing the PSCS service provider. Hence, with the introduction of the PUI, the PSCS subscriber need not ever change the PSCSN, making it a truly personal number.

A PUI identifies a certain FSP. It must contain information about the location of the database containing this FSP and the identity of the FSP within the database. The PUI will be transferred over PSCS service provider boundaries. To allow this, it must have a standardized format. The structure of the PUI could be similar to the structure of the PSCSN. In any case it is assumed here that the PUI consists of 15 digits for world-wide identification of an FSP.

2.2 Database Location

The database storing a PSCSN to PUI relationship may be:

1. a national database in the home country of the home PSCS service provider,
2. an international database,
3. a national database in the call-originating country, or
4. a network database belonging to the home PSCS service provider.

These alternatives must be combined with the PSCS numbering scenarios. Alternatives 1 and 4 require a national structure of the PSCSN. Hence, the national and the international global schemes are possible. Alternatives 2 and 3 require no national structure of the PSCSN to be retained, making the global scheme a possible solution.

Aspects that have to be considered are that alternatives 1, 2, and 4 require international access to the databases, alternative 3 requires world-wide updating procedures, the management of a global database may be a sensitive political issue, and alternative 4 requires a PSCS service provider to manage the PSCSN to PUI relationship long after the subscriber may have changed to another service provider.

3 Reference Network

Figure 1 shows the layered architecture of the existing network of the Deutsche Telekom AG with three layers of transit exchanges. The highest layer consists of

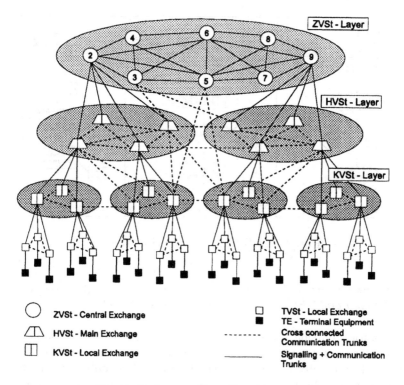

Fig. 1. Reference Signalling Network

8 central exchanges. Below this, 62 main transit exchanges are located, and in the third layer 720 regional exchanges. Connected to these are about 2500 local exchanges. The signalling flows that are the focus of this investigation are strictly hierarchical. Bearer connections are also possible between exchanges of different levels. For the *Intelligent Network* (IN)[3] based solution inside the network the different IN *Functional Elements* (FE), that are needed to support the PSCS functionality, were distributed in this architecture.

4 Simulated Models

Based on the reference signalling network there are some IN *Functional Elements* (FE) subject to location considerations. In the *Centralized Solution* (Figure 2a) all elements are placed in the eight transit exchanges of layer 1.

---

[3] **PUI**: Personal User Identity; **FSP**: Flexible Service Profile; **IN**: Intelligent Network; **FE**: Functional Elements; **SCP**: Service Control Point; **SSP**: Service Switching Point; **SDP**: Service Data Point; **STP**: Service Transfer Point; **SP**: Switching Point; **ISUP**: ISDN Signalling User Part; **IAM**: Initial Address Message; **PE**: Physical Element;

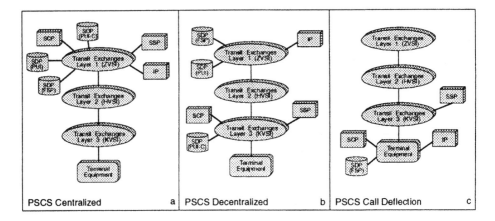

Fig. 2. PSCS Data Distribution Models

In the Decentralized Solution the *Service Control Point* (SCP) and the *Service Switching Point* (SSP) are located in the lowest layer of the transit exchanges (Figure 2b).

The solution outside the network, the Call-Deflection-Based solution, (see figure 2c) does not request any database functions in the network. They are implemented externally in an intelligent terminal. The only network function necessary is the ISDN Call Deflection Service. This service enables a user to forward a call that is waiting to be accepted to a third party. The PSCS user registers at location C informing his intelligent terminal at location B about his current location. The intelligent terminal includes all data that is necessary for registration and intelligent call routing. A caller will first be connected to the location B. There the intelligent terminal analyses the call and spawns the action that is specified in the filter script, for instance, the forwarding of the call to C. The forwarding is performed via the ISDN Call Deflection service or in the intelligent terminal using both B channels of the ISDN line.

5 Message Sequence Charts

The messages of the ISDN User Part (ISUP) are described in ITU-TS Recommendations Q.763 and Q.764 [5]. The messages contain various fields; some fields are mandatory with fixed length, others mandatory with variable length, and some are optional [2, 12].

It is assumed that the PSCS functions are merged into normal ISDN call setup by including new PSCS information in the related messages. The PSCS features require messages to carry information related to authentication procedures, service profile interrogation, call registration, etc. In [7, 3, 10, 11] proposals for the messages required to realize the most frequently used PSCS features are given. The message sequence charts are based on ITU-TS Recommendation Q.1200 on IN [6].

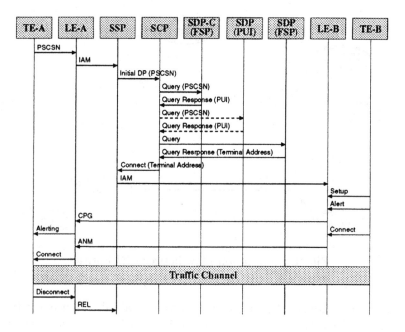

Fig. 3. PSCS Incoming Call

The new messages required by the PSCS service can be classified as either invoke or response messages, and can be divided into three subgroups with different lengths.

A part of the message sequence chart for a PSCS incoming call is presented in figure 3 in order to give an idea of the messages involved. Figure 4 shows the information flows in an IN environment[4]. Additional to IN-CS1 some data cache functions are introduced to minimize the access attempts to the PUI-SDF.

6 Analysis

6.1 Parameters

Based on information from the Deutsche Telekom AG, the models are based on 60% local, 25% regional, 12% long distance and 3% international calls. The result of the connection setup attempt can be a successful call setup, subscriber busy, and no answer.

The models have a network PUI-*Service Data Point* (SDP) and a remote PUI-SDP containing the PSCSN-to-PUI relationships for subscriptions to the PSCS service provider in the considered network and in remote networks, respectively. It is assumed that the fraction of PSCS calls requiring interrogation in the remote

---

[4] **CCF**: Call Control Function; **SSF**: Service Switching Function; **SCF**: Service Control Function; **SDF**: Service Data Function; **SMF**: Service Management Function; **SRF**: Specialized Resource Function; **CCAF**: Call Control Access Function;

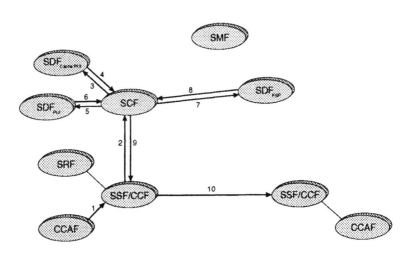

Fig. 4. Information Flow between IN Entities

PUI-SDP is the same as the fractions of international calls. Hence, the fractions of PSCSN to PUI mapping stored in the network PUI-SDP and in the remote PUI-SDP are 0.97 and 0.03, respectively.

In the centralized model the fractions for the storage of the FSPs are assumed to be the same as for the PUI-SDP, i.e. the fraction of FSPs stored in the network FSP-SDP and in the remote FSP-SDP are 0.97 and 0.03, respectively. With distributed FSP-SDPs it is assumed that the FSP is stored in the FSP-SDP co-located with the nearest transit SP with a probability corresponding to the fraction of local and regional calls. Hence, the fractions for FSP interrogation in the nearest FSP-SDP, in another FSP-SDP in the same network and in the remote FSP-SDP are 0.85, 0.12 and 0.03, respectively.

6.2   Service Times

The propagation delay is 3 ms on all signalling links. The subscriber line response time, which starts when the terminating local exchange sends the setup message to the called party and ends when it receives the alerting message, is assumed to be 200 ms [12].

The overall delay in a signalling node is expressed by the cross-office transfer time. It is defined as the time between the last bit of the incoming message leaving the signalling link and the last bit of the corresponding outgoing message entering the signalling link. It includes processing on the required levels in the node and queueing delay and transmission time on the outgoing link.

The cross-office transfer time is specified for *normal*, +15% and +30% signalling traffic load of the *Service Transfer Point* (STP) [5]. The value 100 for relative offered traffic corresponds to the *normal* load, which is the load the signalling point is engineered to handle. The cross-office transfer time assumes 64 kbit/s signalling bit rate and 15 bytes average message length. For longer messages the longer transmission time must be taken into account.

Node	Service time [ms]
Propagation delay	3
STPs (simple)	300
STPs (processing intensive)	470
Subscriber line response time	200
SCP	100
PUI-SDP	50
FSP-SDP	100
remote PUI-SDP	770
remote FSP-SDP	670

Table 1. Service times

The handling of messages in a *Switching Point* (SP) involves the level 4 user functions. The level 4 processing time is different for different SS7 user parts, so each user part specifies the cross-office transfer time applicable to its messages. There are two types of ISUP messages [5]: the *processing intensive* and the *simple messages*. The former arrive at an exchange and require detailed examination before transmission to the next exchange, while the latter require little or no examination or modification before transmission. Of the *ISDN Signalling User Part* (ISUP) messages included in the models, only the *Initial Address Message* (IAM) is a "processing intensive" message.

Table 1 gives the figures for the service times in the signalling network and the service times in the PSCS *Physical Elements* (PEs). The times for the remote PUI-SDP and remote FSP-SDP also include the propagation delay through intermediate nodes in both directions. As the objective of the analysis is to study the PSCS impact on the signalling nodes, and it can be assumed that the PSCS PEs have sufficient capacity to handle the amount of offered signalling load considered in the analysis, these figures are kept constant throughout the analysis.

6.3   Simulation Scenarios

In average there are 5 outgoing calls from a PSCS user and 5 incoming calls to a PSCS user in the period between registration and deregistration. The 5 incoming calls include calls to a busy PSCS user and to a PSCS user not answering the call. The registration periods may be of varying lengths. A user on the move during the day may be registered at a terminal less than an hour before registering at a new terminal. In the evening and on week-ends, however, he will probably be registered for a longer time at each terminal. Other people may be registered at the same terminal for days. Hence, the 5 outgoing calls and 5 incoming calls in a registration period are average values. 1 % of all users are assumed to be PSCS users.

6.4 Simulation Method

The simulation of the models is carried out with the aid of OPNET[5] [1]. Simulations in OPNET are modelled using a hierarchical modelling approach. The highest layer is formed by the network layer, followed by the node layer and the process model. Figure 5 show the distribution of the 8 highest layer exchanges in the OPNET network layer.

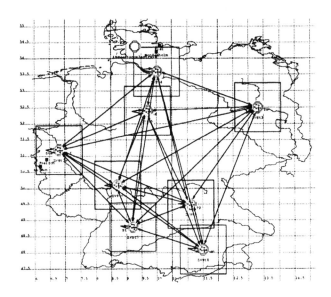

Fig. 5. OPNET Model

The node layer is used to model the network nodes of the network layer. A node model is defined through a set of modules, that communicate with each other using *Packet Streams* and *Statistic Wires*. The modules that can be employed in a node model are *Packet Generators, Processors, Queues, Transmitters* and *Receivers*.

The Round Robin based processor model, as displayed in figure 6, models the mutual influence of different message types on the load in the signalling processors and thus provides a solution to the problem of independent processors for each message type that arises in the usage of the M/D/1 multiple processor model.

In the Round Robin based processor model all messages are processed in the same queueing system in contrast to the previous model. The arriving PDUs

---

[5] a registered trademark of MIL3, Inc., Washington, USA.

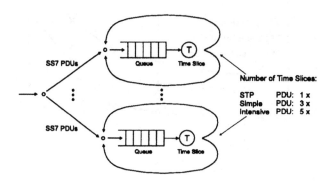

Fig. 6. Round Robin based Processor Model for Switches

are cyclically distributed on the Round Robin systems and are inserted in the queue. Depending on their processing type – Intensive, Simple or STP – the messages have to pass through the whole system one or several times. STP messages have to be processed one time, simple messages 3 times and intensive messages 5 times. This is due to the relation between the cross-office times for those message, being 35 ms : 110 ms : 180 ms in the ITU standard.

In the simulation this processor model provides a very exact approximation of the values prescribed in the standard, while at the same time all message types are now handled by one processor, and thus influence each other, which is an effect that is found in the operation of real switches.

6.5 Simulation Results

An important aspect that must be considered when introducing new telecommunication services is how the quality of existing services is affected. The call setup time is one of the major aspects to consider. For ISDN calls this is the time which starts when the last digit of the B-number is received by the originating local exchange and ends when the A-party gets an indication that the B-party is being alerted.

The focus in this simulation series is on the call setup times for the end-user. Figure 7 shows the mean ISDN to PSCS (PSCS incoming call) call setup time and figure 8 the one for PSCS to ISDN (PSCS outgoing call). The lowest curves show the normal ISDN call setup time as a reference. Many scenarios with different database distributions were simulated and the most important are presented in the figures.

The PSCS outgoing call setup time includes the time needed for PSCS procedures to make sure that the user is allowed to make this call, and the ISDN setup time. The PSCS incoming call setup time defined similarly also includes the time required to locate the called PSCS user.

The centralized solution has the longest setup times and can carry less PSCS traffic than the decentralized solutions. But the centralized version is useful for quick introduction of a PSCS service in the network because only the software

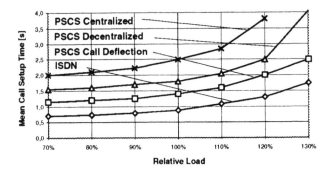

Fig. 7. ISDN to PSCS Call Setup Time (1% PSCS)

in the 8 transit exchanges of layer 1 needs to be modified. The mean call setup time of 3.5 seconds for the busy hour (corresponding to 100% relative load) is acceptable. Shorter setup times are achieved with the distribution of the functions, but this would lead to modifications in each one of the layer 3 transit exchanges (720).

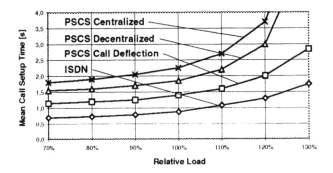

Fig. 8. PSCS to ISDN Call Setup Time (1% PSCS)

First studies of the new German network architecture for 1998 which will have only 2 layers of exchanges have been started. As in the old network concept, the shortest setup times were achieved with the PSCS solution outside the network, that performs the forwarding to the user's location with the ISDN call deflection service. This is due to the fact, that for this solution no "time consuming" central database requests are necessary.

7 Conclusion

The results show that the setup time for ISDN calls is influenced by the PSCS service in varying degree in the different models. In addition, the setup times for outgoing calls and incoming calls of the PSCS service itself varies in the models.

The shortest call setup times are achieved with the Intelligent Call Manager outside the network. More centralized solutions results in longer call setup times and a bigger load in the signalling processors of the higher network layers.

The economic aspects of the distribution of functions versus upgrading of the signalling network must be considered. The decentralized models require multiple PSCS nodes (SCPs, FSP-SDPs, PUI-SDPs) in the network. The same performance may be achieved with a more centralized model if the signalling network is upgraded. The expense related to these alternatives may be different.

The use of a PUI is recommended. The setup time is not significantly longer than without the PUI, and the advantages with regard to network evolution and a truly personal number are considerable.

References

1. Cohain Alain, Geoff Kelsch, Archie Kevin, Nudelman Eric, Finn Russell, Cathey George. *OPNET Modeller: Modelling Manual.* MIL 3,Inc., 2.4.b edition, Nov. 1993.
2. M. Bafutto, P.J. Kühn, G. Willmann. *Modelling and Performance Analysis of Common Channel Signalling Networks.* AEÜ, Vol. 47(5/6), pp. 411–419, 1993.
3. Ulrich Boetzel. *Strategien zur Verteilung von UPT-Funktionen im SS-No.7 Signalisierungs-Netz basierend auf IN-Konzepten.* Master's thesis, Lehrstuhl für Kommunikationsnetze, RWTH Aachen, Oktober 1994.
4. CCITT. *E.164 Numbering plan for the ISDN era.* CCITT, Geneva, 1988.
5. CCITT. *Q.700 Specifications of Signalling System No. 7.* CCITT, Geneva, 1989.
6. CCITT. *Q.1200 Series Intelligent Network.* CCITT, Geneva, 1992.
7. Arne Folkestad. *Strategies for Distribution of UPT Service Logic in an SS-No.7 Signaling Network.* Master's thesis, Lehrstuhl für Kommunikationsnetze, RWTH Aachen, April 1994.
8. ITU-TS. *E.168 Application of E.164 numbering plan for UPT.* ITU, Geneva, 1993.
9. Ulf Jonsson, Stephan Kleier. *Personal Communication - Network Aspects and Implementations.* In *Lecture Notes in Computer Science 851*, pp. 321–331. 2nd International Conference on Intelligence in Broadband Services and Networks (IS&N'94), Aachen, Springer Verlag, Berlin, Heidelberg, New York, September 1994.
10. Thomas Oepen. *Intelligente Netze: Analyse von Realisierungen im zukünftigen Telekommunikationsnetz der DBP Telekom.* Master's thesis, Lehrstuhl für Kommunikationsnetze, RWTH Aachen, März 1995.
11. M. Rüßmann. *Corporate Networks: Analysis of New Mobile Communication Services and Interfaces to the Public Network based on Intelligent Network Technology.* Master's thesis, Lehrstuhl für Kommunikationsnetze, RWTH Aachen, Juni 1995.
12. G. Willmann, P.J. Kühn. *Performance Modeling of Signaling System No. 7.* IEEE Comm. Mag., Vol. 28, pp. 44–56, July 1990.

A Self–Organisation Plane
for Distributed Mobile Wireless Networks

A. O. Mahajan and K. V. Lever

Defence Communications Research Centre, Institute for Telecommunications Research
University of South Australia, Adelaide, Australia

Abstract: Self–organisation functions are capable of providing organisation intelligence in communication networks. Self–organising radio networks support diverse user services in a distributed and dynamic environment. These networks involve such a variety of functions that establishment of a functional guideline or framework is necessary to manage these functions systematically. Such a framework is proposed in this paper in the form of a "plane" of self–organisation functions. Various control mechanisms or protocols can be mapped onto these functions. A practical application of the self–organisation model is demonstrated and other uses of this approach are suggested.

Key words: Radio networks; distributed architecture; intelligent system.

1 Introduction

Intelligent network [IN] concepts provide flexible network services, and are gradually emerging in present telecommunication networks [1], [2]. Although service intelligence and operational intelligence functions are being considered [1], network organisation (self–organisation) functions are not prominent in these networks, since most of them are "static" and pre–installed. With the emerging trends towards providing unlimited network access even to mobile users, it becomes necessary to deliberate about network organisation issues. This is especially needed due to the distributed and dynamic nature of future generation networks. In dynamic networks, in addition to user–node mobility, relay–nodes and control–nodes also operate in a mobile and distributed manner. In this paper, we discuss the concept of self–organisation plane capable of providing a functional framework for supporting network organisation intelligence in communication networks.

Self–organising networks (SONs) [3], [4] have evolved from defence, emergency and disaster relief service requirements for network deployment and use of various radio transmission media (satellites, microwave, VHF/UHF, HF) for service integration and survivability in adverse environments. On the other hand, the need for interworking between civilian telecommunication networks and defence networks is becoming important. Defence networks are slowly migrating towards this path [5]. Therefore, it is important that future efforts in standardising civilian telecommunication networks should consider such self–organisational needs.

We are investigating adaptive protocols and algorithms which are aimed at, but not confined to, a specific defence SON architecture [6]. Because of their variety and complexity self–organisation functions require modelling and visualisation tools in support of the network design process.

This contribution briefly presents in Section 2 a network classification with a mobility perspective, and describes SON characteristics. In Section 3, we define generic self–organisation functions and introduce a functional plane model useful for designing self–organising architectures. Various control techniques in these architectures are mapped to show the completeness of the functional plane. The relationship between this model and implementation of a simulated test scenario is then discussed in Section 4. This section also provides some examples where this self–organisation plane concept is relevant. Our conclusions are presented in Section 5.

2 Self–Organising Networks

2.1 Network Classification

In traditional public networks one can differentiate between user–nodes (or terminals) and the centralised switching nodes where most of the network functions are carried out. Most present mobile networks are also based on centralised operations and user–nodes are well distinguished from other network elements.

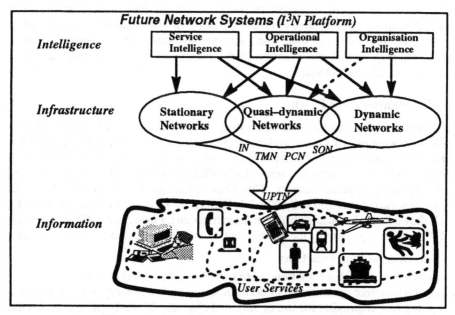

Figure 1: A mobility view of network systems

If unlimited mobility to users is to be allowed with desired services, networks should be adaptive to changing service demands. If the provision of fixed infrastructure is not feasible, deployable network elements (node entities) are an alternative. Networks can be classified on the basis of the nodal mobility allowed for network elements including the user terminals: (1) Static or stationary networks – in these networks all the network elements are fixed; (2) Quasi–static networks – only user–node mobility is allowed in these types of network; (3) Dynamic (mobile, distributed) networks – these networks permit user–node mobility, together with relay–node and control–node mobility, and allow distributed operation. Using this classification SONs are dynamic networks.

Figure 1 depicts a speculative view of universal communication networks which will include three types of intelligence functionalities. Synergy of these functionalities along with the integration of three different network infrastructures will give a versatile platform providing seamless diverse services to users "any place, any time". To emphasise the role of the three forms of intelligence in Figure 1, we coin the acronym I^3N to describe this type of network.

2.2 What is Self–Organisation?

Self–organisation can be defined as the ability of a set of nodes deployed within a certain geographical area to automatically form and maintain a network. The expectation is that various nodes deployed in an arbitrary area will be able to maintain communication whilst exercising practically unlimited and unpredictable mobility. This expectation can be fulfilled by using wireless transmission media and self–organisation capabilities.

SONs can be characterised by their ability to operate in a distributed environment. Therefore, these networks are capable of supporting a varying degree of mobility. These networks use adaptive control protocols in their operations to support dynamic and reconfigurable topologies. Especially such networks may use multi–band transmission links and support different traffic mixes consisting of voice and data services. Data services can be both real–time and store–and–forward with variable bit rates. Inter–operability, high levels of survivability and communications security are other characteristics of these networks.

3 Self–Organisation Plane

SONs involve issues such as diverse services, link dynamics, nodal deployment and protocol adaptability. This functional complexity can not be easily comprehended until we have some mechanism for visualising and categorising the functional requirements of SON architectures. During our search for a suitable network design tool, we devised a self–organisation plane concept. This plane encompasses six generic functions (Figure 2) which are defined in Section 3.1. Further, we show a functional block model which shows interrelationships between these functions. Finally, the sufficiency of this model is verified by mapping the various control techniques used in SON architectures.

3.1 Self–Organisation Functions

Frequency Management Function (*FMF*) : SONs use a variety of radio transmission media, and therefore optimisation of network resources is necessary. *FMF* is required for sensing and monitoring the use of allocated channels. Multi–band communication concepts require an integrated approach while planning and allocating frequencies. Particularly for narrow–band frequency (HF and VHF) systems, channel availability depends on local propagation characteristics. Therefore, frequency management should not be completely based on any centralised database and distributed nodes should be able to replicate this function.

Network Formation Function (*NFF*): This function permits network start–up in a distributed environment. In addition to network initialisation it also involves network planning issues. In contrast to stationary networks where network elements are pre–planned, in SONs network planning becomes adaptive. Here, individual nodes are required to implement control strategies essential for network start–up.

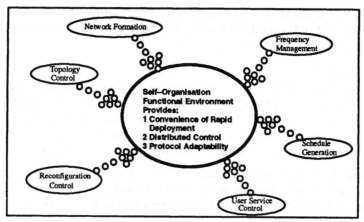

Figure 2: Functions for distributed operations

Topology Control Function (*TCF*): This function implements a desired network topology. In military and rescue operations a flat topology is practical. However, a hierarchical topology is required for managing a large network. Overall connectivity between nodes decides the final topology. Moreover, in radio networks increased reliability and throughput is achievable by controlling link connectivity.

Schedule Generation Function (*SGF*): This function associates traffic links or paths to various network elements. The broadcast properties of radio medium make allocation and use of scheduling rights an important issue. *SGF* is needed in order to achieve fair distribution of frequency channels or time–slots. The network as well as service requirements within a communication region can change from time to time. This unpredictability necessitates a combination of access mechanisms in order to optimise the use of scarce radio transmission resources.

Reconfiguration Control Function (*RCF*): This function is concerned with the issues arising from nodal mobility and node or link failures. The requirements of SON can change during network operation due to such dynamic conditions. In order to retain the communication infrastructure in proper order even under such conditions, *RCF* is necessary.

Although network formation and reconfiguration control functions are similar, reconfiguration protocols are more complex than formation protocols. This is because during network formation nodes are responsible only for organisation of networks: user traffic is expendable. During reconfiguration, however, nodes need to use proper restraints on organisation traffic in order to support as much user traffic as possible. In addition to nodal mobility and node failures other causes of reconfiguration are (a) congestion control due to change in node distribution or change in traffic distribution (b) introduction of new services (c) changes in transmission resources.

User Service Control Function (*UCF*): Obviously, self–organisation functions adapt the provision of network services to meet user expectations. However, the network will have its own limitations on meeting user demands. At the same time network elements will have to divert their resources to cope with resource limitations. Therefore, there should be some mechanism to provide feedback necessary to user related protocol stacks (such as user access control). *UCF* will allow user functions to become aware of the latest network limitations such as change in channel resources (for example narrow–band/broad–band channel availability) and the types of services supported due to changing propagation conditions.

3.2 A Block Approach

In order to model various self–organisation functions for designing a set of protocols, we need a suitable design platform. Various modelling frameworks used for data networks, cellular networks or ISDN networks in current use are not directly applicable to our problems. Due to their centralised architecture and the absence of frequent deployment of network elements, these networks do not use self–organisation.

The OSI 7–layer model is particularly suitable for data networks, but this hierarchical modelling approach is not convenient for a distributed and dynamic environment. Even in the case of fixed infrastructure networks a different approach has been suggested in order to handle mobility issues [7].

In SONs, it is possible that the organisation functions will have to perform tasks which are associated with more than one OSI–RM. Network formation is a process which can be considered as part of the network layer as well as the data link layer. There is little agreement amongst researchers about placement of TDMA scheduling protocols. In reference [8], it has been described as part of the data link layer, whereas reference [9] characterises TDMA policies as a part of the network layer. Similar questions arise when addressing and mobility issues are considered.

We undertook a block oriented approach for modelling the self–organisation functionality. The various functions discussed in Section 3.1 are separated into partitions or blocks located within the same plane. Figure 3 shows the directed interconnections for necessary interaction between various functional blocks.

In the case of *FMF*, timely input is required from network planners in order to incorporate frequency allocation modifications due to geographical area and regulatory control. The initialisation process can only be started if frequency channel distribution is complete and channels are available at each participating node. On the basis of frequency assignments, *NFF* initiates protocols to accumulate neighbourhood connectivity knowledge. Node functional roles can be assigned to various nodes.

Further, the *SGF* block is necessary in order to implement distributed access schemes. In this case *NFF* provides connectivity data essential for time–slot distributions. Once allotment of time–slots or links is completed, user access control can use them for supporting user traffic.

The *TCF* block provides algorithms and protocols necessary for arranging a particular topology structure. Parts of a network may often have to be configured to suit user requirements. For example, for control and telemetry data communication a star configuration is suitable, whereas for conferencing purposes mesh configuration is required.

The *RCF* block has to interact with *TCF* and *SGF* for redistribution of channel resources in the presence of nodal mobility and node failures.

UCF acts an interface between organisation functions and higher level user service functions. In order to maintain a proper balance between user traffic and network traffic, it is important that user service planes adapt to changes in available network resources. For this purpose *UCF* can be treated as a repository for the organisation database. *UCF* entities can interact with higher user level protocols for conveying address and link status changes. This is necessary for updating the routing database.

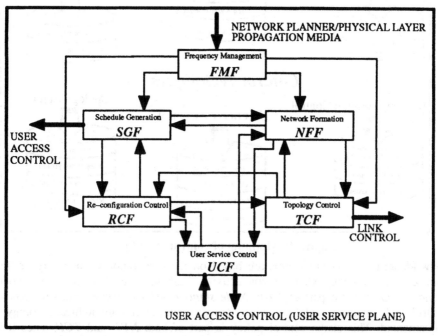

Figure 3: Functional block for a self–organisation plane

We have attempted to show a generic inter–relationship between different self–organisation functions. This functional block has defined an overall structure of the self–organisation plane. One of the objectives of such approach is to ensure some consistency while designing a SON architecture. A specific interfacing between these blocks will differ depending on particular network requirements.

3.3 Mapping of Control Techniques

In order to verify the completeness of the functional block we studied various self–organising architectures from the literature and identified control techniques used in those schemes [10]. Different policies are implemented on the basis of suitability of methods, hardware configurations (such as semi–duplex, duplex, multi–receiver radios) and specific user needs. Figure 4 shows this mapping which we discuss briefly in the following paragraphs.

In multi–channel HF and VHF/UHF networks channel monitoring requires various *Sounding* and *Real Time Channel Evaluation* control techniques. Due to changing propagation characteristics specific *dynamic channel control* policies are also required in such networks. These control mechanisms are categorised under *FMF*. *Channel aggregation* for supporting high data rate services will also be required in radio networks.

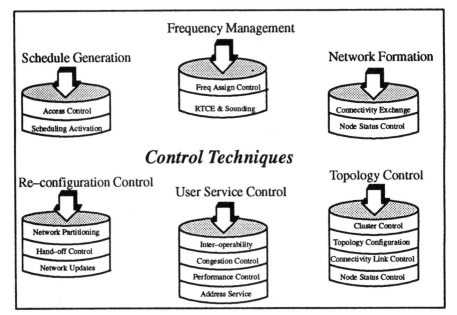

Figure 4: Mapping of control techniques

Network initialisation algorithms are based on the type (node connectivity or link connectivity) and the extent of connectivity database at each node. *Connectivity exchange policies* are part of *NFF*. *Node status control* (for example, selection of control–node, gateway node) procedures are also required if a hierarchical topology is implemented. This status control policy is part of both *NFF* and *TCF*, since it is required at network initialisation and as well as during normal network operations.

Other control mechanisms which are part of the *TCF* block are *cluster control* or *cell structure control* techniques. Local clusters can be formed based on disjoint clusters or overlapping clusters [4]. Similarly, cell structure depends on a particular application. For example, in PCN a multi–storeyed building environment requires a three dimensional cell structure, whereas for transport applications band shape cell structure is needed. *Topology configuration* control is necessary to support different types of services to users. In ALE HF networks different protocols are used to provide net configurations such as star and mesh. The control mechanisms such as *logical link control* and *physical link control* are part of the *TCF*. Connectivity redundancy and link control mechanisms (such as power range control) are required to achieve optimum connectivity.

Due to a mix of service types, networks should implement hybrid as well as dynamic *access control schemes*. This is necessary since performance of asynchronous and synchronous schemes differ depending on traffic distribution and whether the type of traffic is bursty or continuous. Depending on *node* or *link activation*, slot assignment procedures will also differ.

In SON, network elements are also mobile and therefore, there is always a possibility of network partitioning. *Network partitioning* and *automatic link transfer* (*hand–off control*) are two important control techniques in *RCF*.

In order to guarantee service availability to users even during reconfiguration, careful considerations are required to select a *event–driven update policy* or a *periodic update policy*. The selection of update policies depends on dynamic conditions such as link or node failures.

The *UCF* block is mainly concerned with *address service, congestion control, performance control* and *inter–operability control*. *Address service control* is especially required in these networks due to frequent disintegration of clusters. Due to scarce radio resources and resource failures it is possible that nodes are not aware of possible causes of congestion. Special congestion control techniques are required to manage the absence of timely and correct feedback.

The quality of service depends on organising various resources adequately. Both Quality of Service (QOS) and Guarantee of Service (GOS) issues should be considered because the changes in availability of resources are more frequent in SONs. The QOS and GOS parameters are also critical when inter–operability between a fixed infrastructure network and SON is considered. The two networks should neither put additional demand on performance, nor put any restriction on the normal operation of the other. As management of transmission resources is dealt with within the self–organisation block, the *UCF* block can provide appropriate feedback regarding the type and quality of service supported.

4 Applications of the Plane

4.1 Designing a Self–Organisation Architecture

The functional block approach discussed in the previous section can be used to develop a set of protocols for a SON architecture. Figure 5 shows a placement of the self–organisation plane for a distributed radio network environment. This autonomous configuration can be used at each node.

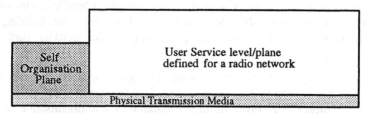

Figure 5: Self–organisation plane used at a radio network node

We have used this approach to segregate organisation related functions from user service related functions. We are investigating adaptive protocols for network formation and schedule generation in a distributed environment NO TAG. One network scenario considered is: (a) narrow–band frequency operation, (b) flat (non–hierarchical) topology, (c) asynchronous channel access mode for network start–up, (d) node activation TDMA access mode for network operation. We used a simulation approach to design and validate the various algorithms for connectivity access, TDMA slot allocation, mobility handling procedures. These procedures are modelled using the OPNET software tool [11]. The corresponding node structure in OPNET is shown in Figure 6.

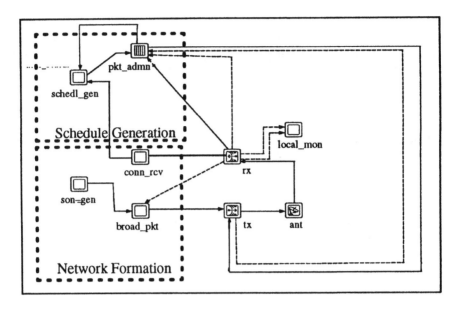

Figure 6: A self–organisation node structure

Separate blocks for network formation and schedule generation allowed us to reduce implementation complexity and develop protocols to specific details. The various control procedures for implementing the asynchronous access mode for start–up, building connectivity database, and distributing time–slots are modelled using various processor modules. Packet generator (son_gen), connectivity packet receiver (conn_rcv), and packet broadcaster (broad_pkt) form the network formation block. The schedule generator module (schedule_gen) implements various slot allocation procedures (organisation slots and user traffic slots). Although our emphasis is on the organisational aspects, the packet administrator module (pkt_admn) is implemented for directing network and user traffic. Various low level functions are: building 2–hop connectivity data, adaptive scheduling, processing of connectivity packets and slot–information packets, monitoring connectivity stability.

Due to the multiplicity of control options being anticipated for SONs, we selected a set of assumptions and test cases in our experiment. This functional approach allowed us to handle the complexity involved in implementation of self–organisation functions. Moreover, basing system realisation on this approach provides insight into performance of adaptive access control for network formation [6]. Having implemented the network formation and schedule generation functional blocks, we believe that the same methodology is extendable to implementing the remaining functional blocks.

4.2 Placement of the Self–organisation Plane within the B–ISDN Framework

In ISDN, B–ISDN networks, the multi–plane Protocol Reference Model [PRM] is used [12]. The approach divides network operation into service management and call control functions. The three planes are *user plane, call control plane* and *management plane*. In PRM different sub–networks can be modelled as strata and sub–network service can be accessed through Service Access Point (SAP). The internal plane structure in these planes is layered based.

If the stratification concepts are extended for distributed radio networks, self–organisation functions will be required in the PRM framework. In order to widen the scope of PRM, addition of the self–organisation plane is necessary. This plane can be a separate plane or part of the management plane as shown in Figure 7.

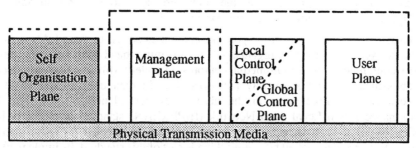

Figure 7: Self–organisation plane with respect to B–ISDN PRM

4.3 Relevance of this Plane to Other Network Systems

Although the self–organisation functional plane described above has been used specifically for reducing complexity of military requirements for distributed, narrow–band deployable networks, this modelling approach should also be useful to other network planners interested in keeping uniformity while integrating various network technologies. Figure 8 shows various network systems which use or are expected to use self–organisation concepts in one form or other [10].

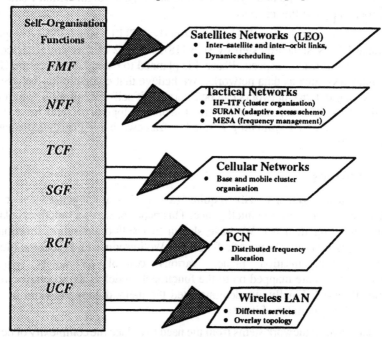

Figure 8: Applicability of self–organisation plane

For example, one Low Earth Orbit (LEO) system (Iridium) will require inter–satellite and inter–orbit link management and dynamic reuse channel assignment in order to improve link connectivity [13]. Simplification might be achieved by use of the self–organisation plane.

Self–organisation concepts are being researched in the cellular network context [14]. Similarly in the case of a PCN scenario, multi–storeyed environments and the presence of different network providers in the same area lead to the implementation of, self–organising frequency assignment procedures [15].

Wireless LAN technologies allow portable computers access to main LAN networks. These networks require self–organisation for (a) multi–bandwidth services (desktop video conferencing and multimedia compressed video) (b) adaptive architectures (establishment of ad hoc overlay networks wherein data transmission is allowed between two or more users in the absence of control–nodes) [16]. The High Performance Radio Local Area Network (HIPERLAN) standard is presently being drafted by ETSI [17]. Here, a layer based approach is selected for radio interface. Although the provision of ad hoc networking is part of this standard, its implementation details are not yet clear.

Catering for adaptive services without changing infrastructure provisions has initiated intelligent service management concepts. With the introduction of various network capabilities such as broad–band, multi–media, and mobile services, IN concepts are being extended to network management issues in order to take into account the addition of these new resources [1]. In the future, telecommunication management and IN are expected to converge on a single platform. In order to augment the similar types of services in a distributed and dynamic environment, self–organisation issues will have to become a part of this process.

The standardisation bodies ITU–TS (formerly CCITT) and ETSI have planned to introduce UPTN standards step by step [18]. Phase 2 of this project is considering mobile communication aspects. Phase 3 is planned to extend and integrate UPTN into other networks such as data networks. We believe that compatibility between civilian networks and defence networks can be improved if self–organisation requirements are considered. In this respect our generic functional approach could be a catalyst for conceptualising and formulating a suitable framework for distributed operations in future UPTN.

5 Conclusions

Future network systems will require (a) service intelligence, (b) operational intelligence (c) organisation intelligence. This paper presents a functional plane model for incorporating item (c). We have shown that self–organisation functions can be broadly classified into six generic functions. We introduced a functional block model of the self–organisation plane. Various control techniques required for self–organisation are mapped on to the functional model to show completeness of the functional model. Finally, we demonstrated the methodology in action by providing some practical examples of its use.

This functional framework stems from the need to reduce the complexity of design steps for developing algorithmic and protocol suites for defence networks. These self–organisational capabilities will also be necessary in future civilian networks. Therefore, we believe that our approach is applicable in these areas also.

6 Acknowledgments

This work was performed with the support of Defence Science and Technology Organisation Salisbury, Australia, under contract CAPO TO 2504.

7 References

[1]   J. J. Garrahan, P. A. Russo, K. Kitami, and R. Kung, "Intelligent Network Overview," IEEE Communications Magazine, Vol 3, pp. 30–36, March 1993.

[2]   B. Jabbari, "Intelligent Network Concepts in Mobile Communications," IEEE Communications Magazine, Vol. 2, pp. 64–69, 1992.

[3]   T. G. Robertazzi and P. E. Sarachik, "Self–Organising Communication Networks," IEEE Communication Magazine, Vol. 24, No. 1, pp. 28–33, January 1986.

[4]   N. Shacham "Organisation of Dynamic Radio Network by Overlapping Clusters: Architecture Consideration and Optimisation," Performance 84, E. Gelenbe (Editor), Elsevier, 1984, pp. 435–447.

[5]   Andrews, F. B. and Kollar, G. I. "The Australian Defence Communications Corporate Plan and its Underpinning Research Program." NATO Symposium on Military Communications Networks Interoperability and Standards, The Hague, The Netherlands, June 1993.

[6]   A. O. Mahajan, A. J. Dadej and K. V. Lever, "Performance of Fixed and Adaptive Access Schemes for Self–Organisation in Distributed Radio Networks," Proc. 7th International Conference on Wireless Communications, Wireless 95, July 1995, pages 20.

[7]   J. Dunlop, P. Cosmini, D. Maclean, D. Robertson, D. Aitken, "A Reservation Based Access mechanism for 3rd Generation Cellullar Systems," Electronics and Communication Engineering Journal, pp. 180–186, June 1993.

[8]   S. Ramanathan and E. L. Lloyd, "Scheduling Algorithm for Multi–hop Radio Networks," Computer Communications Review, SIGCOMM92, pp. 211–222, 1992.

[9]   D. Olsen and N. Dave, "Dynamically Reserved Slot Management (DRSM) for Packet Radio Networks," Proc. IEEE Military Communications Conference, MILCOM 91, 1991, pp.172–176.

[10]   A. O Mahajan, A. J. Dadej and K. V. Lever, "A Self–Organisation Functional Block for Radio Networks," Proc. Australian Telecommunication Networks and Applications Conference, ATNAC'94, 5–7 December 1994, pp. 839–844.

[11]   OPNET Manuals, Release 2.4, MIL3 Inc., Washington, DC, 1993.

[12]   R. Guarneri and F. Zizza, "Evolution of Modelling Techniques for Communication Protocols in B–ISDN," Broadband Communications, A. Casaca (Editor), Elsvier, 1992, pp. 385–397.

[13]   Z. M. Markovic, "Integration of LEO Communication satellites into Third Generation Wireless Networks," Proc. Mobile and Personal Communications Systems Conference, 1992, pp.83–96.

[14]   A. Ishida and J. G. Yoo et al., "Layered Self–Organising Method in Packet Radio Networks," Proc. of 1991 Singapore International Conference on Networks, 1991, pp. 365–370.

[15]   J. C–I. Chuang and A. Ranade, "Self–Organising Frequency Assignment for TDMA Portable Radio in a Multi–Storey Building Environment," Proc. IEEE International Conference on Communications, 1991, pp. 6–11.

[16]   T. Percival, J. O'Sullivan and A. Young, "Wireless Systems at High Bit Rates – Technical Challenges," Proc. Mobile and Personal Communications Systems Conference, 1992, pp. 57–66.

[17]   G. A. Halls, "HIPERLAN: the high performance radio local area network standard," Electronics and Communication Engineering Journal, Vol 6, Iss. 6, pp. 289–96, December 1994.

[18]   G. Arndt, R. Lueder, "Mobility in All Networks," Telecom Report International, Siemens Communications, Vol. 16, No. 2, pp. 8–10, 1993.

Future Hypermedia Retrieval Systems and their Impact on Transfer Systems

Raschid Karabek, Wilko Reinhardt

Technical University of Aachen, Dept. of Computer Science - Communication Systems,
e-mail: raschid|wilko@informatik.rwth-aachen.de

Abstract: Among the most popular distributed applications in today's networks are Hypermedia Retrieval Systems. Due to their rising popularity as well as performance limitations imposed by the underlying communication system the well known 'World Wide Web' is turning a 'World Wide Wait': Response times increase and at peak traffic times many sites are virtually unreachable. On the other hand emerging network technologies like ATM offer a guaranteed Quality of Service (QoS). Future Hypermedia Retrieval Systems will have to be able to reserve bandwidth and request certain timing constraints depending on document media type. Due to the properties of these new media, additional temporal requirements are not only imposed on the presentation (i.e. synchronisation of audio, video and textual information) but also on the transmission of these media streams. There is a desperate need for a uniform QoS architecture and interface for mapping presentation quality to communications requirements to efficiently support the requirements of future distributed multimedia systems and to make optimal use of the QoS capabilities of new networking technologies.

Keywords: Hypermedia Retrieval Systems, World Wide Web, QoS driven Transfer Systems

1 Introduction

The major drawback of almost all commercial multimedia products – except for a few first attempts at distributed services – is that they are only available on permanent storage devices which can only be up-dated by replacing the information carrier. This is especially true with more or less rapidly changing kinds of information such as railroad timetables, mail order catalogues or other types of information systems. The alternative are the network-based (currently non-commercial) hypermedia retrieval systems like the World Wide Web (WWW). Here the user is able to access the information interactively via communication networks. The information itself is presented as multimedia documents ranging from textual information to video clips. Such information systems based on hypermedia structure are becoming more and more popular and find their way from an insider information system for researches to widely used information system, providing all kinds of information.

Commercial information providers are recognising this trend and offering services like on-line newspapers, product information or mail orders. Information providers can easily design a 'homepage' and integrate their information in large networks of links between related documents.

Due to easy structure such an hypermedia system can be seen as the key for all kinds of future multimedia systems. Video on demand, video conferences, contacting a provider instantly by activating a button, all these new services can potentially be integrated. Thus the hypertext browser has the potential of becoming the standard viewing and communications tool of tomorrow.

The most serious problem of today's distributed hypermedia systems is the transmission performances and the variation of the response times. WWW for example is slowly turning into a 'World Wide Wait'. The reasons for that are manifold: On one hand the number of users browsing through the world wide net of information systems increases from day to day. Therefore WWW seems to be the killer application for the current Internet concept. On the other hand WWW suffers from an insufficient communication concept. The complete WWW traffic uses the best effort service without regulating the request from the users. There is no interaction between the hypermedia system and the communication system. The user is allowed to perform all requests without consideration of the current network and server load.

Beside the development of the distributed hypermedia application new network technologies like ATM offer higher transmission speed and - more important - a guaranteed end-to-end performance. From our point of view the solution to the problems of today's distributed hypermedia systems is their adaptation onto the capabilities of the networks. For the retrieval of documents the communication system should reserve adequate network resources to provide the end users a sufficient transmission quality.

The remainder of the paper is organised as follows: Chapter two gives a brief introduction to the communications requirements of Hypermedia Retrieval Systems, their shortcomings and aspects of future systems. Chapter three will take a detailed look at the drawbacks of current data transfer systems, the impact of new network technologies and the necessary changes to transfer systems to make use of the networks new capabilities.

2   Future Hypermedia Retrieval Systems

2.1   Communication Requirements of Hypermedia Retrieval Systems

The hypermedia retrieval systems of today primarily offer means for presenting hypertext documents containing links to other documents at arbitrary locations within the Internet. Multimedia elements such as pictures, audio and video information can be included in documents or displayed separately using resources of the client system. One of the most serious bottleneck of today's hypermedia systems is the insufficient co-operation with the underlying communication system. Modern systems like ATM or the Internet-Service [RFC 1633] environment offer the possibility to guarantee Quality of Services parameters requested from the user. Currently there is no interface bringing together the requirements of the hypermedia systems and the services offered by the networks. Mainly the following properties and requirements can be identified:

- Hypermedia documents contain the full range of possible media types, e.g. graphics, text, video and audio. Each media type has different requirements for the transmission. Isochronous data streams resulting from video and audio sources request high bandwidth, in-time delivery of data and low delay variation. They can accept some bit errors or packet losses, depending on the used compression scheme. On the other hand asynchronous data streams resulting from the transfer of still images, graphics or textual documents need a reliable transmission but can accept small delays.

- Beside the pointers to large documents hypermedia documents also provide links to a number of smaller sub-documents (e.g. in-lines graphics, icons etc.). To retrieve these short documents the system opens a connection, transmits the data and closes the connection.

- Documents are stored on various remote sites, running with different operating system environments and connected to different types of networks. Each hypermedia system that will reach a large acceptances has to run in a heterogeneous computing and communication environment.

- Documents containing videos or audio files should be retrievable from systems not capable to store the file before displaying it. Therefor a streaming mechanism is required that delivers the file packet by packet and displaying the data directly after the sink receives the data. The handling of such documents should be the same like the transmission of live videos, providing the full range of real-time transmission capabilities.

2.2   Drawbacks of Existing Hypermedia Retrieval Systems

The properties of distributed hypermedia documents have several effects on the distribution system providing the transport of the documents. If we consider these properties and requirements the weakness of today's systems become obvious. In the following we shortly summarise the main shortcomings of the hypermedia document retrieval:

- All media types within a hypermedia document are handled in the same way without consideration of their special communication needs. Therefore it is not possible to transmit sub-documents providing connections that fulfil the special needs of the related media type.

- There is no streaming mechanism available that allows the real-time transmission of continuous media to provide immediate presentation and save resources like memory or hard disk capacity at the receiving host system.

- There is no intelligent connection management. For each single element of a document, be it an icon, text or in-lined graphics a transport connection is set up, maintained and torn down. Consequently a large number of often extremely short-lived associations have to be managed by the host systems protocol implementation. Intelligent multiplexing of low rate associations and QoS driven connection establishment are necessary for efficient data transfer.

- But even if QoS support is integrated the currently used hypermedia description languages such as HTML [BeCo93] or MHEG [ISO94] are not capable of specifying the QoS demands of their sub-documents. Both description languages are designed to describe the final presentation of the document without considering the communication requirements. Therefore it is not possible to open adequate communication channels providing optimal support of the sub-documents media type.

2.3 Future Hypermedia Retrieval Systems

Future hypermedia retrieval systems and their document description languages will have to make use of a general hypermedia communications architecture, capable of providing QoS-driven transmission and presentation of all types of hypermedia objects increasing performance and user satisfaction. The main features of such system can be summarised as follows:

- QoS-driven document and sub-document transfer. Each document to be retrieved from a hypermedia system should contain a description of its traffic characteristics. For continuous media this should include the used coding and compression scheme, the rate and the burstiness of the resulting data stream, the required bandwidth as well as the acceptable end-to-end delay and delay variations. For the transmission of discrete media like text or graphics this information should include the size of the file, the file format and the used compression algorithm. If such a document is retrieved from a hypermedia system an appropriate channel is set up that is able to support the QoS requirements described in the requested document.

- These requirements should be mapable onto to the communication system providing resource reservation for QoS-driven data transport. By reserving network resources like link bandwidth, processing power in the intermediate nodes as well as buffer capacities the network is able to provide end-to-end guarantees. This allows the transmission of real time data streams.

- Advanced network environments like ATM or the future Internet Service Architecture should be supported. This also includes the compatibility with today's HTML/HTTP-based systems.

- To provide a full commercialise service billing and security features are mandatory. Information providers will need the possibilities to charge the customers for the service. Depending on the service the information provider or the user should charged for the communication costs.

Although these last two issues are also of increasing importance in this paper we will primarily focus on transport aspects of QoS-driven document retrieval.

3 Transport System Support

The performance limitations of today's systems are largely due to the insufficient support of data-transfer with traffic characteristics typical for hypermedia retrieval systems.

On the server side this is partially due to deficiencies of today's transport protocols. HTTP systems like for instance the World Wide Web establish individual TCP/IP connections for HTTP requests and their responses. If a requested document contains graphical elements such as pictures, icons, coloured lines etc. for each of these elements an individual TCP/IP connection is set up and released. Consequently for every user interaction like clicking a link or accessing a URI [RFC1630] one or more extremely short-lived TCP/IP connections are established and corresponding requests are generated and send. After the reception of the corresponding response the connections are torn down. Figure 3.1 shows the HTTP message exchanges for the retrieval of a typical hypermedia page [BeFi94].

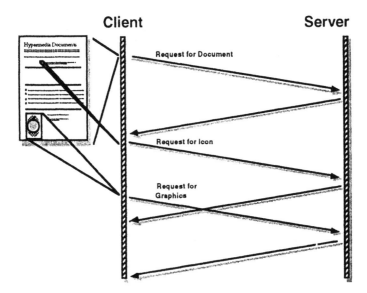

Fig. 3.1: HTTP messages for a typical hypermedia page

This leads to a large number of connection establishments and tear-downs per unit time, especially at highly frequented servers and proxies. Popular WWW-servers today serve up to several hundreds of requests per minute [Lycos95]. Since in each single WWW session several of these 'transactions' (request response pattern) per minute occur, these systems must be able to efficiently manage a large number of extremely short-lived associations. The problems associated with this behaviour are addressed in chapter 3.1.

Furthermore today's transport systems offer no support for the different media and data types contained in a hypermedia document in the area Quality of Service: If for instance the user initiates a large binary transfer (file-, picture-, sound- or video-retrieval), he has no control over the duration of the transfer. Even if the underlying network technology (e.g. ATM) is capable of guaranteeing a certain QoS-Level all today's transport systems offer is a best effort service. Future transport services will have to be able to pass the benefits of QoS capabilities of the underlying system on to the user. In the example mentioned above it would for instance be beneficial if the browser lets the user specify weather he wishes the transaction to be carried out using a (cheap) best effort service or within certain timing constraints.

In addition to the above features future hypermedia systems will offer enhanced services such as real-time-video and -audio (i.e. On-line Customer Support 'Hotlines' referenced from a hypertext document) or on-line presentations, games and chats. The timing constrains associated with real-time communications which network technologies like ATM can meet are not supported by today's transport systems. Chapter 3.2 will address these issues.

3.1   Transaction Performance of Today's Transport Systems

Future transport systems will have to be able to support a large number of short-lived transport connections ('transactions') per unit time. To evaluate the transaction performance of TCP/IP based transport systems of today the following measurement scenario was set up: Two SUN IPX workstations running under SunOS4.1.3 and connected via Ethernet acted as client and server. In addition to TCP/IP as transfer protocol the XTP3.6 protocol was examined. This protocol incorporates several advanced design features such as more efficient connection management, user selectable error detection and recovery mechanisms (go-back-n, selective or no retransmission), user selectable flow control (window- or rate-based) [StDe92]. The duration of a single transaction is similar for TCP/IP and XTP (ca. 7 ms). When the number of consecutive transactions was increased these results change dramatically. While the transaction duration remained nearly constant for XTP at ca. 7 ms it rose for TCP/IP up to more than 25 ms. The individual duration for 10000 consecutive transactions is shown in figure 3.2.

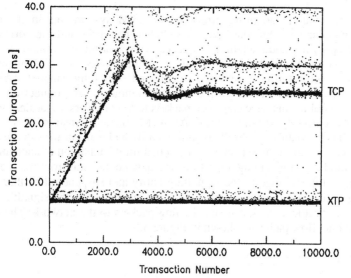

Fig. 3.2: Transaction duration for TCP/IP and XTP

The reason for this behaviour lies in the so called TIME-WAIT state of the TCP state machine: After a connection is released the connections state machine enters into the so called TIME-WAIT state. It stays in this 'zombie state' for four minutes before the context is finally destroyed [RFC793, RFC1337]. This mechanism prevents the interpretation of delayed duplicated connection establishment messages arriving after the release of a connection as requests for a new connection. In end-systems with a high transaction rate such as an hypermedia document server a large number of TCP connections in the TIME-WAIT can be expected. The corresponding timer management leads to the high CPU load encountered in the measurements [DaKa95].

These measurements show that improved transport protocols with efficient transaction support are necessary for efficiently support of distributed high performance client-server applications.

3.2   Transport Systems for Networks with QoS Capabilities

Modern network technologies like ATM provide advanced communication services by allowing the application to specify traffic characteristics of a connection and to receive corresponding QoS guarantees. In addition to their performance deficiencies today's transport systems do not adequately support the capabilities of these increasingly important network types.

Hypermedia retrieval applications could benefit from this support in the following ways:

- better overall performance,
- reduced request-response cycle duration by transporting (low bandwidth) requests over a guaranteed channel,
- transfer duration guarantees for large binary transfers,
- new services with stringed timing requirements (real-time services).

The main problem in adapting protocols to ATM system lies in the different signalling mechanisms. Standard packet switched networks and the corresponding protocols employ in-band signalling: all necessary signalling information is included in the header of a data packet. ATM on the other hand uses Out-of-Band signalling: All signalling information is sent via a dedicated signalling channel [ATMF94]. This poses new problems in protocol implementation. A connection oriented transport protocol for instance can issue packets to lower layers, regardless whether a connection is currently in place or not. An ATM adapted transport protocol on the other hand has to contact the ATM signalling entity and wait for the establishment of a connection prior to sending packets. The actual implementation of such an adapted protocol would, upon receiving a request for a new connection from the user, send a connection setup message to the signalling entity and suspend all further protocol processing until a response from the network is received (via the signalling entity). Only then the protocol stack can start issuing packets to the network. The different control and data flow paths are shown in Figure 3.3.

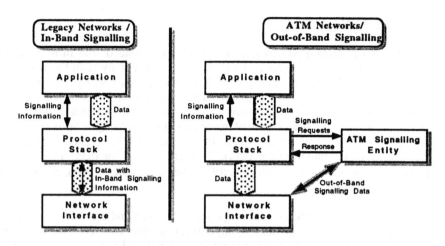

Figure 3.3: Signalling and Data Flow

The second necessary modification is an enhancement of the interface between the protocol and the users applications. Today such interfaces typically offer two primitives for connection establishment and tear-down (open, close) and two primitives for data transfer (read, write). There is no provision in this scheme for the negotiations of QoS guarantees between user and network. For an ATM adapted protocol a means must be provided to specify the required QoS prior to connection setup. Further primitives are needed for QoS re-negotiation and multicast/group management.

Based on these extensions Hypermedia Retrieval Applications can establish dedicated data connections for different traffic types. The following table sums up examples of traffic types and their QoS demands:

Traffic Type	QoS demands
control information and low bandwidth textual data	low bandwidth, moderate delay constrains
high bandwidth file retrieval	high bandwidth depending on user specified delay constrains
best effort file retrieval	available bandwidth service, no delay constrains
real time audio/video streams	stringent delay constrains

An ATM-adapted version of the XTP transport protocol is currently being realised at our institute. On top of this communication platform a hypermedia retrieval system based on a modified HTTP/HTML-system is being implemented. Initial results and experience with this implementation will be the subject of a following paper. A communications architecture incorporating dedicated transport/network protocols for different network technologies and corresponding interworking mechanisms is discussed in [Kara95].

4 Conclusion

Hypermedia Retrieval Systems have proven to be one of the most widespread distributed applications today and will become increasingly popular in the future. Currently used systems are hardly suitable for the demands set by tomorrow's users. Transmission performance for hypermedia retrieval systems as well as for other new applications like multimedia data bank access, multimedia mail, video on demand etc. is insufficient. With the advent of networks supporting Quality of Service and performance guarantees such as ATM this can not just be blamed on the lack of capabilities of the underlying network. Protocol support for these applications and protocol adaptation to advanced networking techniques becomes increasingly important.

The service requirements for future hypermedia systems and the necessary changes in these systems as well as the underlying transfer systems have been identified in this paper. The presented adapted hypermedia retrieval and transfer system for ATM networks is the first step towards a high performance distributed computing environment.

5 References

[ATMF94]   THE ATM FORUM: '*ATM User-Network Interface Specification - Version 3.1*', ATM Forum, September 1994

[BeCo93]   T. BERNERS-LEE, D. CONNOLLY: '*Hypertext Markup Language - HTML*', Internet Draft 1993.

[BeFi94]   T. BERNERS-LEE, R. T. FIELDING: '*Hypertext Transfer Protocol HTTP/1.0*', Internet Darft 1994

[DaKa95]   P. DAVIDS, R. KARABEK: '*Transport Protocols and Client Server Applications*', XTP Forum Research Affiliate Annual Report 1994, XTP Forum, Document XTP 94-104, pp. 100-102 and TRANSFER - XTP Information, Jan./Feb. 1995, Vol. 8, No. 1, pp. 11-13

[ISO94]   DIS 13522-1 - Information Technology - Open Systems Interconnection - Coding of Multimedia and Hypermedia Information (MHEG) - Part1: MHEG Object Representation Base Notation (ASN.1), 1994

[Kara95]   R. KARABEK: '*A high performance transport service for ATM LANs*', *accepted for the Broadband Islands* '95 Conference, Dublin, September 1995

[Lycos95]   CARNEGIE MELON WWW-SERVER 'LYCOS': '*Accesses per Day*', http://lycos.cs.cmu.edu/hourly.html

[RFC793]   J. POSTEL: '*Transmission Control Protocol*', RFC 793, September 1981

[RFC1337]   R. BRADEN: *'TIME-WAIT Assassination Hazards in TCP'*, RFC 1337, May 1992

[RFC1630]   T. BERNERS-LEE: *'Universal Ressource Identifier in WWW (URI)'*, RFC 1630, Juni 1994

[RFC1633]   R. BRADEN, D. CLARK, S. SHENKER: *'Integrated Services in the Internet Architecture: an Overview'*, RFC1633, 1994

[StDe92]   T. STRAYER, B. DEMPSEY, A. WEAVER: *'XTP - The Xpress Transfer Protocol'*, Addison-Wesley Publishing Company, 1992

TMN Systems Implementation Issues

Session 5a Introduction

Chairman - Richard Lewis

Cray Communications Ltd
Caxton Way, Watford Business Park
Watford, Herts WD1 8XH
Tel : +44 1923 258019
Fax : +44 1923 258890
e-mail: rl@cray-communications.co.uk

The three papers in this session are based on work performed by three of the Project Line 2 RACE projects which continue through the final year of RACE - ICM, PREPARE, and PRISM. The papers report the valuable implementation experience the projects have gained applying the results of their own research, together with appropriate RACE and external results, to network and service management. This experience is further described below, and expanded upon in the three papers which follow. However, before summarising the papers it is worth reviewing the progression from RACE I to RACE II, and from the start of RACE II to the present day (the final RACE IS&N Conference) in order to see the papers in their proper context.

RACE I, whose main objective was "to review the options", while constrained to a certain extent by established standards and norms, was quite free to review and use a wide range of methods and technologies. In those early days, RACE I TMN projects operated, quite rightly, as standalone projects with limited inter-dependency (though interworking was encouraged, for example though the GUIDELINE Project) and without necessarily being constrained by the need to interoperate with real network technology. This freedom was a great advantage which allowed the projects the flexibility to make good progress. The RACE I TMN projects produced very significant results, which were used as a starting point by the RACE II management projects, and with some results being incorporated in de jure and de facto standards by external bodies. The RACE I results therefore directly and indirectly influenced the RACE II projects, providing a good foundation for continued research.

RACE II, whose main objective was "to prepare for IBC deployment", made early reference to RACE I work and to existing standards. Significantly, during the period of RACE II, external activity in the area of IBC management has accelerated, with the Network Management Forum and the ATM Forum being particularly influential towards industrial progress, while ETSI and Eurescom are influencing the Public Network Operators. Now IBC related products, particularly network level equipment, are becoming more readily available, and so the later work in RACE II and the developing and developed standards can be validated against real equipment, in a real operational context. This has obliged the projects to conform to these available technologies and reduced project's experimental freedom. On the positive side, projects have been able to validate the relevance and performance of their results against commercial and near-commercial technology. The three papers in this section report on different aspects of this experience.

The first paper, from the PRISM Project, describes the implementation of a telecommunications service prototype (VPN) and associated management services which have been specified according to the Services Management Reference Configuration. The implementation is achieved making use of an OSF Distributed Computing Environment (DCE) platform. The paper reports how the architecture is adapted to conform to constraints imposed by the platform.

The second paper, from the PREPARE Project, applies TMN principles to the management of ATM Virtual Paths over the European ATM Pilot network comprising a number of ATM cross-connects. The management system is constrained on the one hand by the TMN Architecture considerations (Network and Network Element levels) and on the other hand by the need to make use of commercial network management system. Further constraint is imposed by the physical network, in the form of commercial cross-connects which present standard information models to the management system.

The third paper, from the ICM Project, describes a TMN Operation System (OS) which monitors the performance achieved by a real ATM network. Again, the implemented system is constrained by TMN and network equipment standards. The OS, Performance Verification, is further constrained by the need to inter-operate with other TMN OSs, for example Routing.

As identified above, the TMN projects are becoming more and more constrained by emerging standards and by the latest equipment being deployed in the networks. This is likely to apply even more in ACTS. However, these constraints are not universal. There remain a large number of management issues which are still unresolved, where researcher's talents and imagination can be applied. Some of these, which will furnish the challenges of the future, are identified in the papers which follow. I look forward to their resolution in papers at the ACTS conferences!

VPN on DCE: From Reference Configuration to Implementation

Jean-Paul Gaspoz, Constant Gbaguidi
Swiss Federal Institute of Technology (EPFL)
Telecommunications Laboratory, CH-1015 Lausanne

Jens Meinköhn
GMD-FOKUS, Hardenbergplatz 2, D-10623 Berlin

Abstract. This paper presents the results of the implementation of a Virtual Private Network telecommunications service prototype and associated management services on a distributed platform. The specification thereof has been carried out according to the RACE II project R2041 PRISM Service Management Reference Configuration. The implementation of the service prototype and its management has been achieved on an OSF DCE (Distributed Computing Environment) Platform.

Keywords: service management, virtual private network, DCE, ODP

1 Introduction

Due to liberalisation trends in the telecommunications market, e.g. driven by the Open Network Provisioning directive of the European Community as well as by market pressure, and the introduction of Integrated Broadband Communications (IBC) technologies, a large number of complex telecommunications applications is expected to appear in the near future. These applications will run on integrated computer networks and will provide customised innovative and flexible multimedia services to users in a competitive deregulated arena.

Among the large number of emerging services, several reasons led a number of RACE II projects to select VPN as the IBC service to demonstrate their work. In particular, VPN is a very good example for representing increased service flexibility vis-à-vis customers. Indeed, when customers purchase a VPN service they want the VPN to represent their own virtual network as if it were a real private network. Moreover, they also want to get rid of the burden to design and maintain a dedicated physical network. For this purpose, VPNs represent a class of services providing virtual networking facilities with different levels of outsourcing and functionality so that the complete range of customer requirements can be met. On the other hand, although VPN is considered to be a major IN service, the IN paradigm has so far been predominantly voice-based and concerned with services that have their origin in telephony rather than in future data and multimedia services. Currently existing standards work reflects this IN view of VPN [Q.1211]. The investigation of VPN services within RACE has a wider scope and is focusing on VPN as a data service and on the requirements for managing such a service in an IBC environment. Some RACE I work was devoted to this, for example [CFS-D721a] and [CFS-H409], and it is being further developed within RACE II.

Service management is becoming more important as the service market expands and competition between service providers increases. The separation of the network from the services means that the services must be managed separately from the network and a distinction must be made between network management and service management. The management of end–to–end services in an open service market requires co–operation between those involved in providing and managing the service, which is possible only if there are open, standardised management interfaces, a clear partitioning of management responsibilities between all entities concerned, and a common understanding of the management information and of the operations on this information. The work on service management in RACE II is intended to show how this can be achieved in the IBC environment.

The PRISM (Pan–European Reference Configuration for IBC Services Management) project, initiated within the RACE prógramme is currently developing a comprehensive framework for designing such management systems. This framework, termed Service Management Reference Configuration (SMRC), provides methods and concepts for going from an overall enterprise model, which relates the roles of each actor in a particular management scenario, down to detailed specifications of management service functionality and the management information necessary in each domain to fulfil the management needs.

In this context, the objectives of the work presented in this paper are threefold: firstly, it aims at validating the relevance of the concepts and methods advocated by the SMRC via a simple but real implementation in order to gain insights as to the applicability and practical value of the SMRC. Secondly, it attempts to consider the specification of a service and its management in an integrated manner, according in this respect to the Information Networking Architecture (INA) approach [Rubi94]. Thirdly, it offers the opportunity to evaluate the limits of a currently available distributed implementation and execution platform when trying to realise an Open Distributed Processing system (ODP).

The remainder of this paper is organised as follows: section 2 provides an introduction to the main concepts and methods put forward by the SMRC. Section 3 is structured according to the ODP viewpoints. It comprises the specification of a VPN service and its associated management services as well as its implementation on the DCE platform. This paper concludes with summary remarks and some issues required to be resolved by future extensions of this work.

2   The Service Management Reference Configuration

2.1   What is the SMRC?

The SMRC defined by PRISM can be seen as a generic blueprint for the specification and design of service management systems. In particular, a methodology for the stepwise decomposition of complex management services into less complex components is provided. The SMRC includes considerations of user requirements, structure, information, communication, external access, security and non-functional constraints which have an impact on the specification of management functionality.

ODP is used to structure the SMRC, but further concepts are needed for aspects of customer service management, for external access to TMNs, for service management security, for issues related to the boundary between service and network management layers and for the requirements capture which results in specific features of management services. Also, it addresses the information modelling for both managing and managed part of the management system. As such it is using inputs from TMN and OSI management.

The scope of the SMRC is to establish a generic framework for the development of TMN systems for managing service and network environments in the context of IBC. The SMRC is composed of the SMRC Framework, the Abstract Architecture and the Concrete Architecture (Fig. 1).

The *SMRC framework* is used as a meta–model for abstract architectures to be applied by system architects and for concrete architectures to be applied by implementors.

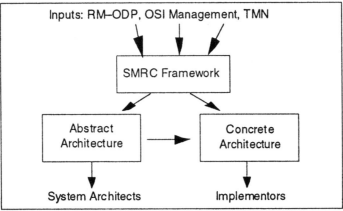

Fig. 1: The SMRC

The *abstract architecture* describes generic management services in relationship to specific actors and domains, leading to the definition of reference points and interfaces. The abstract architecture does not consider any available platform for supporting the communication, distribution and execution needs of the management services.

The *concrete architecture* does take into account the platforms available on which the management services will actually run. It obeys the concepts, rules and methods given by the meta–model and elaborated by the abstract architecture, maybe adding additional rules and concepts. The result of the concrete architecture is to be used by implementors to implement a management system. The process of moving from the SMRC framework to abstract and concrete architectures is illustrated in figure 2.

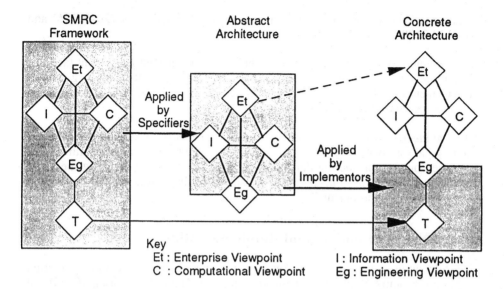

Fig. 2 : Applying the SMRC to abstract and concrete architectures

The scope of the work presented in this paper covers both the definition of the abstract VPN architecture, that is the specification of the enterprise, information and computational viewpoints as well as the realization of the concrete architecture on DCE.

2.2 Using ODP for the SMRC

Due to the distributed and highly complex nature of the resulting management systems for service management, the application of ODP (Open Distributed Processing [RM–ODP]) concepts as a framework for structuring their definition is appropriate. The ODP Reference Model is a framework of abstractions helping to position services relative to one another as well as to guide the selection of appropriate models of services. The ODP Reference Model encompasses the viewpoints by partitioning the concerns to be addressed when describing the ODP system. PRISM is not concerned with the specification of software modules and their implementation, so the technology viewpoint has not been covered. The following gives a brief description of the determined viewpoints:

The *Enterprise* viewpoint is directed at the needs of the users of the management services and can be modelled in terms of required management functions, TMN-domains, TMN-actors and their roles. It covers business and management policies with respect to access rights and actor's roles. Additionally non-technical aspects covering regulatory and non-functional requirements such as Quality of Service may be considered as well.

The *Information* viewpoint describes the structure of information elements that define the information needs required from the enterprise viewpoint. It defines rules stating the relationship between information elements and the forms in which information and information processing is visible to the users. The information model is achieved in accordance with the structure of management information (X.720–X.725) and TMN M.3100.

The *Computational* viewpoint describes how the management processing facilities functionally or logically perform the processing task. Rules for grouping management functions into blocks are described as well as interactions between these blocks regardless of their distribution.

The *Engineering* viewpoint identifies a set of concepts for describing the physical entities and for positioning the computational building blocks on these entities. This is carried out by considering the communication needs and by defining the TMN interfaces in terms of concepts such as capsules, channels and their supporting communication environment.

3   VPN Specification and Implementation

The telecommunications service retained for specification and implementation purpose is a simple VPN service basically offering a Closed Used Group (CUG) and a Private Numbering Plan (PNP) facility. Management aspects addressed are the ability for an authorized manager to add/remove users to/from the CUG, and to change their privileges and restrictions as well [PrismD3]. Moreover, the principal security issue considered is the access control to the service and to the management facilities and information.

The following subsections present the way all those features are treated using the methods and concepts advocated by the SMRC. In turn, the enterprise, the information, the computational and the engineering viewpoints are considered. As the latter needs provision of a distributed platform we also experimented the mapping between the SMRC and a real distribution framework, namely OSF DCE (Open Software Foundation - Distributed Computing Environment).

3.1   The Enterprise Viewpoint

The main topics relevant to the enterprise viewpoint are the requirements upon the service, the identification of the actors involved in providing, using, subscribing to or managing the service, the interactions between these actors, and the scenarios of service usage. The actors considered in the case of a VPN are the service user, the service provider, the customer, who subscribes for the service, and the network provider [CFS-D721b].

The scenarios of usage are based on the architecture illustrated in figure 3. This shows the organization of a corporate telecommunications network in sites connected by a VPN. Therefore, the VPN deals only with inter-site communications, while the CPN is responsible for intra-site interactions. The scenarios of usage considered include:

- access control to the service, using password and username verification at log-in time;

- management facilities (adding or removal of a user, modification of users' privileges), performed by the customer or a delegated manager;

- service facilities, such as call processing or password modification. Any person authorised by the customer to use the service can access these facilities.

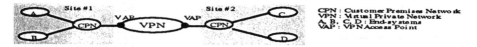

Fig. 3 : Organization of a Corporate Telecommunications Network

These scenarios deal with both management and service issues, mainly by considering two kinds of users, simple users and the manager. Within the simple users group, a distinction is also made in terms of privileges and restrictions. Hence, some users are not allowed to originate (respectively receive) calls going (respectively coming from) outside the VPN. Therefore the CUG feature, one of the core VPN concepts, has been thoroughly expressed.

3.2   The Information Viewpoint

The information viewpoint specifies information bearing entities and information processing activities. These use the information objects to give the processing details of the operations identified in the enterprise viewpoint. The main mapping rule from the enterprise to the information viewpoint is that any operation from the scenario cannot be effected, unless the information objects are exhaustive enough to enable that operation. Therefore any operation that cannot be supported by the information objects set is illegal.

To comply with this rule, object models have been developed for the VPN features of interest in this study, namely CUG and PNP. To conciliate readability and formalism, both a graphical notation -OMT (Object Modelling Technique [Rumb91])- and textual ones -GDMO (Guidelines for the Definition of Managed Objects [X.722]) / GRM (General Relationship Model [X.725]) have been used during the specification. For the sake of brevity, GDMO and GRM specifications have not been reproduced in the paper.

The CUG facility restricts access to the VPN service to authorized users registered in the group only. Every such user is assigned with both a logical and a physical address. The mapping between these addresses is performed based on the private numbering plan. In this respect, a PNP realizes the promise of a seamless, yet virtual, private network, as users need not be aware of physical addresses. Instead they use logical addresses to interact with each other. We choose to present the most relevant feature, that is the CUG facility. The corresponding object model (fig. 4) exhibits a set of *Policies* applied to the group (billing and security policies) and the way a *user agent* is viewed as a *CUG member*, having some *Privilege* and/or *Restriction* or neither. A *Privilege* object is concerned with the permission for receiving and originating off-net calls, that is, calls which have some endpoints not connected to the VPN. A *Restriction* object is concerned with the permission for receiving and originating on-net calls, i.e., calls whose endpoints are all connected to the VPN. To cover call aspects the information model of network resources has also been designed. For the sake of brevity, this model has not been reproduced here.

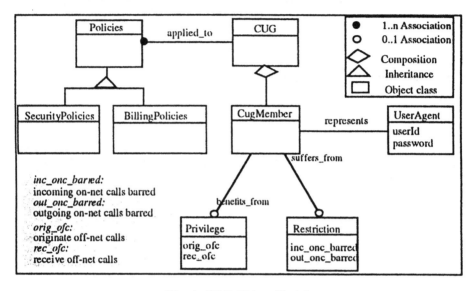

Fig. 4: CUG Object Model

3.3 The Computational Viewpoint

The objective of the computational viewpoint is to define the object *interfaces* necessary to deploy the VPN functionality, both in terms of service and management thereof. Computational objects (COs) are accessible via their *interfaces* only. Access rules for interfaces may be different from one interface to another, even though both interfaces are supported by the same CO. Also, service and management interfaces are properly separated. Therefore, service and management aspects may be modeled in a consistent and uniform way. To deploy the VPN functionality, each operation from the scenarios outlined in the enterprise viewpoint is realized by means of interactions between computational objects. As a result the overall functionality is distributed over these objects via their interfaces (fig. 5).

To have an overview of the roles taken by each represented entity, note that *CPN* conveys the user requests to the VPN, precisely to the *AccessSession* object. This latter is created at log-in time. Its creation ensures that the user is authorized to access the VPN service. The *VapDirectory* object gives the VAP (VPN Access Point) information related to any registered *CPN*. Moreover, the *Attachment* object creates a relationship between a user and the VAP he is attached to, while the *Ownership* object deals with the relationship between a user and his related *AccessSession* and *VpnSession* objects. Indeed this latter is created in order to manage the logical connections involved in the interactions between the users. Each logical connection is characterized by its carried information type (voice, video, etc.) and the required bandwidth. Network connections aspects are relevant to the

PubNetwork entity. One should note that we did not detail call aspects, especially connection management. Indeed, due to the short time scale of this study, we mainly restricted the scope to the service access control and the service management.

Users are represented within the VPN by many objects. Hence, the *CugMember* object is the CUG view of the user, the *UserAgent* is concerned with access control aspects (password checking) and the *UserAddr* entity maps the logical address of the user (his private name or that of his equipment) to his public address. The *DialList* entity permits to find out the *UserAddr* entity related to a given logical address. This entity has been introduced to make the deployment of the VPN functionality easier.

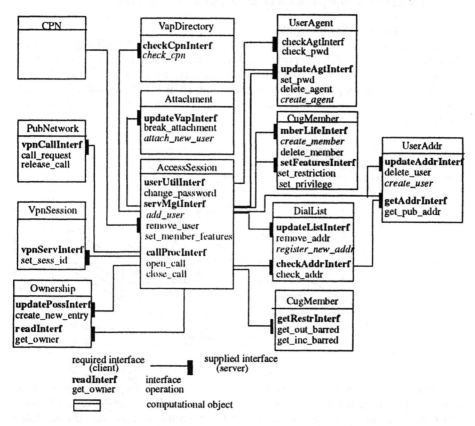

Fig. 5: VPN Computational Object Type Diagram

An example how computational objects represented in figure 5 interact to provide the required system behaviour is highlighted on the same scheme using italic style. First, the CPN CO asks for adding a new user (*add_user* request), supplying the CPN this one should be attached to, and the information needed to create the various objects mentioned above. Then, the consistency of the supplied parameters has to be checked, particularly the target CPN (*check_cpn* request to VapDirectory object). If this is correct, then the different instances representing the new user can be created (*create_user, create_agent, create_member*, and so forth). Obviously, if the new user has been already registered as a person authorized to use the VPN service, an error message is returned to inform the manager performing the registration operation.

3.4 The Engineering Viewpoint

The engineering viewpoint reveals the control and transparency mechanisms deployed within the distributed environment to distribute the designed system. The rationale for using such an environment is to save applications and services designers from knowing about those mechanisms. Therefore it is worth presenting the distribution framework as well as equipments used in the purpose of this work.

The Distribution Framework
The platform chosen for this work is OSF DCE version 1.0 [OSF92]. Based on industrial standards, DCE supports the development and execution of distributed applications on networked computer systems (fig. 6) [Ros93]. In addition to providing a consistent programming environment, DCE is independent of operating systems and protocols and offers a wide range of built-in security services and mechanisms.

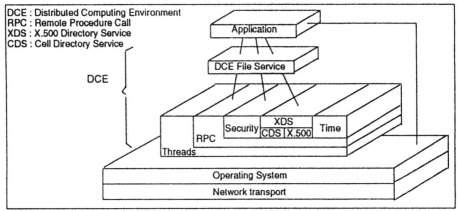

Fig. 6: Overview of DCE

The programming environment includes facilities such as *threads* for concurrent programming, *RPC* for designing client/server models, *DCE File Directory Service* and *Cell Directory Service* for managing file migrations, and *Distributed Time Service* for synchronization issues. The application transparently accesses all those facilities.

The Results
As DCE only realizes concepts pertaining to the ODP computational and engineering viewpoints, the practical lessons learned especially concern these two viewpoints. In particular we investigated the way a computational model built using the SMRC can be mapped onto the computational and engineering concepts offered by DCE.

The very first learning was the impossibility to use an object-oriented language such as C++ with our release of DCE. Instead the use of the C programming language to implement clients and servers did not make it possible to fully exploit the benefits provided by the use of an object oriented approach in the specification and design phases, in particular genericity and inheritance. In order to represent an object in a suitable way, we used a structure type to record object attributes, while the object's interfaces were thoroughly represented as DCE servers. Likewise, in order to have persistent objects, whose lifetime does not depend on any running process, the structure types have been put into files. Thus, a user is registered only once in a file and its removal does not rely on the application runtime but only on the service manager who can decide of adding or removing a user. We found no other means for creating persistent objects, because DCE lacks supporting tools in this area.

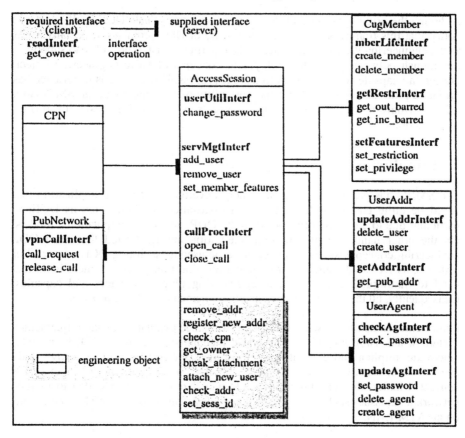

Fig. 7: VPN Engineering Model on DCE

The accurate application of SMRC tends to give rise to fine grained computational objects. As a consequence, the number of these objects can be important. On the other hand, DCE 'objects' are rather large grained processes. Consequently, the memory consumption is high when using DCE, and then fine grained processes must be avoided. We first tried to implement the computational model using a one-to-one mapping between computational objects (COs) and basic engineering objects (BEOs) and in implementing each of these as a DCE process. As a result, we found that each generated process was nearly 2.5 Mbytes large. Therefore, for reasons of efficiency and memory constraints, several objects whose services where exclusively used by each others have been grouped and their respective interfaces realized by means of internal actions of the resulting composite object. Conceptually, this operation consisted in grouping several tightly coupled COs into one basic engineering object (many-to-one mapping) and in implementing each BEO as a DCE process. Consequently, the number of implemented clients and servers was actually smaller than the original number of computational objects. The resulting VPN engineering model on DCE is shown in fig. 7. As notations are not distinct from those used in fig. 5, it is worth mentioning that the basic engineering objects depicted are implemented as processes in DCE.

Although the original purpose of this work was no to evaluate DCE itself, we could notice that this framework does not grant many of ODP promises, especially as far as the engineering viewpoint is concerned. In particular, DCE provides no support to distinguish between the ODP deployment concepts of basic engineering object, cluster and capsule which all map to a DCE process. However, this experience has provided promising insights as to the portability of concepts from the SMRC to a real distribution framework and it represents a stimulating basis for further investigations.

4 Conclusion

The SMRC has proven very useful to help develop a distributed application that could be quite easily implemented on a distributed platform such as DCE, in particular due to the similarities between the ODP and DCE computational models. On the other hand, the present implementation has illustrated that most ODP engineering concepts were too advanced to be used as such in a DCE context. For instance, as DCE only supports access and location transparencies there is no real need to distinguish between basic engineering objects, clusters and capsules. Consequently, all these concepts are equally implemented as DCE processes.

Future extensions of this work include performance evaluation of the implemented system when using the different binding mechanisms offered by DCE, namely, automatic, implicit and explicit binding. In the same way, a more extensive study and use of the security mechanisms offered by DCE is certainly of interest in a VPN context. Finally, the distributed application realized will be extended to include network level issues such as the establishment and management of ATM connections on an ATM LAN.

5 References

[CFS–D721a] "Virtual Private Networks", RACE CFS D721, Issue C, Jan. 1993

[CFS–D721b] "IBC VPN Services", RACE CFS D721, Issue D, January 1994

[CFS–H409] "Customer Query and Control", RACE CFS H409, Issue C, December 1992

[OSF92]   Open Software Fundation: Introduction to OSF DCE, Prentice Hall, 1992

[PrismD3]   "VPN and UPT Service Management: First Case Study Report, Volume 2: VPN Case Study", PRISM Deliverable 3, March 1993

[PrismD8]   "Reports on Selected Areas of the Service Management Reference Configuration", PRISM Deliverable 8, Sept. 1994

[Q.1211]   "Introduction to IN Capability Set 1", CCITT Study Group XI, Working Document XI/4-38, Geneva, March 1992

[RM–ODP]   Basic Reference Model of Open Distributed Processing Parts 1–5, ISO/IEC/JTC1/SC21/WG7, #N431, N6079, N432, N433, N343

[Ros93]   Rosenberry, W., D. Kenney and G. Fischer, "Understanding DCE" -, O'Reilly & Associates, Inc., May 1993.

[Rubi94]   Rubin, Harvey and Natarajan, N., "A Distributed Software Architecture for Telecommunications Network", *IEEE Network*, Jan-Feb. 1994

[Rumb91]   Rumbaugh, J. et al., *Object-Oriented Modeling and Design*, Prentice Hall, Englewood Cliffs, New Jersey, 1991

[X.722]   "Guidelines for the Definition of Managed Objects", ITU Rec. X.722, Sept. 1991

[X.725]   "Information Technology - Open Systems Interconnection - Structure of Management Information - Part 7: General Relationship Model", Draft ITU Rec. X.725, Jan. 1994

ATM Public Network Management in PREPARE

Wolfram Kisker

IBM European Networking Center
Vangerowstr. 18
D-69115 Heidelberg
Germany
Tel. +49-6221-59-4422
Fax. +49-6221-59-3300
E-mail. woki@vnet.ibm.com

Jürgen M. Schneider

IBM European Networking Center
Vangerowstr. 18
D-69115 Heidelberg
Germany
Tel. +49-6221-59-4205
Fax. +49-6221-59-3300
E-mail. jms@heidelbg.ibm.com

Abstract: The Asynchronous Transfer Mode (ATM) technology is the basis for the Integrated Broadband Communications (IBC) environment of the envisaged information society. In this paper we present results from the application of the ITU-T TMN framework and associated ETSI standards for ATM management within the PREPARE[1] project. In particular, we describe the design of the TMN architecture for the public ATM network part of the PREPARE 1995 Demonstrator testbed and the design and implementation of the ATM network management layer operations system (OS).

1   Introduction

The Asynchronous Transfer Mode (ATM) technology is the basis for the Integrated Broadband Communications (IBC) environment of the envisaged information society. All major public network operators (PNOs) and also private carriers in Europe are currently preparing the introduction of ATM in the public network sector through ATM pilot networks and associated usage trials. With this growth and international interconnection of public ATM networks there is an increasing demand for standardized ATM network management solutions. This includes the provisioning of standardized ATM network management services, management information models and applications conforming to a standardized management framework.

As the management framework for public ATM networks the **Telecommunications Management Network (TMN)** recommendations from ITU-T (M.3000 series) has been widely accepted. But how to define the functional TMN architecture of ATM networks ? Which ATM management information models

---

[1]This work was partially supported by the Commission of European Communities (CEC) under project R2004 PREPARE of the RACE II program. The paper does not necessarily reflect the views of the PREPARE consortium.

and management services can be used ? What are the early application and implementation experiences ?

The PREPARE project 1995 testbed and demonstrator is focussing on ATM networks as the basis of an integrated broadband services infrastructure and its cooperative end-to-end management by several network operators and service providers [BjSch94]. Initial results from this work with respect to the questions above are presented in this paper. This includes

- the application of existing standards,
- the design of a suitable TMN architecture for ATM networks,
- the definition of the requirements on the ATM network management system,
- the specification of the interfaces of the ATM network management system, and
- the realization of ATM network management application behavior.

2   PREPARE 1995 Demonstrator Testbed

The focus of this paper is on the public ATM network part of the PREPARE 1995 Demonstrator testbed. Within this testbed, permanent ATM virtual paths (VPs) provided by the European ATM pilot network are used to interconnect several ATM cross-connects (XCs) at Copenhagen, Berlin and London. Attached to these ATM cross-connects are private ATM LANs, hubs and end-user terminal equipment at these different sites (see Figure 1). The end-to-end ATM infrastructure is managed and controlled by several cooperating management systems from different providers. Another aspect of the Demonstrator is the management of multimedia teleservices, such as multimedia conferencing (MM conferencing) and multimedia mail (MM mail), and the integration of teleservices management with the ATM end-to-end infrastructure management.

The realization of the permanent VPs between Copenhagen, Berlin and London within the European ATM pilot network is transparent and cannot be managed by PREPARE. In order to be able to manage the public ATM network domain the PREPARE project has placed additional ATM cross-connects at the edge of the permanent VPs of the European ATM pilot network. There are two ATM cross-connects at Berlin that form one public ATM network (PNO domain), managed by an associated TMN (see Section 3). A second public ATM network is constituted by another two ATM cross-connects (one in Copenhagen, one in London), managed by a second TMN system. The purpose of this configuration is to facilitate the demonstration of cooperating ATM public network domains.

The service provided by the public ATM network part is an ATM virtual path (VP) service. The ATM LAN switches at the different sites are using the public VPs setup between them for transparent end-to-end signalling in order to setup ATM virtual channels (VCs) between end-systems.

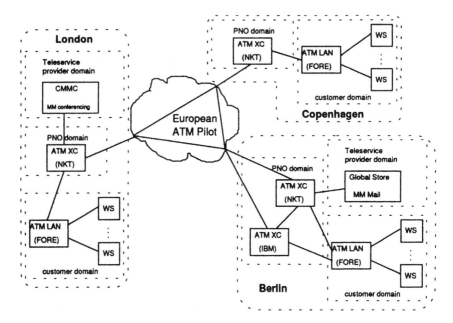

Figure 1. PREPARE 1995 Demonstrator Testbed

3   TMN Architecture for ATM

The management architecture of the public ATM networks in the PREPARE 1995 Demonstrator testbed has been designed according to the ITU-T TMN recommendations [M3010]. The functional TMN architecture distinguishes between operations system function blocks (OSFs) at the

- **network element management** layer (NE-OSF),

- **network management** layer (N-OSF), and the

- **service management** layer (S-OSF)

of an ATM network. In the physical TMN architecture all of these OSFs are physically instantiated as separate operations system building blocks (NE-OS, N-OS, S-OS) with interoperable TMN interfaces between them as shown in Figure 2.

3.1   ATM Network Element Management Layer

For each of the individual ATM network elements (ATM cross-connects) there is an agent application running at the network element management layer. This agent application controls the software and hardware resources of the ATM network element and provides functions to manage these resources via a local operator interface or from

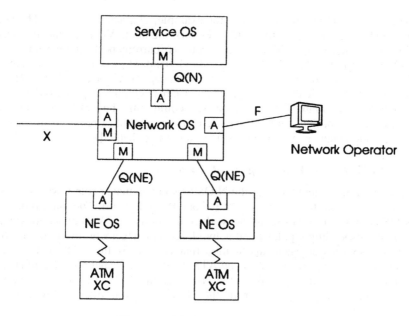

Figure 2. ATM TMN Architecture

remote. The communication between the agent and the network element hardware is achieved in a vendor-specific way.

The individual ATM network element agents represent NE building blocks (realizing TMN network element functions (NEFs)) if they are directly connected to the N-OS via Q management interfaces. In cases where the network element agent does not directly support a Q management interface means of Q-adaptation can be employed to connect the network element management to the TMN. Such Q-adaptor functions (QAF) are needed, for example, if

- the network element agent does not support the required Q protocol stack (CMIS/CMIP, OSI etc. [Q.961][Q.962]), or
- the management information model used by the network element agent is different from the Q management information model.

Both the NKT and IBM ATM cross-connects in the PREPARE 1995 Demonstrator testbed are managed using Q-adaptation. A physical Q-adaptor system which is realized by an operations system building block (NE-OS) exists for each of these ATM network elements as shown in Figure 2. A Q3 protocol stack is used between the NE-OSs and the N-OS, which is CMIS/CMIP over OSI presentation, session, transport and X.25. The Q3 network element management information model used here is a subset of [ETS 52210] as needed to support equipment management and ATM VP services from the network element point of view.

The network element management information model described in [ETS 52210] has been derived from [M3100] and refined specifically for ATM. Therefore, there are managed object classes representing, for example, characteristics of ATM access points (ATM interfaces) in the ATM network element, VP trail and connection termination points and ATM cross-connections, current performance data of the network element etc. Overall, this recent ETSI standard proved to be adequate and applicable in the context of network element management for the PREPARE 1995 Demonstrator, although agreement on the precise behavioral interpretation of some of the attributes and managed object classes could only be achieved after some discussions in the project. In addition, there are quite a lot of relationships and side effects that implementers have to be aware of (compare [Ritt95]).

3.2 ATM Network Management Layer

At the network management layer the N-OS acts in a manager role towards all ATM network elements in the ATM network (of one PNO domain). The main function of the N-OS is to manage end-to-end network connections (i.e., VP connections) and to maintain corresponding topology and network level management information. To this end, the N-OS runs an agent application that is used by the S-OS at the service management layer to provide end-to-end ATM services of the network. Optionally, the management information stored by the N-OS agent can also be used by network operators and peer N-OSs. Overall, the N-OS has four different management interfaces:

1. The interfaces to the network elements (Q(NE)) are used to control the network elements, whereat, in this design, each of them reflects one ATM cross-connect.
2. The interface to the service management layer (Q(N)) provides the capability to manage the ATM network as a whole, the concept of network connections and performance information.
3. The interface to the network operator (F) provides the control and coordination of all network elements and network connections within this domain via a graphical user interface.
4. An interface to peer N-OSs (X) may optionally exist, but has been excluded for the purpose of concentrating resources in PREPARE.

The Q protocol stack employed between the N-OS and the S-OS is the same as between the N-OS and the NE-OSs. For the F-interface to the workstation (WS) building block two approaches are being followed: one uses the Q3 interface to the NE-OSs, the other is using CMIS/CMIP over TCP/IP to interwork with the N-OS agent. With respect to the management information model used for the N-OS agent there was no standard available for ATM networks. Therefore, an ATM management information model for the network management layer view was developed within PREPARE for the N-OS. The realization of the N-OS and the design of the ATM network layer management information model is further detailed in Section 4.

3.3 ATM Service Management Layer

At the service management layer, the S-OS runs as a separate physical system. Its function is to facilitate the usage and management of the services provided by the ATM network. To this end, it maintains management information about individual service contracts with customers, the end-to-end virtual paths through the ATM network they are using, quality of service, accounting and billing information etc. This management information is stored by an agent application in the S-OS providing the service management interface for external requests. To complete external requests the S-OS acts as a manager towards the N-OS agent and cooperates with peer S-OSs to provide one-stop shopping.

3.4 ATM TMN Architecture Characteristics

The main objectives met by this TMN architecture design are:

- clear separation of management functions and services between functional layers
- precise and unambiguous specification of management information available at different functional layers
- modular, parallel and independent implementation of TMN building blocks
- performance gains through parallel processing
- application and efficiency of security policies and mechanisms
- extendability and scalability of the overall TMN system.

All of the physical building blocks of this TMN architecture (i.e., OSs, QAs, WSs etc.) are currently being implemented by different partners of the PREPARE project consortium.

4 ATM Network Management Layer

The network management layer presented here has been designed to address the functional requirements of the management scenarios defined for the PREPARE 1995 Demonstrator. Based on these requirements general management functions are defined for configuration and limited fault and performance management capabilities. These management functions are aimed at providing a simple interface for handling virtual paths (VPs) of an ATM network. In addition the service management layer will be notified about faults of objects related to specific VPs and it can request to monitor information about transmitted cells related to specific VPs. Within this section we describe the defined management functions and the information model for the ATM network management layer and we discuss mapping issues between the network management layer and the network element management layer.

4.1 Management Functions

This section describes the management functions for managing virtual paths through an ATM network over several ATM cross-connects.

- **Create a VP Subnetwork Connection**
  This function provides the possibility to setup a virtual path in an ATM subnetwork consisting of several ATM cross-connects. The virtual path is identified by the start point, by the end point and the required quality of service. Start and end point are located at the edge of the subnetwork and interconnected with subnetworks of other domains. Start and end points are specified by topological points or more detailed by specific virtual path identifiers. The routing and bandwidth reservation between the start and end point is done by the N-OS.

- **Activate a VP Subnetwork Connection**
  An existing VP Subnetwork Connection with a reserved bandwidth is enabled for data transmission.
- **Deactivate a VP Subnetwork Connection**
  An activated VP Subnetwork Connection is disabled for data transmission.
- **Modify a VP Subnetwork Connection**
  The attributes of a VP Subnetwork Connection (e.g., bandwidth) are modified dynamically.
- **Delete a VP Subnetwork Connection**
  An activated or deactivated VP Subnetwork Connection is released.
- **Notification of topology changes**
  If the topology of the subnetwork is changing, for example if a new link is added between two ATM cross-connects, there is the possibility to receive notifications about these changes.
- **Notification of faults**
  If there is a fault in the subnetwork, which effects specific VP Subnetwork Connections, there is the possibility to receive notifications about this fault.
- **Get accounting data**
  This function allows to retrieve information about transmitted cells of specific VP Subnetwork Connections.
- **Setup of basic configuration**
  To get the basic configuration, the N-OS must be initialized. E.g., ATM cross-connects can be grouped into subnetworks; the links between the ATM cross-connects can be initialized, etc.

4.2 Information Model

Presently, there is no standard for the network management layer existing which provides an ATM specific management interface to the service management layer by managing several ATM cross-connects that are interconnected to form (sub-) networks. To define such a management interface the managed object class library from [ETS 43316] was profiled. This library provides generic managed object classes for the network management layer view.

To build the information model 8 classes were derived from the NA43316 to add ATM specific information. Two new classes (*currentData* and *interDomainLink*) were derived from *top* defined by [X.721] to provide specific information which is not available from the generic model. The inheritance tree and the naming tree of the resulting ATM network management information model are shown in Figure 3 and 4. For the full GDMO/ASN.1 specifications we refer to [PNIHB95]. The key managed object classes are briefly described below:

1. **ATM Subnetwork (atmSN)**
   The ATM Subnetwork represents a logical collection of ATM Topological Points, VP Network Connection Termination Points (CTP) and VP Subnetwork Connections between them. Each ATM Subnetwork can be devided into interconnected ATM Subnetworks and ATM Internal Links. By this partitioning several abstraction levels of the ATM network management layer can be created, where at the lowest abstraction level an ATM Subnetwork represents one ATM cross-connect of the network element management layer.

2. **ATM Internal Link (atmIL)**
   The ATM Internal Link represents the relationship between two ATM Subnetworks. It connects two ATM Topological Points and defines the maximum available bandwidth and current available bandwidth between them.
3. **ATM Topological Point (atmTP)**
   An ATM Topological Point is a logical collection of VP Network CTPs. It is an access point to ATM Subnetworks at the same level within this domain or to ATM Subnetworks from other domains. ATM Subnetworks from the same domain are connected by an ATM Internal Link; ATM Subnetworks from other domains are connected by Interdomain Links.
4. **VP Network CTP (vpNCTP)**
   The VP Network CTP terminates a VP Network Connection and a VP SubNetwork Connection. The identifier of the VP Network CTP is the VP identifier of the virtual path link represented by the VP Network Connection.

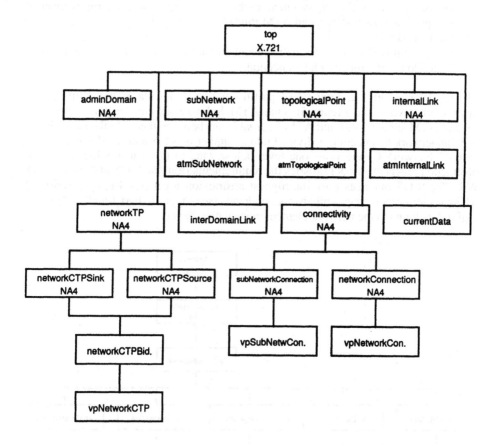

Figure 3. ATM Network Management Information Model Inheritance Tree

5. VP Network Connection (vpNC)

The VP Network Connection connects two VP NetworkCTPs between two ATM Subnetworks. It represents a virtual path link between two ATM cross-connects and defines the peak cell rates allocated for this virtual path link.

6. VP Subnetwork Connection (vpSNC)

The VP Subnetwork Connection connects two VP Network CTPs within an ATM Subnetwork. At the lowest abstraction level (where the ATM Subnetwork represents an ATM cross-connect) the VP Subnetwork Connection represents an ATM cross-connection. At all other levels a VP Subnetwork Connection is assembled from a number of VP Subnetwork Connections and VP Network Connections from one level below.

7. Interdomain Link (IDL)

The Interdomain Link describes the capacity between two ATM Topological Points of different administration domains. It is also used to describe ATM related restrictions at ATM Topological Points to external subnetworks, which are not part of the cooperative management architecture (e.g., to describe the permanent VPs provided by the European ATM Pilot).

8. Current Data

The Current Data holds information about the transmitted cells and discarded cells related to a VP Subnetwork Connection.

The model supports the concept of partitioning to divide the network in several abstraction levels (see Figure 5). One ATM Subnetwork can be partitioned into several ATM subnetworks and ATM Internal Links between them. At the same time a VP Subnetwork Connection is divided into a number of VP Subnetwork Connections and VP Network Connections one level below. Due to this partitioning in several abstraction levels the model satisfies the requirements from the S-OS and the network WS. The S-OS manages only the highest abstraction level (see Figure 5, Part 1), whereas the N-OS manages all abstraction levels (see Figure 5, Part 1 and 2; Part 3 of Figure 5 belongs to the network element management layer).

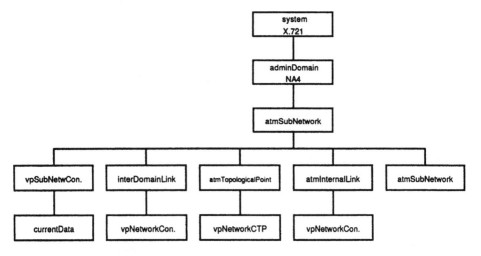

Figure 4. ATM Network Management Information Model Naming Tree

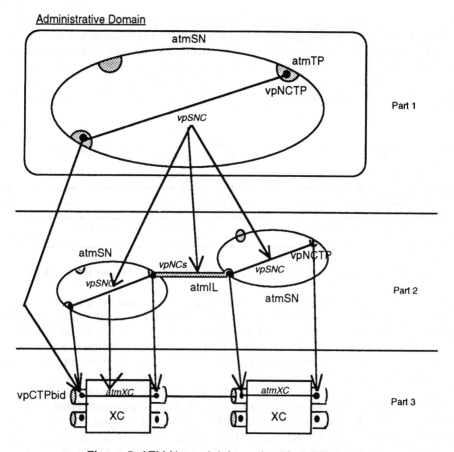

Figure 5. ATM Network Information Model Concept

For each ATM Topological Point there is an ATM Topological Point with the same resource reference in all abstraction levels below. The level of an ATM Topological Point depends on the level of the link to which it has a reference. For example an ATM Topological Point which has a reference to an Interdomain Link must be visible in all levels. The same principle is valid for the VP Network CTP.

4.3 Mapping between Network and Network Element Management Layer

As described in Section 4.2, the hierarchical modelling of an ATM network with several levels of ATM Subnetworks and Internal Links at the network management layer ends at the network element management layer, where an ATM Subnetwork represents a single ATM cross-connect network element. Therefore, there is a requirement for a mapping from the lowest abstraction level created at the network management layer down to the network element management layer and to the individual ATM network elements.

As described in Section 3 the standardized information model NA52210 from ETSI is used at the network element management layer. The NA43316 contains a conditional package *neAssignmentPointer*, which can be included in network management layer managed object instances to provide a reference from the network management layer to the network element management layer. A GDMO attribute is used to store the information about the mapping relationship as depicted in Figure 5 and listed in Table 1.

Network Management Layer (N-OS)	Network Element Management Layer (NE-OS)
ATM SubNetwork (lowest level)	atmNE (XC)
ATM Topological Point	atmAccessPoint
VP Network CTP	vpCTPbidirectional
VP Subnetwork Connection (lowest level)	atmCrossConnection (atmXC)

Table 1. ATM Network and Network Element Management Layer Relationships

The behaviour of the mapping is described by an example for the setup of a virtual path. The S-OS has a view of the highest abstraction levels with all ATM Topological Points which have a link to other domains. To create a virtual path from one ATM Topological Point to another, it just requests the action *createVpSNC* defined in the ATM Subnetwork class. If the ATM Subnetwork is not part of the lowest abstraction level (there is no mapping to the network element management layer), the N-OS uses the two ATM Topological Points as input and starts the route finding algorithm. The algorithm returns a list of ATM Topological Points. The N-OS chooses a VP identifier of unused VP Network CTP objects for each ATM Topological Point. If a VP Network CTP is created in the N-OS, a corresponding vpCTPbidirectional object is created in the NE-OS from the N-OS. In the next step the N-OS creates the VP SubNetwork Connections and VP Network Connections between the VP Network CTPs. If the ATM Subnetwork is part of the lowest abstraction level, the N-OS creates an appropriate *atmCrossConnection* object in the NE-OS.

5   Implementation Experiences

Today, there are still little experiences in implementing and operating layered TMN systems such as the one for ATM networks outlined in Section 3. The decision to instantiate physical OS TMN building blocks at each of the network element, network and service management layers within the PREPARE project proved to be very efficient in terms of distributing management functionality and focussing management application semantics to the layer requirements. In addition, the clear definition of interfaces in the TMN architecture, including CMIS/CMIP service and protocol stack, as well as information models specified in GDMO, facilitated parallel, independent application development by different partners of the PREPARE project consortium. The TMN architecture, as well as the ATM network management layer information model design, allows to define several abstraction levels through the definition and creation of subnetworks. At the lower levels, in particular, components of the physical TMN architecture such as NEs, QAs, NE-OSs, and N-OSs can be replicated and associated with a variable span of management control.

The TMN approach also proved to be very well suited as a framework for network management integration. For example, the management of ATM network elements from different vendors (such as the ATM cross-connects in the PREPARE 1995 Demonstrator testbed) or the permanent VPs provided by the European ATM pilot can be integrated into a common TMN system without major conceptual difficulties. The mechanisms to achieve this management integration for the PREPARE public ATM network are abstraction through information modelling, mediation and Q-adaptation.

First international standards for ATM management, such as [ETS 52210], are available and ready for usage. As indicated in Section 3, we found this ETSI standard applicable for the network element management layer and implemented most of its managed object classes. It is important, however, to correctly understand its function at the network element management layer and to understand its relationship to the management and connectivity model via trail and connection termination points and cross-connections as defined by [M3100]. At a first glance, some important relationships between attributes and managed object classes, and side-effects of management operations can be easily overlooked [Ritt95].

Furthermore, our experience is that any attempt to exploit the [ETS 52210] model to derive management information which applies to the ATM network as a whole should be carefully considered both in terms of efficiency as well as semantics. With a growing size of the ATM network the management information from one ATM network element used as the basis for ATM network wide management information (such as the status of one ATM cross-connection supporting an end-to-end VP trail) is quickly out-of-date.

The generic class library defined by [ETS 43316] provides some useful managed object classes for the network management layer view. Some of these managed object classes are used in the inheritance hierarchy of our ATM network management layer information model. However, the full set of classes defined in [ETS 43316] still needs further discussion and improvement.

As soon as the actual implementation is approached it becomes very clear that the specified information models just provide a functional and informational framework within which the actual realization of the ATM N-OS takes place. Whereas the set of possible managed object instances, their attributes, actions and notifications syntaxes are precisely specified, the precise meaning of the managed object behaviour and their correct implementation still needs to be clarified through verbal discussions and interoperability tests.

The setup of a virtual path described in Section 4.3 is a good example for the fact that with the specification of the TMN architecture and the ATM management information models at the network element and network management layers the main realization problems of the ATM N-OS are not solved. The endpoints of a requested virtual path on one side and a number of ATM cross-connects with well defined ATM Internal Links between them on the other side are the input for the algorithm to find a route. The main effort involved for the setup of a virtual path is the implementation of the algorithm to find an optimal routing and path selection according to different criteria such as shortest path, tariffs, time-of-day etc.

In addition, the functional and informational framework defined for ATM network management does not by itself provide solutions for a number of classical network management problems, such as, for example:

- route selection
- automatic problem detection and fault recovery by, for example, re-routing or activation of backup resources
- bandwidth reservation and quality of service guarantees
- statistical multiplexing, dynamic bandwidth allocation and adaptation.

To realize the ATM N-OS acceptable solutions and implementations of adequate algorithms to tackle these network management problems are required. These algorithms require non-trivial mappings between the N-OS and several NE-OSs.

Finally, the practical implementation of layered TMN systems requires powerful management platform support. The design of management services and information models has to be supported by GDMO browsers and "application composers". The main parts of management applications realizing agents, Q-adaptors, mediation devices, managers, etc. need to be generated from GDMO and ASN.1 compilers. The remaining logics of management application programs should be filled in by the use of high-level, object-oriented application programming interfaces (API). The X/Open XMP/XOM API, for example, is inadequate for this purpose. For the testing of management application programs and platform interoperability the platform communications infrastructure needs configurable protocol stacks and MIB browsers that are easy to use.

6   Conclusions and Outlook

While the first pan-European ATM networks are currently being setup and operated a lot remains to be done as far as their network management is concerned. The PREPARE project works with a real pan-European ATM testbed and the newest evolving standards for public ATM network management. Our first results from the application of ITU-T recommendations (TMN) and ETSI standards have been presented in this paper. A TMN architecture for ATM networks and an information model design for the ATM network management layer have been discussed. Some early implementation experiences have been described.

The TMN technology is complex, but powerful and sufficiently mature to be deployed. Additional standards for the ATM network and service management layers are needed. Our work proves the feasibility of applying the TMN concepts to ATM and provides guidelines for the realization of an operational TMN system for ATM networks. Based on the prototypes, the work will be continued to enhance management functionality and to derive additional results with respect to performance and scalability of the TMN approach.

References

[M3010]   ITU-T Rec. M.3010, Principles for a Telecommunications Management Network

[M3100]   ITU-T Rec. M.3100, Generic Network Information Model

[X721]   ITU-T Rec. X.721, Structure of Management Information: Definition of Management Information

[BjSch94]   L.H. Bjerring, J.M. Schneider, End-to-end Service Management with Multiple Providers, Proc. 2nd IS&N Conference , Aachen, 1994, Springer LNCS 851

[ETS 52210]   ETSI NA 52210, B-ISDN Management Architecture and Management Information Model for the ATM crossconnect

[ETS 43316]   ETSI NA 43316: TMN Generic Managed Object Class Library for the Network Level View

[Q961]   ITU-T Rec. Q.961, Lower Layer Protocol Profiles for the Q3 Interface

[Q962]   ITU-T Rec. Q.962, Upper Layer Protocol Profiles for the Q3 Interface

[Ritt95]   A. Ritter, Design und Implementierung von Testfällen für Netzmanagementobjekte, Diploma Thesis (in German), University of Dortmund, 1995

[PNIHB95]   PREPARE Public Network Implementers Handbook, 1995

Acknowledgements

A number of people from the PREPARE project have contributed to the design of the ATM network management layer services and information model. In addition, the implementation of the N-OS described in this paper has been tested together with PREPARE people. In particular, we hereby acknowledge the contributions of L. H. Bjerring (KTAS), U. Harksen (NKT), C. Lorenzen (NKT) and S. M. Pedersen (KTAS). Our special thanks for the help we received on the ETS 52210 information model through discussions and managed object test cases go to B. Bär, A. Ritter and O. Zimmermann.

Management Services for Performance Verification in Broadband Multi-Service Networks

Panos Georgatsos
ALPHA Systems S.A., Athens, Greece, Tel: +30.1.48 26 014-16,
Email: panos@alpha.ath.forthnet.gr
David Griffin
FORTH-ICS, Heraklion, Greece, Tel: +30 81 39 17 22, Email: david@ics.forth.gr

Abstract. This paper presents a practical management system for performance monitoring and network performance verification to support the larger goals of performance management systems. We show how the rich and powerful features of OSI systems management can be used in a hierarchical manner in the TMN, to achieve sophisticated performance monitoring and performance verification without imposing a large communications overhead in the TMN and hence in the managed network itself.

1 Introduction

The intelligence provided by network management has been widely recognized as an important aspect of telecommunications networks. The coexistence of different services, with potentially widely differing requirements on performance and quality of service, on the same broadband networking infrastructure, imposes the need for effective management schemes. ATM technology itself provides several degrees of freedom for multiplexing traffic that if not managed properly can prove disastrous in terms of network provisioning and performance.

The ITU-T introduced the Telecommunications Management Network (TMN) [9], as a means of provisioning the required management intelligence and have distinguished between the management and control planes in the operation of communications networks [6], [7]. The TMN relies on the OSI systems management concepts and functions, developed by ISO and ITU-T (cf. X.700 series recommendations), to model the network and service resources at various levels of abstraction and the communication of management information. Utilizing these concepts, TMN suggests a hierarchical management architecture enabling separation of concerns and encapsulation of lower level functionality. TMN therefore, implies a hierarchical distribution of management intelligence and it can be regarded as a separate network, logically distinct from the network being managed.

The majority of the management systems deployed today are concerned with network configuration and network monitoring. There is a trend [1]-[5] to increase the intelligence of management functions to encapsulate human management intelligence in decision making management components to move towards the automation of the monitoring, decision making and configuration management loop.

Performance management has been identified as one of the major management areas by ISO and ITU-T. The aim of the performance management functions is to guarantee that the network meets the required performance targets of the range of services that it supports. The on-going work in performance management related areas (routing, VPC bandwidth management, etc.) justifies the need for efficient network performance management systems.

The actual efficiency of a performance management system cannot be established unless reliable measures on network performance are provided. Furthermore, network performance assessment is required for identifying undesirable trends in network performance and triggering the appropriate management functions so that the necessary corrective actions to be taken. In [2] the need for network performance verification capabilities has been emerged as an essential part of a management architecture for VPC and routing management.

Taking into account the multi-service environment, network performance should be evaluated and assessed on the basis of measures relating to the individual performance characteristics of the bearer connection types (or classes of service, CoSs) supported by the network. Although there are practical implementations of network monitoring and performance evaluation systems, the problem of network performance assessment from a network management perspective and taking into account the different performance characteristics of the multi-class environment, has not been fully addressed.

Within the above framework and recognizing the emerging need for enhanced automated network management systems, this paper concentrates on the problem of network performance assessment within the overall context of network management.

A TMN approach is followed. The paper defines an appropriate management service, the Performance Verification management service, for network performance assessment. The paper describes the Performance Verification management service in the context of the performance management functional area and proposes a specific approach and algorithms to network performance verification. The management service is decomposed and mapped to the TMN architectural framework.

By taking advantage of the rich and powerful features of OSI systems management, the paper shows how the calculation of the required measures can be made for all CoSs over all source destination pairs, introducing minimum overhead into the management system. Specifically, by pushing the monitoring functionality down to the NEs, polling from the network management layer is avoided; the Performance Verification management service residing in the network management layer receives only the emitted threshold crossing notifications indicating unacceptable network performance. Requirements on supporting monitoring objects are identified as well as enhancements on the 'classical' OSI metric and summarization monitoring objects [16], [17].

The paper is organized as follows: Section 2 introduces the Performance Verification management service, describing its objectives and its interactions with other management services/components and the network. Section 3 describes the main functional aspects of the Performance Verification management service and outlines a specific approach for network performance verification. Section 4 presents a TMN compliant management architecture for the Performance Verification management service. Section 5 elaborates on system design aspects, presenting efficient means for realizing the specified management functionality on the identified architecture. Finally, section 6 presents the conclusions and highlights aspects of future work.

2 The Performance Verification management service

This section describes the scope of the Performance Verification management service in the framework of ATM network management, specifying its objectives and its relationship with other management services/components, the network and the TMN users.

The managed environment is assumed to be a public ATM network supporting a wide range of services made up of network bearer service classes. The term CoS denotes a network bearer service class. Different CoSs are distinguished according to their bandwidth requirements and performance targets. The view taken is that there is a range of network bearer services of different quality (in terms of performance targets) and costs to support the AAL services. This view is in accordance with the views of the ATM Forum [11], where they explicitly recommend the augmentation of the AAL service classes with a range of quality service classes.

Within the above environment, the objectives of the Performance Verification management service are:

- to ensure that the network meets the performance targets of the different CoSs supported by the network,

- to warn performance management related components when network performance has dropped below the acceptance levels per CoS, so that corrective actions can be taken,

- to analyze customer complaints with respect to the quality of the network services they use.

The scope for network performance verification is necessitated from the fact that the network supports multiple CoSs of guaranteed performance. In the case of a network providing connections without a guaranteed performance, the scope for network performance verification and performance management in general is limited to performance monitoring.

Although the Performance Verification management service can be regarded as a management service in its own right, it could well be taken as a management service component of other management services in the performance or other management functional areas (e.g. billing in cases where the network accounting policy allows for compensations when the quality offered by the network is outside the specified levels). In [2] the above management service has been proposed as a management service component of an overall management system architecture for VPC and routing management for networks supporting guaranteed performance connection classes. In particular, this management service was responsible for verifying network performance and for emitting quality of service alarms in case of undesirable trends in network performance.

The Performance Verification management service is beneficial to the network operation in an indirect manner in the sense that it is beneficial to the management system that runs on the top of the network. Specifically, the Performance Verification management service quantifies the performance of the network management system, providing therefore an indisputable measure of the efficiency of the management system.

The Performance Verification management service should not be confused with the performance monitoring services of the network. The performance monitoring capabilities of a network are responsible for calculating and supplying measurements concerning various network entities like nodes, VPCs, VCCs. The Performance Verification activities are making use of the network monitoring capabilities, requesting the measurement of the statistics required for network performance evaluation and verification. In the same sense other network management components - in the performance or other management functional areas (e.g. accounting) - are making use of the network monitoring capabilities for the purpose of applying their intelligence. Therefore, the functionalities of the Performance Verification and Performance Monitoring management services are quite distinct.

The Performance Verification management service belongs to the performance management functional area. Figure 1 shows the relationship of the Performance Verification management service with other performance management service, management functional areas and the TMN users.

The boundaries of the management responsibility of the Performance Verification management service are shown in Figure 2, which depicts the interactions between the management and control planes from the point of view of the Performance Verification management service.

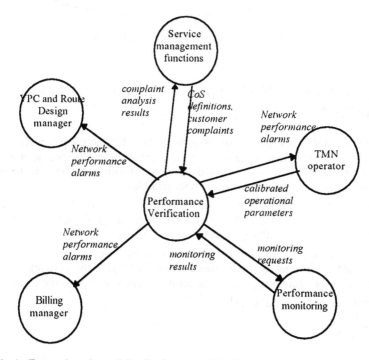

Fig.1: Enterprise view of the Performance Verification management service

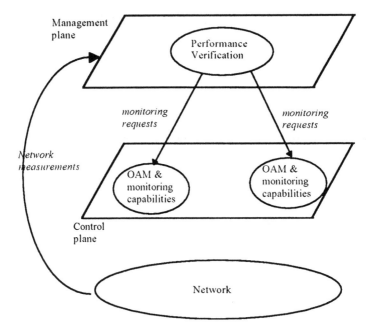

Fig 2: Management and Control plane interactions.

3 Performance Verification functional aspects and approach

This section highlights the main aspects of the functionality involved in the Performance Verification management service and proposes a specific approach for achieving its objectives. As implied by its objectives, the functionality of Performance Verification management service includes

- evaluation and verification of network performance with the purpose to notify other management services of network unacceptable performance and to analyze customer complaints with regard network performance.

Evaluation and verification of network performance is done on the basis of measurements concerning the performance of the CoSs that the network supports. It is assumed that the following parameters define the performance of the CoSs:

- rejection ratio (blocking probability),
- cell delay,
- cell loss ratio,
- jitter.

The above parameters are those which are directly influenced by the activities of the VPC and route design managers [2]. Other performance parameters may be associated with network CoSs, such as set-up delay; but because the role of Performance Verification is to feedback performance analysis results to the performance management system, to influence its future behaviour, these parameters are outside the scope of Performance Verification.

These parameters have been widely accepted as meaningful connection performance parameters and they are in accordance with the performance parameters defined by ATM Forum [11]. The jitter refers to the variance in the cell delay within a connection; it is equivalent to the 2-pt CDV (two point cell delay variation) defined by ATM Forum [11]. It is assumed that for each CoS there is an upper bound (performance target) defining the range of acceptable values. Network performance is within acceptable levels if these bounds are preserved within certain confidence limits.

Measurements of rejection ratio statistics could be done directly from appropriate raw data (e.g. counters of connection requests and rejections) available at the management interfaces of the network elements. The measurement of cell related statistics (delay, loss ratio, jitter) requires the existence of measurement instruments at the source and destination end-points or the existence of OAM (operation and maintenance) capabilities at access switches including appropriate management interfaces.

Based on measurements on the above parameters, Performance Verification evaluates network performance and by comparing it with the maximum allowable values (performance targets) verifies whether network performance is within acceptable levels or not. Evaluation and verification should be done in the scope of the total population of CoSs and source-destination pairs. Therefore the measurements should cover all (or a representative portion of) the source-destination pairs and supported CoSs, to increase the validity of the produced results. On the other hand, the verification process should introduce minimum overhead to the management system for collecting the required network information and it should not be sensitive to transitory situations.

For a given CoS, network performance evaluation and verification could be done in two modes:

- Per source-destination (s-d) pair, or
- Network-wide (statistically averaged over all possible s-d pairs by sampling)

Network-wide performance estimates could be taken either exhaustively (over all possible s-d pairs) or following sampling techniques in cases where the number of all s-d pairs is huge. In the latter case, sampling per strata should be pursued, to guarantee homogeneity of the population strata.

The proposed approach to network performance verification adopts the first mode, since it is considered more useful to the management services requiring network performance deterioration alarms. Indeed, the routing management systems construct and manage routes of guaranteed quality for all source-destination pairs for a given CoS; therefore, as a means for quantifying routing management efficiency, network performance estimates per s-d pair are required. This view was adopted in [2], where a hierarchical management architecture for VPC and routing management was proposed. Network-wide performance estimates may hide unacceptable performance situations for some s-d pairs. Moreover, network-wide performance estimates inevitably require the use of polling for retrieving the necessary sample values which subsequently creates a communications overhead in the management system. As it will be shown in the following sections, network performance verification per CoS and per s-d pair, paradoxically enough, although all s-d pairs are involved, does not require as great a management overhead as in the network-wide case; in fact it creates minimal management overhead.

For a specific s-d pair and a given CoS the essence of the proposed approach is to calculate the following probabilities as a means of ensuring and therefore verifying that the network performance is within acceptable levels:

$$\text{Prob}[^{(i)}R_{sd} <= {}^{(i)}T_r] >= A \qquad (1)$$

$$\text{Prob}[^{(i)}R_{sd} > {}^{(i)}T_r] < 1 - A \qquad (1a)$$

where:

$^{(i)}R_{sd}$ denotes a measurement of a CoS related performance parameter R (rejection ratio, cell delay, cell loss ratio, jitter) for CoS i and source destination pair s-d. For rejection ratio, $^{(i)}R_{sd}$ is calculated as an instantaneous value or as a moving average estimate. For the cell related performance parameters (cell delay, cell loss ratio, jitter), $^{(i)}R_{sd}$ is calculated as the arithmetic average over a specific number of connections.

$^{(i)}T_r$: is the maximum allowable value of performance parameter R for CoS i.

A: is the confidence level.

Formula (1) says that it is almost certain (with a confidence level A) that the network performance with regard the performance parameter R for connections of CoS i between the source-destination pair s-d will be within the acceptable levels (performance target) specified for that CoS. Violation of condition (1) or (1a) means that the network performance with regard the performance parameter R for CoS i between the source-destination pair s-d has fallen below the acceptable level. In such cases, network unacceptable performance alarms should be triggered in order the necessary corrective actions to be taken by the appropriate management services.

The calculation of the previous probability can be done continuously, or during specific verification periods, with dynamic duration and inter-period times, depending on network state. For a given source access node, the above probability measure could be calculated either exhaustively (over all possible s-d pairs) or for a sample of destination nodes following sampling techniques -in cases where the number of the s-d pairs for all destination nodes is huge. In the latter case, sampling per strata should be pursued, to guarantee homogeneity of the population strata.

Note that by taking the Probability of the event $ER = [^{(i)}R_{sd} \geq {}^{(i)}T_r]$, transitory fluctuations in network performance are not taken into account. The measurement of the probability is taken at a given window, within the verification interval; then the probability is approximated by the frequency (ratio) of the times of occurrences of the event ER over all the observations that were made within this window. The choice of the length of the probability window should be done taking into account the nature of the measurements (instantaneous values or moving average type of values or number of connections over which the cell related performance parameters are calculated) for the performance parameter S. These are left as design options.

Table 1 summarizes the proposed approach outlining the main steps of an algorithm fulfilling the objectives of the Performance Verification management service. The details of the Performance Verification algorithm is not in the scope of the paper, since the paper deals more with architectural and design issues.

Retrieve performance targets [i]T_r for all CoS related performance parameters R.

    For all access source nodes

        For all CoSs, determine destination nodes over which network performance will be verified

            For each CoS related performance parameter, initiate appropriate monitoring activities for the selected s-d pairs for monitoring the probability of *unacceptable* performance (see (1a)).

        Collect threshold crossing notifications.

            If one received, wait for a specific time period and collect any other notification that may come within this *time* period. At the end of the period send the collected notifications (network unacceptable performance alarms) to the interested management services.

Schedule the next verification interval.

Table 1. Performance Verification functionality.

It is worth noting that the Performance Verification management service operating with the previously described functionality creates *minimal overhead* into the management system. As it will be shown in section 5, all the required monitoring activities may be pushed down at the QA (Q3 adaptation) level providing the management interface of the network elements. Only the notifications resulting from threshold crossings are forwarded to the management system. The number of notifications depends apart from the performance of the network, on the sensitivity of the measures. By regulating appropriately the measurement characteristics of the connection related performance parameters and the probability window, the tradeoff between the validity of the measurements and their sensitivity can be managed.

4 Functional architecture

In this section we place the Performance Verification Management Service in the context of an architectural framework that will allow design and implementation of the management functionality introduced in the previous sections. We have adopted the TMN approach [9] recommended by the ITU-T. By following the methodology of Recommendation M.3020 [10], Management Services are decomposed into Management Service Components (MSCs) which are in turn decomposed into Management Functional Components (MFCs). Expanding this methodology, the derived MFCs are mapped to the hierarchical layers of the TMN [9] and to the TMN function blocks of the TMN functional architecture.

The Performance Verification Management Service is decomposed into:

- a *performance verification* MSC, performing the functionality of the Performance Verification management service as outlined in the previous section. This MSC is responsible for controlling the required network monitoring activities, analyzing the measured statistics against performance targets, and for analyzing and validating customer QoS complaints against measured performance. Following the functional analysis presented in the previous section, this MSC is further decomposed into the following MFCs:

- a *performance analysis* MFC which is responsible for: determining which performance parameters in which network resources and on which connections should be monitored in order to estimate network performance; comparing measured performance against target performance to identify whether the network is not meeting its performance obligations; and raising appropriate alarms to trigger a re-configuration of the network to resolve performance degradation.

- a *customer complaints analysis* MFC which interfaces to the customer complaints functions in the service level; compares customer complaints on quality degradation to the measured performance in the network; analyses whether the customer complaints are justified; initiates additional performance monitoring probes in the case where complaints arise on specific portions of the network which previously were not being monitored explicitly.

- a *performance management control* MFC which is responsible for: interfacing to the network performance monitoring functions (see below); controlling the initiation and termination of monitoring activities such as logs and alarm thresholds; specifying the attributes of monitoring activities. These activities are performed at the request of the *performance analysis* and the *customer complaints analysis* MFCs.

- a *performance monitoring* MSC which is responsible for retrieving and reporting on the necessary data from the network elements to support performance monitoring. Performance monitoring is performed on two aspects of the network: the network resources and the network CoSs supporting customer calls. The performance monitoring MFC is involved in the collection of raw data as presented by the network elements and also for transforming this data by summarization, averaging, and statistical analysis to more comprehensive forms (e.g. rates, probabilities, averages) according to the requirements of the *performance verification* MSC. The performance monitoring MSC is decomposed into:

  - a *current load model* MFC which is responsible for collecting, storing and reporting on data concerning the traffic on specific network resources (e.g. nodes, links, VPCs).

  - a *connection performance model* MFC which is responsible for collecting, storing and reporting on data concerning network CoSs. This may often be achieved by interacting with the OAM capabilities of network switches, or by means of network monitoring tools.

- a *connection-type model* MSC which stores the performance targets associated with each CoS the network supports. It provides the performance verification MSC with the cell loss, delay, delay jitter, connection rejection ratio, etc. targets. It is decomposed into a single MFC:

  - a *connection-type model* MFC

- a *network model* MSC which stores information about the available network resources and their configuration, including network topology and resource capacity. This MSC is used by the performance verification MSC to identify where appropriate monitoring probes should be placed. It is decomposed into a single MFC:

  - a *network model* MFC

The *performance verification* MSC is specific to the Performance Verification management service, its functionality and algorithms have been derived in the previous section. The *performance monitoring* MSC and the derived *connection performance model* MFC may also be regarded specific to the Performance Verification Management Service; but the *current load model* MFC is more generic and may be used by other management services [2], [3]. The *connection-type model* and the *network model* MSCs are not specific to the Performance Verification Management Service, they are supporting MSCs of a more general use in management systems and may be used by a number of management services. They even may be regarded as MSs in their own right. Note that elsewhere the *network model* MFC may be a sub-component of a configuration management MS.

Figure 3 shows the allocation of MFCs to OSFs and their allocation to the TMN architectural layers. The ICF and MF function blocks [9] have been included to guarantee access from the network layer to the network element layer information models, and to enhance the network element specific information model, respectively. The actual choice of whether the physical counterparts of the QAF, MF or ICF are implemented as appropriate in a NE, QA, MF or a Network Element Level OS is left open according to the capabilities of the underlying network elements. However, it is assumed that a management component, for each network element (or a set of network elements) supplying a Q3 interface will exist below the network management layer.

The functionality of the Performance Verification management service has been placed at the network management layer following the directives implied by the decomposition of the logical TMN architecture. Performance verification is concerned with collecting performance data from a number of network elements, collating that information and comparing performance data obtained from more than one network element. For these reasons, it needs to have a global view of the network, and therefore must exist at the Network Management Layer. However, the performance monitoring functionality is distributed over the network management and element management layers. This distribution allows frequent data collection for a single network element to be carried out close to the source of the data.

A hierarchical system architecture is an important consideration in fulfilling the objectives of the Performance Verification management system. By virtue of the proposed hierarchical system architecture, the management overhead for acquiring the required network statistics may be minimized, as monitoring activities can be delegated down the hierarchy as close as possible to the network elements themselves where the raw performance data is generated.

Based on the identified architecture, and by making use of the rich facilities of OSI management, the following section proposes an efficient means for the design and implementation of the functionality of the Performance Verification management service.

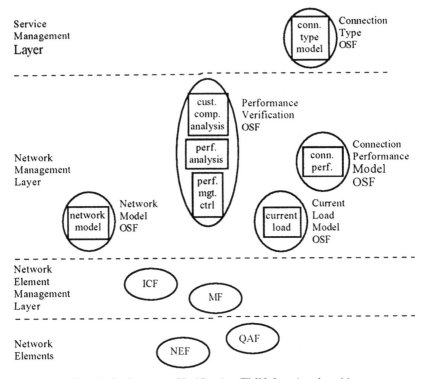

Fig. 3 Performance Verification TMN functional architecture.

5 Design aspects

By following the TMN approach which uses OSI systems management concepts, performance monitoring is achieved by virtue of the OSI systems management functions (SMFs). In particular, event reporting [12], alarm reporting[13], log control [14], test management [15], workload monitoring [16] and measurement summarization [17] SMFs are used. These are standard facilities of OSI systems management, pertinent to any Q3 interface, and demonstrate the advantage of adopting the TMN approach for implementing the Performance Verification management service. It is believed that these facilities provide a rich and powerful set of generic management tools, which - when properly used - may considerably reduce the cost of developing and of operating the management system.

The hierarchical TMN architecture and the OSI SMFs imply and furthermore facilitate distribution of the required monitoring activities over the network management, network element management and network element (management interface) layers. This distribution enables frequent data collection for a single network element to be carried out close to the source of the data. The results of monitoring activities in the network elements are then forwarded to the higher management layers either on request or at exception, at threshold crossing instances. This design consideration ensures that the management communications overhead is as small as possible as the bulk of the data in the management plane will be transferred locally.

Furthermore, by adopting the per s-d verification approach (see section 3), rather than the network-wide one, decomposition at the functional level is achieved, in the sense that individual network element performance measures do not need to be further summarized, in higher management layers; network performance is assessed individually on a s-d basis. By extending the measurement summarization functions to include probability calculations, as described in section 3, the calculation of the required performance measures (see (1), (1a) in section 3) can indeed take place at the network element (management interface) level, incurring no polling cost to the management system whatsoever.

Specifically, when the Performance Verification OSF requests a particular performance measure, a monitoring activity is created in either the Current Load Model OSF or the Connection Performance Model OSF. In turn, these later OSFs delegate element level monitoring activities to the network element (management interface) layer (in an element level OSF, MF, QAF or NEF, according to the actual implementation or the type of elements to be monitored). The monitoring activities and data retrieval between the Performance Verification OSF and the Current Load Model/Connection Performance Model OSFs is achieved by the creation of monitoring activities in the form of metric and summarization objects. By creating thresholds to identify degraded performance conditions, and by using event and alarm reporting, asynchronous, rather than synchronous polling communications are achieved, reducing both the communications overhead, and the processing overhead required in the Performance Verification OSF. In turn, the interaction between the Current Load Model/Connection Performance Model OSFs and the underlying OSFs/MFs/QAFs/NEFs is achieved by the same mechanisms, allowing asynchronous communications on exception and therefore reducing the communications load between the Network Management Layer and the Network Element Management Layer. Figure 4 illustrates the proposed design approach.

Only if the Network Elements themselves do not support the required SMFs is synchronous polling required between the lowest level management functions and the elements themselves. This means that high load communications inherent in polling mechanisms is limited to local area communications, reducing the load in the rest of the TMN and the underlying managed network. So increasing the capacity for revenue earning traffic.

By distributing the functions required over the Network Management and Network Element Management Layers and by utilizing the powerful framework of OSI management, the architecture of the TMN hierarchical system can be designed to avoid the management communications overhead inherent in centralized systems by pushing management intelligence and frequently used management functions as close as possible to the network elements.

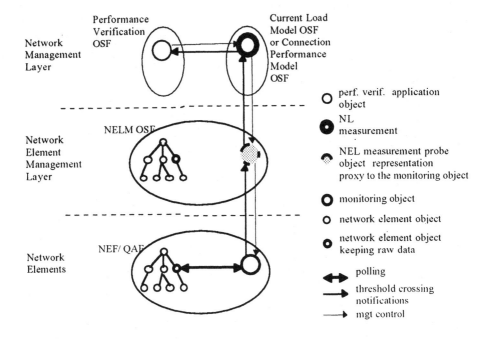

Fig. 4: Performance Verification management system design.

6 Conclusions and future work

In this paper we dealt with the issue of network performance verification in broadband multi-service network. Recognizing the emerging need for enhanced automated network management systems, the paper concentrated on the problem of network performance verification within the overall context of network management.

A network management approach (management service definition, algorithms and architecture) was followed based on TMN and OSI management principles. The paper defined an appropriate management service, the Performance Verification management service, for network performance assessment. The Performance Verification management service was described in the context of the performance management area, as well as its interface with management components in other management functional areas, the network and the TMN users.

The introduced Performance Verification management service is beneficial to the network operation in an indirect manner in the sense that it is beneficial to the management system that runs on the top of the network. Specifically, the Performance Verification management service quantifies the performance of the network management system, acting as an indisputable measure of the efficiency of the deployed management systems.

The paper elaborated on the parameters characterizing network performance and on how monitoring of these parameters can be achieved within the following constraints:

- management overhead to be introduced for collecting the required network information
- validity and sensitivity of the required measurements.

The functionality of the Performance Verification management service was analyzed and a specific approach was proposed. A TMN-compliant management system architecture fulfilling the objectives of the Performance Verification management service was presented. By taking advantage of the rich and powerful features of OSI network management, the paper proposed an efficient design approach for implementing the identified functionality. The proposed design approach, although covering all s-d pairs and CoSs, introduces minimum overhead into the management system. Specifically, by pushing the monitoring functionality down to the network elements, polling from the network management layer is avoided; Performance Verification, residing in the network management layer, receives only the emitted threshold crossing notifications, indicating unacceptable network performance.

Requirements on supporting monitoring objects to cover the required performance monitoring needs were drawn and enhancements on the 'classical' OSI metric and summarization monitoring objects were identified.

A Performance Verification system has been prototyped following the architecture and design principles presented in this paper.

Future work, includes extensive experimentation for quantifying the performance of the Performance Verification management system design in terms of the overhead introduced to the network management system. Relative comparisons with alternative network performance verification approaches and designs, is another important dimension of future work. Other aspects of future work include further research for enhancing the identified functional components e.g. the customer complaint analysis functionality. Certain aspects of this work are currently being undertaken in the RACE II ICM project.

7 Acknowledgments

The work described in this paper has been carried out by the authors in the course of the Integrated Communications Management (ICM) project (R2059) in the framework of the RACE II programme. The RACE II programme is partially funded by the Commission of the European Union. The authors wish to acknowledge Peter Baxendale (of University of Durham), Andy Carr (of Cray Communications) Kostas Kassapakis (of ALPHA Systems S.A.), George Mykoniatis (of National Technical University of Athens) and Bruno Rossi (of Monetel/ASCOM) for implementing the proposed management system components.

8 References

1. K.Geiths, P.Francois, D.Griffin, C.Kaas-Petersen, A.Mann *"Service and traffic management for IBCN"*, IBM Systems Journal, Vol.31, No.4, 1992.

2. D.Griffin, P.Georgatsos *"A TMN system for VPC and routing management in ATM networks"*, Integrated Network Management IV, Proc. of 4rth intern. symposium on

integrated network management, 1995, ed. Adarshpal S. Sethi, Yves Raynaud and Fabienne Faure-Vincent, Chapman & Hall, UK, 1995.

3. P.Georgatsos, D. Griffin, *"Load Balancing in Broadband Multi-Service Networks: A Management Perspective"*, Proceedings of the Third Workshop on Performance Modelling and Evaluation of ATM Networks, July 1995

4. M.Wernic, O.Aboul-Magd, H.Gilbert *"Traffic management for B-ISDN Services"*, IEEE Network, Sept.1992.

5. G.Woodruff, R.Kositpaiboon *"Multimedia traffic management principles for Guaranteed ATM Network Performance"*, IEEE J. Select. Areas of Comm., Vol.8, No.3, July 1992.

6. ITU-T Recommendation I.320 *"ISDN protocol reference model"*

7. ITU-T Recommendation I.321 *"B-ISDN protocol reference model and its application"*

8. ITU-T Recommendation I.150 *"B-ISDN asynchronous transfer mode functional characteristics"*

9. ITU-T Recommendation M.3010 *"Principles of a telecommunications management network"*

10. ITU-T Recommendation M.3020 *"TMN interface specification methodology"*

11. ATM Forum, "ATM User-Network Interface Specification", Version 3, Sept.1990.

12. ITU-T Recommendation X.734, *"Event Report Management Function"*(ISO/IEC 10164-5)

13. ITU-T Recommendation X.733, *"Alarm Reporting Function"* (ISO/IEC 10164-4)

14. ITU-T Recommendation X.735, *"Log Control Function"* (ISO/IEC 10164-6)

15. ITU-T Recommendation X.745, *"Test Management Function"* (ISO/IEC 10164-12)

16. ITU-T Recommendation X.739, *"Workload Monitoring Function"* (ISO/IEC 10164-11)

17. ITU-T Recommendation X.738, *"Measurement Summarization Function"* (ISO/IEC 10164-13)

Technologies for Service Engineering

Session 5b Introduction

Chairman - Rick Reed

TSE Ltd., 13 Weston House,
18-22 Church St., Lutterworth, UK
email: rickreed@tseng.co.uk

It is now generally accepted that computer technology and telecommunication technology are converging and the term "Information Technology" created a few years ago is appropriate to apply to the merger. Full integration will probably only come about as Asynchronous Transfer Mode becomes widely available, but all the other enabling technologies, including service engineering, need to be in place to support the merger.

Telecommunications services engineering is a specialised field within software engineering in general. There some characteristics of the software for telecommunication services that require special attention such as:

- services operate between terminals and across networks belonging to different parties, so that the software needs to be implemented in different places in a heterogeneous environment;
- the telecommunication system is on a huge scale - the number of terminations, the area covered, the throughput requirements;
- the expectations of reliability and interoperability based on the success of the ordinary telephone system.

The technologies required for service engineering are therefore those needed for software engineering plus additional features to handle the special requirements of telecommunication services.

The creation of services needs customised software environments. This is partly to satisfy the need to support the languages of the telecommunication domain: GDMO, ASN.1, MSC, SDL, TTCN. Three of these languages GDMO (Guidelines for the Definition of Managed Objects), ASN.1 (Abstract Syntax Notation One), TTCN (Tree and Tabular Combined Notation) have their roots in OSI Standardisation and therefore come from the boundary between telecommunications and computing. MSC (Message Sequence Chart) and SDL (Specification and Description Language) originate from the ITU sector. The environments also need to be customised to allow the created services to be deployed on the operational networks.

There are other special requirements such as the study of service feature interaction or the creation of service scripts if the Q.1200 series IN architecture is used as the basis of modeling or implementation. One paper in this section deals with these issues.

It seems obvious that database technology should be used to store the data for services, but it is not certain the current or emerging database implementations are capable of meeting the requirements for services. For example, object based databases seem to be appropriate to service requirements, but this is new technology and the performance may be inadequate.

The question to be answered is how special the data bases need to be for service use. These issues are investigated in two papers in this section.

Although the services architecture has not yet evolved to a point that it is stable, at least a *de facto* standard is required for interoperability between the equipment of different parties. Obviously the architecture has a major impact on the technologies, but fortunately there is enough consensus on work already done that it is low risk to assume that systems will be Open Distributed Processing (ODP) compliant. Therefore, a candidate for the service software interface definitions is therefore the CORBA interface definition language (CORBA IDL) and the interfacing of this with IDL is covered in a paper in this section.

Although a standard for service architecture is in the future, pilot and demonstration systems can be built today utilising available networks that have some capabilities of the integrated broadband communication networks of the future. The final paper in this section describes just such a demonstration.

At the level of interfaces standards are essential. For the technology to create and deploy services, the benefit of standards is less obvious: service creation environments do not have to be similar - they just need to be capable of producing service software components that can be integrated with each other and meet the standard interfaces. There is a case for competing technologies, but if there is consensus for a more unified approach then there can be economies of scale and it is easier to trade tools and components. The role of standards in this scenario is to provide an independent definition of the agreed technique.

An SDL Based Realisation of an IN Service Development Environment

Conor Morris, John Nelson,
University of Limerick, National Technological Park, Limerick, Ireland.

Abstract: This paper describes a computer-aided service development environment (CSDE) which supports the graphical specification of Intelligent Network (IN) services for service designers using the ITU-T's Service Independent Building Block (SIB) methodology. The CSDE presented here consists of a textual/graphical SIB editor, a translator, which translates the SIB description of a service to SDL (ITU-T's Specification and Description Language), an SDL compiler, and a graphical based simulation/debugging environment. The SIB editor and simulation/debugging environment provide the service designer with a sophisticated man-machine interface. To illustrate the functionality and effectiveness of the CSDE, the Universal Personal Telecommunications (UPT) feature Access, Identification and Authentication (AIA) is used as an example.

1 Introduction

Intelligent Networks (IN) are being developed [COMM95] in which service designers can easily and rapidly create telecommunication services according to the customers' needs. Two primary goals of service development under the Intelligent Network are rapid service creation using new software technologies and the minimisation of service development costs through switch vendor independence. To achieve the above goals the ITU-T (formally CCITT) has adopted a phased standardisation process towards the target IN architecture, defining a capability set for each phase. Each IN capability set (CS) is intended to address requirements for one or more of the following : service creation, service management, service interaction, network management, service processing, and network interworking. The ITU-T has developed the IN conceptual model (INCM) to provide a framework for the design and description of each IN CS.

One important aspect which requires further investigation before there is development and deployment of wide ranging IN based services, is the availability of service specification and development tools. The tools used for service creation, mainly proprietary, consist of editors, compilers, debuggers and execution environments. This paper addresses service development by describing a computer-aided service development environment (CSDE) which supports the specification of IN services using ITU-T's Service Independent Building Block (SIB) methodology. The system is targeted at service designers.

Before introducing the CSDE, an overview of the INCM will first be given, followed by a more detailed look at the SIB methodology for the creation of services. The description of the CSDE consists of a system overview and software architecture. The sibEditor tool is presented along with the simulation/debugging environment. There then follows an outline of a major component of the CSDE, the SDL modelling of relevant planes of the INCM. By modelling the appropriate planes i.e. the INCM Distributed Functional Plane (DFP) and Global Functional Plane (GFP) using SDL, and inserting different chains of SIBs, the SDL environment can be used to simulate the behaviour of services in the IN. Finally, an example specification of a service, the Universal Personal Telecommunications (UPT) feature, Access, Identification and Authentication (AIA) is presented.

2 Overview of the INCM

The concepts for the Intelligent Network are embodied in the IN Conceptual Model (INCM). The INCM, shown in Figure 1, is structured into four planes : a service plane, a global functional plane, a distributed functional plane, and a physical plane. A brief description of each plane follows [Q1201].

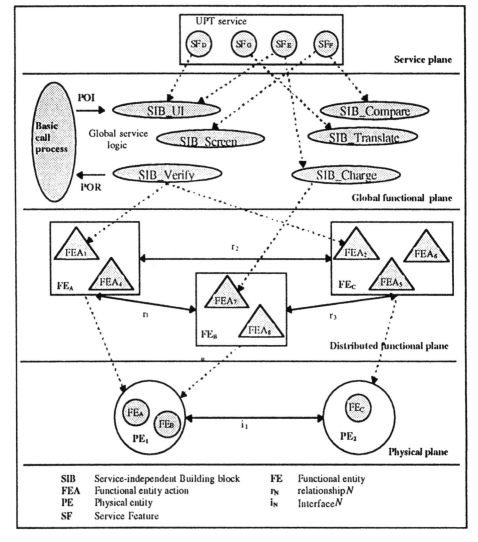

Figure 1 The IN Conceptual Model [Q1201]

2.1 Service plane

This plane is of primary interest to service users and providers. It describes the services and service features from a users perspective, independent of how the service is implemented.

2.2 Global Functional Plane

This plane is of primary interest to service designers. The network is viewed as a single structure from this plane. It includes the description of the units of service functionality, referred to as service independent building blocks (SIBs), independent of how the functionality is distributed in the network. SIBs can be combined with global service logic (GSL) on the global functional plane to realise services and service features on the service plane. The GSL can be defined as the "glue" that defines the order in which SIBs will be chained together to accomplish services.

2.3 Distributed Functional Plane

This plane is of primary interest to network designers and providers. It describes the functional architecture of an IN structured network in terms of units of network functionality, referred to as "functional entities", and the information flows between the functional entities, referred to as "relationships". The functional entities are described independently of how the functionality is physically implemented in the network. SIBs on the global functional plane are realised on the distributed functional plane by a sequence of functional entity actions together with the information flows between them.

2.4 Physical Plane

This plane is of primary interest to network operators and equipment providers. It describes the physical architecture alternatives for an IN-structured network in terms of potential physical components in the network, referred to as "physical entities", and the interfaces between physical entities. One or more functional entities from the distributed functional plane may be realised in a physical entity on the physical plane, and one or more relationships from the distributed functional plane may map onto an interface on the physical plane.

2.5 The SIB methodology

In order to understand how SIBs can be used to specify services the components of the GFP must be understood. The components contained in the GFP are:

- SIBs which are standard reusable network wide capabilities used to realise services.

- basic call process (BCP) SIB which identifies the normal call process from which IN services are launched, including points of initiation (POI) and points of return (POR) which provide the interface from the BCP to the global service logic.

- global service logic (GSL) which describes how SIBs are chained together to describe service features. The GSL also describes interaction between the BCP and the SIB chains.

By definition, SIBs, including the BCP are service independent and cannot contain knowledge of subsequent SIBs. Therefore, the GSL is the only element in the GFP which is specifically service dependent.

The SIB

A SIB is a standard reusable network-wide capability residing in the global functional plane used to create service features [Q.1203]. SIBs are of a global nature and their detailed realisation is not considered at this level but can be found in the distributed functional plane (DFP) and the physical plane. SIBs are also independent of the underlying levels. The SIBs are reusable and can be chained together in various combinations to realise services in the service plane. SIBs are defined to be independent of the specific service and technology on which they will be realised. A SIB has one logical starting point and one or more logical end points.

In order to describe service features with these generic SIBs, some elements of service dependence is needed. Service dependence can be described using data parameters which can be tailored to perform the desired functionality. Data parameters are specified independently for each SIB and are made available to the SIB through the global service logic. Two types of data are required for each SIB, dynamic parameters referred to as, call instance data and static parameters which are called service support data (SSD).

Call instance data defines dynamic parameters whose value will change with each call instance. They are used to specify subscriber specific details like calling or called line information. Service support data defines data parameters required by a SIB which are specific to the service feature description. When a SIB is included in the GSL of a service description, the GSL will specify the SSD values for the SIB.

Basic call process

The basic call process (BCP) is responsible for providing basic call connectivity between parties in the network. The BCP can be viewed as specialised SIB which provides basic call capabilities, including: connecting a call, disconnecting a call, retaining the call instance data for further processing.

IN supported services are represented through the use of chains of SIBs connected to the BCP SIB. The interface points between the BCP SIB and the chains of SIBs are described as points of initiation and points of return :

A point of initiation (POI)	is the BCP functional launching point for the SIB chains
A point of return (POR)	identifies the functional point in the BCP where the SIB chain terminates.

A graphical illustration of the POI/POR/BCP functionality is shown in Figure 2.

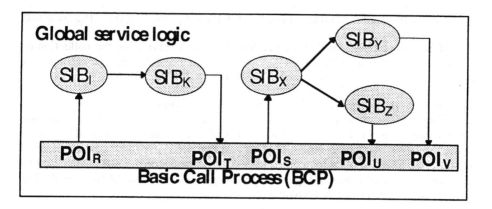

Figure 2 GSL. example

Global Service Logic (Gsl)

The GSL can be defined as the "glue" that defines the order in which SIBs will be chained together to accomplish services. Each instance of global service logic is potentially unique to each individual call, but uses common elements :

- POIs and PORs

- SIBs

- logical connections between SIBs, and between SIBs and BCP interaction points

- input and output data parameters, service support and call instance data defined for each SIB.

Based upon the functionality of these common elements, global service logic will "chain together" these elements to provide a specific service. GSL ensures that all relevant call instance data is maintained throughout multiple SIB chains until termination of each call instance. Figure 2 illustrates the global service logic concept. To avoid complexity SIB data parameters are not shown.

Global service logic in the GFP views the basic call process as a single resource. With this in mind, the following are identified as necessary interactions between global service logic and the BCP [Q.1203] :

Communications from BCP to GSL:

i)   logical start for SIB chains - which is represented by POIs
ii) call instance data, which is required by SIB chains for processing IN services.

Communications from GSL to BCP

i)   logical termination for SIB chains - which is represented by PORs

ii)   call instance data that have been defined by one or more SIBs on a SIB chain.

Therefore, the components that are needed to define a service are the collection of SIBs (including their service support data and call instance data), and the definition of how they are interconnected (to each other and to the POIs and PORs).

Global functional plane view of a service feature

Referring to Figure 3, when a service feature is required, it is invoked by a triggering mechanism from the BCP. The chain or list of SIBs which describe the service feature must be obtained by the global service logic in order to process the service feature. The global service logic must have access to each of the SIB chains used to describe the service features. Figure 3 illustrates both the graphical based and table based representation of the SIB chain.

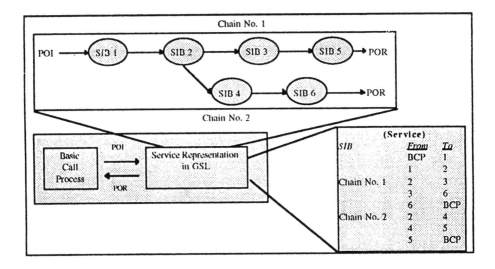

Figure 3 GFP view of a service showing SIB graphical and table based representation

CS-1 SIBs

The following list of SIBs (see Table 1) have been identified [Q.1213] to support the list of targeted CS-1 services and service features identified in [Q.1211]. (The SIBs highlighted in bold have been implemented in the current version of CSDE).

SIB	Description
Algorithm	Applies a mathematical algorithm to data to produce a data result.
Charge	Determines special charging treatment for the call.
Compare	Performs a comparison of an identifier against a specified reference value.
Distribution	Distribute calls to different logical ends of the SIB based on specified parameters
Limit	Limit the number of calls related to IN provided service features.
Log Call Info.	Log detailed information for each call to a file.
Queue	Provide sequencing IN calls to be completed to a called party.
Screen	Perform a comparison of an identifier against a list.
Service Data Management	Enables end user data to be replaced, retrieved, incremented, or decremented.
Status Notification	Provide the capability of enquiring about the status of network resources.
Translate	Determine output information from input information.
User Interaction	Allows information to be exchanged between the network and a call party.
Verify	Provide confirmation that information received is syntactically consistent.

Table 1 CS-1 defined SIBs

3 CSDE Description

In this section the Computer-aided Service Development Environment (CSDE) is described. The system overview includes the objectives and requirements. The software architecture is then presented. The SIB editor and the associated simulation and debugging facilities follow. Finally, the resources and platform used in the development of the CSDE are overviewed.

3.1 System Overview

The objectives of developing the platform are as follows :

- To investigate the SIB methodology for the creation of services.

- To analyse at the use of SDL [SDL88, SSR89] for the modelling/implementation of the middle two planes of the INCM, the GFP and the DFP.

- To evaluate the use of a multi-media platform for service verification/validation.

As a result of the above objectives the following requirements were identified:

- There must be a notation, or description language, to express service ideas. This must be formal enough to be unambiguous and allow for automatic processing.

- There must be an editor for the manipulation of the service descriptions.

- There must be a translator to translate the high level service descriptions to a form which can be executed on a target platform.

- It must be possible to simulate a service description in order to observe the dynamic behaviour arising from the static service description.

- The CSDE should be developed with an open architecture in mind. Tools from different vendors should supported.

From Figure 4 it can be seen that CSDE consists of six main components:

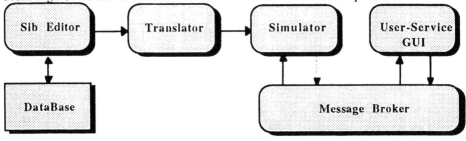

⟶ Communication Path

⟶ Generated Output

Figure 4 System Model

SIB Editor	The service designer is provided with a powerful textual/graphical editor which allows him/her to chain a number of SIBs together in order to realise a service.
Database	A object-oriented database was developed to store SIB-instance descriptions and SIB chain information.
Translator	Translates the SIB chain description of a service, outputted from the SIB editor, to SDL/PR[1].
Simulator	The SDL/PR is analysed to check if it conforms with SDL-88 and an executable form of the service is generated.
User-Service GUI	An X-Windows front end is based on top of the simulation. By using the multi-media capabilities of the work-station the service can be verified and validated.

---

[1] SDL has both a graphical representation, SDL/GR, and a textual representation, SDL/PR.

Message Broker   A messaging mechanism is used to connect the simulations generated by SDT (see Software Architecture) with the external simulation/debugging environment

Figure 5 identifies how the CSDE could fit into the overall service development process. A new service usually starts its life as a concept, or idea, by the marketing department. By identifying the customer requirements for the service concept, the marketing department can pass on high level service descriptions to the service designer. The service designer can then enter a detailed service description into the service development environment, and produce a detailed SIB representation of the service for the service implementor. The service development environment allows the service designer to design, analyse, validate, and test a potential new service.

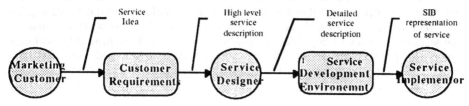

Figure 5 Position Of The Csde In The Service Development Process

On the other hand the CSDE could allow service designers to capture their ideas and investigate the service thoroughly before committing to detailed development and marketing.

3.2 Software Architecture

Figure 6 illustrates the flow diagram of the service development environment. The initial high level service description is used as input to the sibEditor. Using the sibEditor the service designer generates an SDL/PR representation of the service. This SDL/PR description acts as input to SDT. The service specification written in SDL is linked to the IN system also written in SDL. The two SDL systems are written in isolation. The service specification specifies the behaviour of the service, while the IN system describes the behaviour of the IN environment which is acted upon by the service. The IN system is written to ensure the service signals acting upon it are consistent with the service. SDT (the SDL supporting toolset used is TeleLogic's SDT [SDT93]) integrates the SDL description of the service with the SDL description of the IN system to produce an executable model of the service. This model can then be used for simulation and debugging purposes.

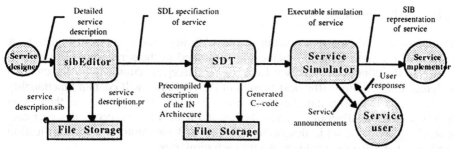

Figure 6: Flow Diagram Of The Service Development Environment

Figure 7 outlines the major components involved in the generation of the executable simulation of the service. The major components are as follows:

SDT Analyser
The SDT analyser examines the service specification created with the SIB editor. The syntactic analyser performs a syntactic analysis according to the SDL-88 definitions. The semantic analyser checks that the specification obeys the static semantic rules of SDL. The dynamic analyser is used for validating the SDL specification with respect to dynamic aspects from a static point of view. The dynamic analyser predicts dynamic errors and reports illogical use of the SDL language.

C-code Generator/ Compiler Linker
SDT automatically translates the SDL/PR representation of the service to C-code. The C-code is then compiled by a standard C compiler and linked with: a precompiled SDL description of an IN architecture, a precompiled simulator library (for simulation purposes the runtime library RealTimeSIMulation is used. This means that will be a connection between the clock in the executing program and the wall clock, i.e. time provided by the operating system), an Abstract Data Type (ADT) library; to form the executable simulation of the service.

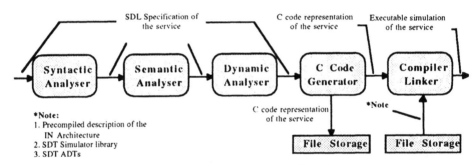

Figure 7 Flow diagram of the SDT tool set

3.3   Resources used

The resources used were a SUN Sparc 5 workstation (24 Mb RAM), running the Solaris 2.3 operating system, TeleLogic's SDT 2.3, Standard Motif library for the design of the GUI's, SparcWorks (ver. 2.0), SparcCompiler C (ver 2.0), SparcWorks Professional C++.

3.4   SIB Editor

A screenshot of the current version of the sibEditor tool is given in Figure 8. A number of the most common menus are illustrated in the screenshot. The current version of the editor offers the service designer limited functionality for the creation of dynamic descriptions (the next version of the sib Editor will contain functionality for the generation and manipulation of dynamic service descriptions), i.e. there is no facility to generate diagrams such as the one in Figure 14. The service designer can, however, generate a static description of each SIB and indicate the identification number of the SIB-instance to invoke for each result type.

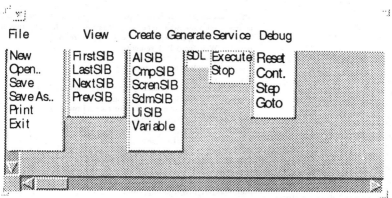

Figure 8 Screen-shot of the SIB editor

The static descriptions for each SIB-instance are entered using dialog boxes. An example of a compare SIB-instance is shown in Figure 9. Two separate file formats are provided by the tool. One specifies how the service descriptions are loaded and saved by the sibEditor tool, while the other specifies the translated format to be inputted to SDT for compilation. Figure 10 gives an example of the output generated for the SIB-instance configuration shown in Figure 9. Referring to Figure 10, the file format on the left is based on a new sib description language developed, while on the right the file format is based on SDL88. The SDL/PR description of a SIB-instance can be split into three sections. Again referring to Figure 10, the first section sets up the parameters for the Compare SIB and outputs a signal to the Compare SIB process. The next section receives the result from the Compare SIB process and makes a decision, based on the result, as to which SIB should be called next. The third section is used for debugging purposes. The sib description language to SDL/PR translator is integrated with the sibEditor tool.

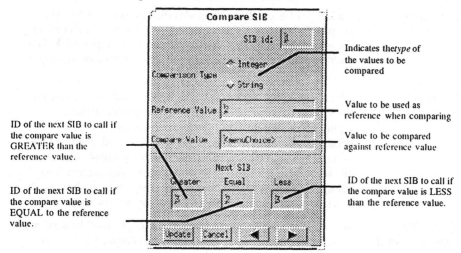

Figure 9 Example configuration of a Compare SIB

```
SIB_1:task compare_cid!what_type := int,compare_cid!int_value := 2,
compare_ref_value!what_type := int,compare_ref_value!int_value := menuChoice
output SIBCmp_params(compare_cid,compare_ref_value);
nextstate Wait_for_SIB_1_response;
state Wait_for_SIB_1_response;
input SIBCmp_Result(result,error_type);
decision result;
(greater_than): output Compare_SIB_Completed(1,result,3);
nextstate Wait_for_SIB_1_debug_instr;
(less_than): output Compare_SIB_Completed(1,result,3);
nextstate Wait_for_SIB_1_debug_instr1;
(equal): output Compare_SIB_Completed(1,result,2);
nextstate Wait_for_SIB_1_debug_instr2;
enddecision;
state Wait_for_SIB_1_debug_instr;
input step;
join SIB_3;
state Wait_for_SIB_1_debug_instr1;
input step;
join SIB_3;
state Wait_for_SIB_1_debug_instr2;
input step;
join SIB_2;
```

```
SIB_TYPE: CMP
SIB_ID: 1
Location: 0 0
compare_value: STATIC_INT.2
reference_value:
DYNAMIC_INT.menuChoice
next_sib_greater: 3
next_sib_less: 3
next_sib_equal: 2
```

SIB instance description using the service description language

Equivalent description of the above SIB instance SDL/PR form

Figure 10 Example of the SDL/PR generated by a Compare SIB Instance

3.5 Simulation environment

An overview of the simulation/debugging environment is shown in Figure 11. The environment consists of four main components.

Keypad GUI	An image of a telephone keypad is displayed on screen. The keypad demonstrates similar functionality to that of an ordinary telephone; by using the mouse, the user can simply click on a button on the keypad and achieve the same effect as pressing a button on a real telephone keypad.
Debugging window	When simulating a service, the designer is able to step through the service description one SIB at a time and view the input/output parameters of the currently active SIB. The debugger also allows the designer to insert break points in the description, simulate the service as far as the break point and allow the designer to view the input/output parameters. The debugging facility allows the service designer to analyse a service in detail and identify any errors in the description.
Executable simulation of the service	A C-code representation of the SDL description of the service is generated by the C-Code Generator, and compiled to form an executable simulation.

SDT PostMaster   SDT PostMaster is a messaging mechanism used to connect the generated simulations by SDT with the external simulation/debugging environment. A C program generated by SDT (executable simulation of the service in this case) can communicate with any application connected to the PostMaster that sends messages according to a defined format.

Announcements to be made to the user are sent as text messages from the executable simulation to the keypad GUI, via the PostMaster. These text messages are interpreted by the keypad GUI, and by incorporating a text-to-speech converter, the announcements are played on speakers attached to the workstation. The user can respond to these announcements by 'dialling' digits on the keypad. As the user interacts with the service, debugging information is presented to the service designer.

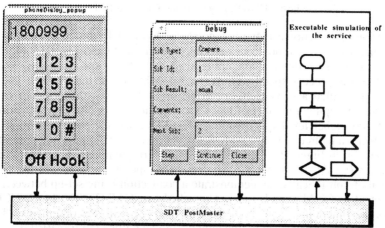

Figure 11 Overview of simulation/debugging environment

3.6   SDL Model of the GFP and DFP

The SDL model of the IN system concentrates on the middle two planes of the INCM, the DFP and the GFP (see Figure 12). There are two channels (Debug_Arch and DFP_User) connecting the two blocks to the surrounding environment. The DFP Block contains the IN architecture ([GARR93],[LAUER94]), while the GFP contains the SIB specifications. All user interaction takes place at the DFP level (i.e. all signals to and from the user to the IN system are carried over the DFP_User channel). The GFP contains only the specifications of the currently implemented CS-1 SIBs (see Table 1), it does not contain the GSL. The GSL (i.e. SDL/PR description of the service) is slotted into the GFP when the executable simulation of the service is being generated by SDT. The channel Debug_Arch is used to carry debugging information to and from the external debugger.

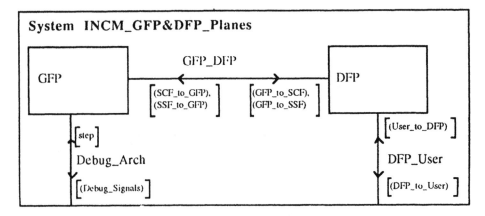

Figure 12 SDL model of the GFP and DFP

The Distributed Functional Plane is represented by the DFP block. As can be seen from Figure 13, there are no processes specified in this block. The DFP block is a partitioned block specification containing lower-level block specifications, which in turn contain the process specifications (i.e. specification of the FEAs). The DFP block can be called a subsystem specification as it allows the designer to gain overview by grouping a number of block specifications (i.e. block specifications of the FEs) into higher level specifications. The FEs , and the communication paths between them, are directly in line with the model of the DFP architecture presented in [GARR93]. From Figure 13 it can be seen that an originating (CCAF block) and a terminating (CCAF_2 block) CCAF has been specified. This allows for the simulation of non-IN calls, i.e. demonstrate a connection being set-up between user-A (attached to CCAF) and user-B (attached to CCAF_2). The block CCFSSF specifies the functionality for both the CCF and the SSF. A combined CCF/SSF specification is chosen to keep in line with the model outlined in [GARR93]

4 Example Service

The UPT [UPT] feature Access, Identification and Authentication (AIA) is used as a test case for the CSDE (the high level specification of the chosen feature is company proprietary). UPT AIA is an essential feature by which the UPT user contacts UPT facilities and verifies his/her identity to be the one claimed. A UPT user must invoke this feature before invoking other UPT features (such as InCall Registration).

Associated with each service description is both a dynamic and a static description. The dynamic description describes how the SIBs are chained together in order to realise a service, i.e. it is a description of the global service logic (GSL) in the GFP. It also describes the interaction between the basic call process (BCP) and the SIB chains. The implemented feature requires over seventy SIBs for a full description. An excerpt from this SIB chain description is shown in Figure 14. Each circle in the diagram represents an instance of a SIB. The type of the SIB-instance is indicated by the identifier in the bottom half of the circle, while the unique identity of the SIB-instance is represented by the number in the top half of the circle.

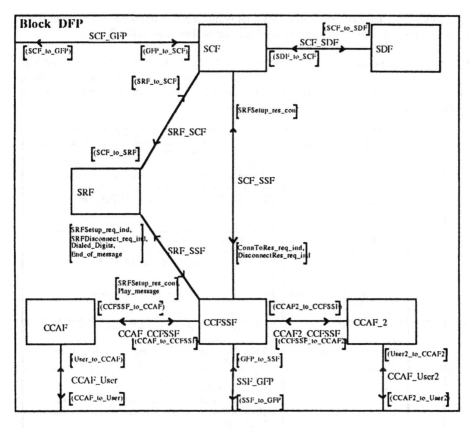

Figure 13 SDL model of the Distributed Functional Plane (DFP)

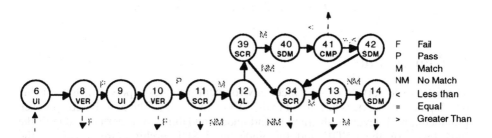

Figure 14 Excerpt from SIB chain description of UPT AIA

The dynamic description illustrates the dynamic behaviour of the service, it indicates the sequence in which the SIBs are invoked. For example, referring to Figure 14, SIB 13 screens the user's profile to see check if the serviceOnline flag is set to TRUE. If serviceOnline is not set to TRUE (a result of NM, no match, is return from the SIB) then SIB 14 will be invoked, otherwise if serviceOnline is set to TRUE (a result of M, match, is return from the SIB) then some other SIB will be invoked (not shown in diagram). However, the dynamic description does not tell us exactly what SIB 13 is doing, it just indicates the type of the SIB. Therefore a static description, which describes what each SIB does, along with its parameters, is required. An excerpt from the static description of the implemented service is shown in Table 2.

The static description is of tabular format. The first column indicates the unique SIB identity number. The second column indicates the type of the SIB-instance. The third column gives a prose description of what the SIB-instance does, while the fourth column details the parameters of the SIB-instance.

ID	Type	Description	SIB Parameters
9	UI	Play message: Please enter PIN.	Message = "Please enter your Personal Identification Number. Press # after your input ", Repetition_req = 1, Repetition_Type = count, Max_repetitions = 3, Collect_info = 1, Collection_var = PIN_C
10	VER	Verify that a PIN number has been entered*.	Element_value := DYNAMIC_CHAR.PIN_C
11	SCR	Check if UPT number corresponds to a valid user profile.	Table_name = "Subscribers.db", Key_name = UPTnumber_I, Key_value = DYNAMIC_CHAR.UPTnumber_C, Item_name = UPTnumber, Item_value = DYNAMIC_CHAR.UPTnumber_C
12	AL	Set goodUidSeen_I := TRUE.	Element = goodUidSeen_I, Action = increment, Offset = STATIC_INT.1
13	SCR	Screen the user's profile , check if serviceOnline flag is set.	Table_name = "Subscribers.db", Key_name = UPTnumber, Key_value = DYNAMIC_CHAR.UPTnumber_C, Item_name = serviceOnLine, Item_value = STATIC_INT.1
14	SDM	Set serviceOnline flag in uer's profle	Table_name = "Subscribers.db", Key_name = UPTnumber, Key_value = DYNAMIC_CHAR.UPTnumber_C, Item_name = serviceOnlIne, Action = replace, Element_value := STATIC_INT.1

Table 2 Static description of each SIB instance

5 Conclusions

The SIB based service development environment provides a powerful tool for the development of services according to the SIB methodology. The SDL based approach provided a high level of abstraction and understandability. Furthermore, due to the modularity of the CSDE and the underlying SDL architecture, many of the components could be reused in a service implementation environment (i.e. to create service logic in a physical IN architecture). Also, as the SDL specifications for new capability sets become available they can be easily incorporated.

However, a number of issues can be identified with the use of SIBs in particular related to the complexity of SIB specifications, the SIB chaining concept including the use of shared data, and the limited functionality of the CS-1 SIBs. An initial investigation of the object oriented methodologies suggest that a similar level of specification could be obtained more simply, and with more extendibility and reusability by adopting an object based approach.

6 Acknowledgements

We would like to acknowledge the support of GTE Laboratories (Waltham, Ma, USA), in particular Peter O'Reilly, for their support in this project.

7 References

[COMM95] Peter Phillips, "I.N. to the future", Communications International, January 1995.

[GARR93] Jams J. Garrahan, Peter A. Russo, Kenichi Kitami, and Roberto Kung, " Intelligent Network Overview", IEEE Communications Magazine, March 1993.

[LAUER94] Gregory S. Lauer, "IN Architectures for Implementing Universal Personal Telecomunications", IEEE Network, March/April 1994.

[Q1201] ITU-T Q.1201 Recommendation, "Principles of the Intelligent Network Architecture".

[Q1203] ITU-T Q.1203 Recommendation., "IN Global Functional Plane Architecture"

[Q1211] ITU-T Q.1211 Recommendation, "Introduction to CS-1".

[Q1213] ITU-T Q.1213 Recommendation, "Global Functional Plane for CS-1".

[SDL88] CCITT Z.100 (1989) 'Specification and Description Language SDL.'

[SDT93] SDT 2.3 Reference Manual Volumes 1 and 2, Telelogic, Sweden, 1993.

[SSR89] Roberto Saracco, J.R.W. Smith, Rick Reed, "Telecommunications Systems Engineering using SDL", North-Holland, 1989.

[UPT] ETR/NA-70201 "Universal Personal Telecommunications: General service description", ETSI.

Performance Evaluation of Database Concepts for Personal Communication Systems

Arndt Ritterbecks, Damian Lawniczak, Norbert Niebert

Ericsson Eurolab Deutschland GmbH,
Ericsson Allee 1, D-52134 Herzogenrath, Germany
{eedari, eeddal, eednni}@eed.ericsson.se

Abstract. The purpose of the RACE project Mobilise is the definition and characterisation of future advanced telecommunication services. Within this project, the PSCS (Personal Service Communication Space) service concept has been developed which is based on UPT (Universal Personal Telecommunication) ideas. Special emphasis is put on the user's point of view on telecommunication and the support of personal mobility independent of the involved networks. For the establishment of such a service, the service data and logic has to be accessible from everywhere. However, PSCS does not consider an implementation of a database and leaves this to the underlying telecommunication network. It is the aim of this paper to evaluate how far the database concepts developed for the Universal Mobile Telecommunication System (UMTS) are able to support PSCS [1]

Keywords: Personal Communications, UPT, UMTS,
Personal Service Communication Space, Distributed Database

1. Introduction

Future telecommunication systems will be characterised by personal mobility and service personalisation as defined in service concepts like UPT or PSCS [9]. To provide personalisation and mobility for their users, they have to deal with a large amount of data which needs to be searched and maintained under real time conditions. To keep the traffic load in the network and the data search/access times as low as possible, a very efficient database system is needed. Database concepts as used today for instance in GSM have to be evolved further in order to fulfil the high demands of such services.

Within the RACE project MoNet (Mobile Networks), a distributed database concept for UMTS has been developed to support user and terminal mobility which promises well for the management of PSCS data.

---

1. This paper is based partly on the work undertaken in RACE project R2003 "Mobilise" which started in January 1992. Partners of the Mobilise consortium are Ericsson Eurolab Deutschland GmbH, Cap Gemini Innovation, Dutch PTT research, Telia research, RWTH Aachen, Ericsson Telecommunicatie B.V., empirica, Cap Sesa Telecom, Vodafone, Ascom Tech and GIE Cofira. The views represented in this paper are those of the authors and do not necessarily represent the view of the consortium.

The aim of this paper is to investigate whether this database concept is capable to handle data of advanced personalised services (here: PSCS), to propose necessary extensions and modifications of both data concepts, and finally to determine its usability for PSCS on the basis of performance analyses.

In the following, summaries of both the PSCS service concept and the UMTS database concept will be given. Afterwards, proposals for required extensions and modifications of the UMTS database concept and results of performance evaluation will be presented.

2. Personal Communication - *the concept for PSCS*

PSCS focuses on advanced *personal communication*, especially in the context of innovative, and possibly complex, multimedia services across networks of varying and increasing bandwidths.

Personal communication supports every day tasks such as getting and staying in touch with other people, accessing message and information servers, or performing tele-trading services, such as tele-banking or shopping. This places emphasis on another aspect of personal communication: the organisation of communication—the ability of users to decide for themselves when, how and with whom (or which system) they wish to communicate, as originators as well as recipients.

The Personal Service Communication Space (PSCS) is a service concept designed to fulfil the demands of personal communication.

2.1 Functional Entities in PSCS

The following functional entities are recognised as essential for the PSCS description:

- PSCS service framework,
- Application service,
- Connectivity, and
- Access element.

The *PSCS Service Framework* (PSCS SF) provides location and personalisation support services in an application service independent way and offers a platform for service and application interoperability. The main components of the PSCS SF are the actual framework service (Service Agent), performing the PSCS network functionality, and the distributed flexible service profile which handles network data for services, subscriptions and users of the PSCS environment. The service profile is a PSCS provider managed part of the Flexible Service Profile concept, which describes as a whole *all* service, subscriber and user related data existing in the PSCS system.

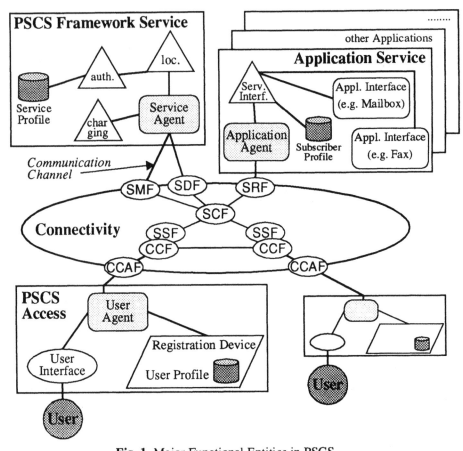

Fig. 1. Major Functional Entities in PSCS

The *Access Service* provides services which are necessary for connection establishment and connection control with the connectivity element and with the PSCS framework service. In addition the access element contains a part of the Flexible Service Profile which is located in the access equipment, and a User Agent supporting locally the end-user in organising his personal working environment.

PSCS can be accessed on a range of terminals from simple telephones to highly integrated PDAs or computer terminals. The PSCS service provider issues the PSCS subscription by means of a *smart card*.

The term *Application Service* describes a functional element which performs a non-PSCS specific service function and provides it to the other entities in the system.

The *Connectivity Service* is concerned with aspects of transmission, switching, routing and internal storage, e.g. buffering of queues. Two aspects of connectivity have to be distinguished and regarded as independent: the signalling (responsible for the connection establishment, control and disconnection) and the communication connection itself.

2.2 The PSCS Enterprise Model

The PSCS conceptual design starts at an enterprise model stage. Enterprise modelling is needed for PSCS in order to put the service features into the right context, to identify on-line and off-line (contractual) relationships and to relate the services to the domain interfaces. Requirements are placed on a framework for personal communication from all the different "stake holders" involved in the deployment and operation of PSCS: End-User, Subscriber, PSCS Service Provider, Network Provider and Access Provider. These "PSCS players" are characterised by their roles and the mutual relationships between them (see Figure 2).

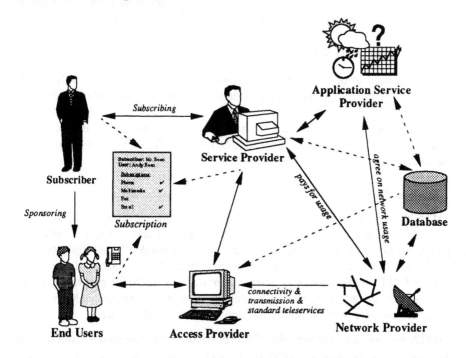

Fig. 2. The PSCS (Personal Service Communication Space) Enterprise Model

Every entity in this model is linked via a contractual relationship to each other. The *Network Provider* provides the connecting telecommunication infrastructure including signalling and basic tele-services like the telephone service. Specialised services (e.g., weather forecasts) are business of *Application Service Providers*. The entire PSCS market is managed by *Service Providers* who resell the mentioned services. The Subscriber subscribes to these services and pays for their usage by the End-User whose rights he defines in the PSCS *Subscription*. Besides through his own terminal, an End-User can access the network via *Access Providers*. The access may consist of the terminal equipment or/and an access network. The *Database* (in most cases established by the Network Provider) takes care for the availability of necessary data for the PSCS service in any point of the network.

2.3 Personalisation in PSCS

The PSCS design represents a complete personal communication environment. An important aspect of personal communication is personalisation, i.e. adapting the telecommunication services to specific user needs. In PSCS, this personalisation is achieved by means of a database containing a personal record per user which "describes his communication space".

PSCS was designed having an IN infrastructure in mind. UMTS similarly capitalises on an IN approach. The distributed database proposed for UMTS may host a service like PSCS, providing the SDF interface to it. The next chapters introduce the database concepts of UMTS and explain how PSCS can be implemented on top of it.

3. UMTS Database Concepts

UMTS is intended to be a global mobile communication system which will include most of today's and future telecommunication systems. A well promising database concept has been developed to enable terminal and personal mobility within and between networks. In contrary to the existing telecommunication databases of current implementations, this database is intended to be open to third parties, for instance telecommunication services like PSCS.

3.1 Database Structure

In UMTS, the highest level of data organisation is the domain level. Each network provider establishes one domain which covers the supported area. Of course, one geographical region can belong to many domains if it is provided by multiple network operators. In a domain, the smallest organisation units are DATA nodes which store the data of the UMTS entities which are registered or roaming inside the geographical area the DATA nodes cover (see Figure 3). No numbering scheme exists to establish a fixed relation between user number and the position of his data in the network.

The DATA nodes as a whole form the UMTS distributed database which maintains and distributes the various data records of the respective entities. Although all data nodes can physically communicate with each other directly, logical data organisation and search operations are managed by means of a search tree. *Directory (DIR) nodes* link the DATA nodes (which form the leaves) together to this search tree. DIR nodes contain only directory data, which means references to other nodes and necessary information about the contents of the DATA nodes beneath them to be able to direct a query to the right node. In addition to normal DIR node functions, the *Root DIR Node* of a DDB tree is the only one which is able to communicate with Root Dir Nodes of other domains and handle queries coming from there.

Queries coming from outside first reach the Query Handling Functional Entity (QH-FE) which analyses and decomposes them into simpler requests, understandable by the Directory Service (DS) FEs. Decisions on where to direct a query may also be taken here. DS FEs manage the distribution and retrieval of the UMTS data which finally will be administrated locally by data (DA) functional entities (see Figure 3).

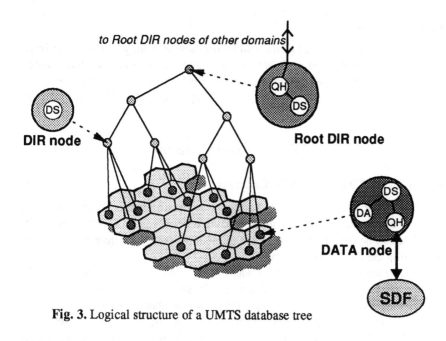

Fig. 3. Logical structure of a UMTS database tree

3.2 Data Types

UMTS databases have to contain and manage all the data needed to support the UMTS service. They start one or more data records for each of the UMTS entities. These are:

- Service Provider
  - Service Provider Data

- Subscriber
  - Subscriber Profile Data

- Terminal
  - Terminal Data

- User
  - Subscription Data
  - Registration Data
  - Routing Data
  - Session Data

Whereas the entities Service Provider, Subscriber, and Terminal are represented by only one data record, special attention is put on the User whose data is managed through multiple data records. However, only the *Subscription Data* record which contains all static user data is stored permanently in the network. *Registration Data* and *Routing Data* are created temporarily on each service registration and have only the lifetime of that service. Registration Data contains the part of Subscription Data which is mandatory for the service it represents. Routing Data is used to query the user state in each registration in order to perform the required checks and avoid routing calls to detached users. Routing Data includes location information about the user. The Session Data record, with the lifetime of one session only, includes a complete replication of the part of Subscription Data used for session setup purposes. It is mainly used by UMTS procedures.

3.3 Data Distribution

Because the overall performance of the UMTS database is mainly shaped by efficient data distribution, MoNet developed four different algorithms (static, resident, visited, resident & visited) using data fragmentation, duplication or none of them. For data retrieval, the database concept provides the techniques chaining, referral, chaining/referral and passing. [7]

4. Evaluation of the UMTS Database Concepts

4.1 Benefits of the UMTS Database Concepts

The flexible and user individual search directory tree construction is advantageous, for instance, for users roaming with a regular behaviour between two (or more) areas. In this way, data can be found very fast although the locations of the querying node and the node storing the data can be far apart.

A suitable example that describes the advantage of UMTS database solutions is the following: A person 'A' with a UPT subscription in Sweden ($\rightarrow$ personal mobility) is currently visiting Germany and borrows there a mobile phone with local GSM subscription ($\rightarrow$ terminal mobility). With this terminal he registers to the UPT service at his home UPT service provider. Afterwards, he roams to Spain without registering again at his UPT provider and is called from there by one of his friends 'B'. The current system would route the call of user 'B' to the UPT provider of 'A', who forwards the call to the german GSM provider from where it will be directed to Spain and finally reach the target, user 'A'. In a UMTS system, the location information about user 'A' would be automatically updated in his resident node, visited node and in directory nodes of the involved databases. Now, the UMTS database in Spain would direct the call from user 'B' to B's parent node where the location information about user 'A' would be found immediately.

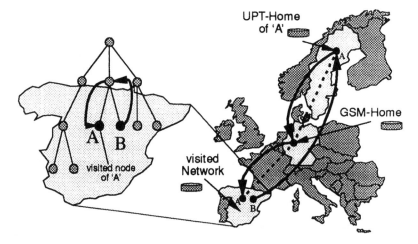

Fig. 4. Effects of transparently implementing UPT on top of GSM

Another example for the "personalisation" of the database the possibility to choose an individual data distribution method for each user. Even the elements which have to be copied from Subscription Data into Routing Data can be determined individually for each user, e.g., depending on the user behaviour.

4.2 Extension, Modification & Adaptation for PSCS

Although the UMTS database concepts seem to be quite suitable for the handling of data of advanced communication services like PSCS, some extensions are necessary. However, it is not intended to convert the UMTS database into a PSCS database by simply implementing all necessary PSCS database features into the UMTS concepts. This would violate the object of being a generic database which is able to support any service. PSCS specific database operations like the support of the PSCS Identification Module (PIM) have to be performed by PSCS functions. Furthermore, it must be taken into account, that the UMTS database concepts are still under development.

MoNets interpretation of services like PSCS or UPT is slightly different from the Mobilise one. Mobilise defines PSCS as a 'service that manages services', running on top of the underlying telecommunication system. In comparison, MoNet wants to real-ise these 'service managing services' on system level. Applied to the database, this would mean that UMTS would treat PSCS as a usual service. For that reason, UMTS has to distinguish between ordinary services and managing services by permitting lat-ter one to operate on the UMTS level in the database and to use higher database com-mands like service registration/deregistration.

By the definition of PSCS, Mobilise considers a more detailed composition of future telecommunication than UMTS. The introduction of the additional players Application Service Provider and Access Provider improves the possibility for commercial exploi-tation of telecommunication. Although the implementation of data records for both players into the UMTS database seems to be quite sensible for a moment, it is not rec-ommended. This is for the simple reason that in that way UMTS database would become too 'PSCS specific'. Instead, it is suggested to generalise the UMTS Service Provider Data record so that it can include data of any provider. The conversion of the data of a PSCS provider to a general UMTS Provider Data record and vice versa would be business of PSCS functions.

The UMTS database concepts offer very advanced methods of data distribution and its allocation, even among different networks. Nevertheless, neither a data model nor low level database functions which are responsible for storing data on hard disk and retrieving it from there, have been considered yet. Until now, all UMTS data elements are only specified by their names, without knowledge about the internal structure. Since the properties of data in future communication systems and services cannot be predicted with regard to type, structure and size, both the data model and the low level of the database must be quite flexible on this matter without wasting resources. It can be assumed that relational data models, as used today in many databases, will not be suitable for the handling of such data. Furthermore, the database must guarantee fast access to comply with the demands of telecommunication actions like call establish-ment for which very short access times are mandatory.

The elaborated data model for the UMTS database is based on the *TriBase* database concept which is recommended for the physical level of the database. TriBase has been designed for the management of data for future telecommunication systems and services, particularly PSCS [13]. The basic idea of TriBase is to reduce the term "database" to an effective mechanism of storing and retrieving flexible data structures with the focus on fast retrieval of data. TriBase stores only the required data, what means that

Fig. 5.
Proposed UMTS
Data Model

there are no empty fields like in relational databases if there is no appropriate data item for this player. Each data item is described by a so called *Triple* which consists of three elements: the *Owner*, the *Attribute* and the *Value*. Owner and Attribute are used for the identification of the respective data element and Value stores the essential data. Whereas Owner and Attribute have well defined data types, the Value is treated by TriBase as a 'black box' and can therefore store data of any type. This agrees outstandingly with the above mentioned demand of data type independence and can be seen as a first step towards security.

PSCS data records consist of hierarchically arranged groups of data elements. To apply this hierarchy to UMTS, the TriBase data model has been extended by the *Pointer Data* part, which contains names of other Triples.

It is true UMTS considers partial and flexible copying of Subscription Data into Registration Data, but nothing has been mentioned about the way in which this can be performed or how to mark a data element that has to be copied. For that reason, another part, the *Control Data*, has been added to the UMTS data model with the function to store optional information about the data element like the mentioned marking or access rights. [10]

Until now, only complex database commands (e.g., service registration, which creates a whole Registration Data record) have been taken into account by the UMTS database concepts. For the interworking between the database and services, the possibility to request individual data items which are part of, for instance, the Registration Data record have to be implemented.

5. Performance Evaluation

The evaluation of the performance of a UMTS-like database applied for personalised and mobile applications is a difficult issue. On the one hand, a broad range of possible design alternatives has to be investigated and, on the other hand, parameters describing users behaviour are difficult to be validated, since the future mobility and call distribution patterns are not known in sufficient details. The choice of key performance criteria has as well impact on the preferred solution. In following all those issues will be discussed in short and a summary of results achieved so far will be given.

5.1 Alternatives, users behaviour and performance criteria

As alternatives will be understand in following those database options, which have essential impact on database performance. Two classes of alternatives can be differentiate:

- structural and
- data distribution.

Among the first class such issues as height and size of the directory tree are of major impact. In the second class, the user data distribution on the data nodes and the pointer structure in the directory nodes are of major importance.

It is obvious, that not all procedures will cause the same effort in the network. For an evaluation, these procedures should be chosen, which have the strongest impact. Two procedures were chosen, namely database update and database searching. Both are very complex and require from the database sophisticated protocol support.

In order to be able to evaluate the chosen procedures, a comprehensive model describing actions, which initiate those procedures has to be described. In our case this model will describe user actions, which initiate both procedures:

- location update initiates the update and
- cal set-up initiates the searching.

A flexible description of users behaviour on sufficient level of details is a very complex task. An example of possible solution is given in [12].

After the alternatives are systematically described, main procedures are chosen and users behaviour is modelled, in the last step before evaluation the key performance parameters have to be chosen. Here only those parameters are considered, which are strictly dependant on the choice of the alternative, that means issues as e.g. data duplication for reliability reasons are not considered, since it influences the effort in the same way in all alternatives. The following parameters were finally chosen to judge about the performance of the database:

- node load,
- link load, and
- total delay in database.

5.2 Evaluation results

The preliminary results achieved so far indicate following conclusions. From the performance point of view (as defined in previous section):

- The data distribution strategy, which is based on static pattern, that means the user data is mainly stored in the home location, is in most cases the preferable solution; the reason for this situation is the fact, that users move and call mainly close to their home location.

  In case, that users are roaming extensively, it is beneficial to copy the required data into the visited data node and keep track only of that data - similar as it is done nowadays in the GSM system.

- As for the directory structure, it is beneficial in all cases to work with a flat directory. The higher the directory, the higher is the effort required for update and searching in the database. This conclusion is independent of the users movement and call patterns.

In order to draw more detailed conclusions about the impact of personalisation data on the database performance additional investigation have to be performed. Issues as e.g. replication of data among network and ID-card or user adapted data distribution have to taken into account.

6. References

1. *"Network Aspects (NA); Universal Personal Telecommunication General Service Description (Version 1.2.0)"*, ETSI Technical Reports ETR NA-70201, ETSI, Valbonne, France, 8 July 1992.

2. Mobilise—R2003, Deliverable D18, R2003/EMP/CT1/DS/P/018/b1, *"PSCS Concept. Definition and CFS—Final Version"*, December 1994.

3. Mobilise—R2003, Deliverable D20, R2003/ETM/CT2/DS/P/020/b1, *"PSCS Specification and CFS: Architectural Framework—Final Version"*, December 1994.

4. H. Mitts, *"Universal Mobile Telecommunication System - Mobile access to Broadband ISDN"*, Proceedings of Broadband Islands 1994, North Holland, June 1994.

5. D. J. Goodman, *"Personal Communications"*, Proceedings 1994 International Zürich Seminar on Digital Comm., Lecture Notes in Computer Science, Springer, March 1994

6. MoNet—R2066, Deliverable 61, R2066/PTT_NL/MF1/DS/P/061/b1, *"Implementation aspects of the UMTS database"*, May 1994

7. S. Kleier, C. Görg, *"Personal Communication System Realizations: Performance and Quality of Service Aspects on SS-No.7"*, Proceedings IS&N '95, Lecture Notes in Computer Science, Springer, October 1995

8. Patricia E. Wirth, *"Teletraffic Implications of Database Architectures in Mobile and Personal Communications"*, pp. 54-59, IEEE Communications Magazine, June 1995, Vol. 33, No. 6

9. N. Niebert, E. Geulen, *"Personal Communications - What is Beyond Radio?"*, Proceedings IS&N '94, Lecture Notes in Computer Science, Springer, September 1994

10. A. Ritterbecks, *"Adaptation and Implementation of UMTS DDB Concepts for PSCS"*, Master's thesis, Lehrstuhl für Kommunikationsnetze, RWTH Aachen, Juli 1995

11. D. Lawniczak, *Modellierung und Bewertung der Datenverwaltungskonzepte in UMTS*, dissertation, Lehrstuhl für Kommunikationsnetze, RWTH Aachen, Juni 1995

12. J. Sauermann, N. Klußmann, *"A new flexible and high Performance Mechanism for Storing and Retrieving Data for personalised telecommunication services"*, Proceedings International Switching Symposium ISS '95, Berlin

Next Generation Database Technologies for Advanced Communication Services

Vera Goebel, Bjørn H. Johansen, Hans C. Løchsen, and Thomas Plagemann
University of Oslo, Center for Technology at Kjeller (UNIK)
Granaveien 33, P.O.Box 70, N-2007 Kjeller, Norway
Email: {goebel, bjornj, hansl, plageman}@unik.no

Abstract. Advanced telecommunication services have to be carefully designed in order to meet the various Quality of Service requirements of a probably very high number of end-users. Due to the enormous data management problems in telecommunications the use of database technology provides an efficient solution for several software components. However, the "ideal" database management system for telecommunication data management seems out of reach. It is the aim of this paper to show where and how next generation database technologies could be used to implement and maintain advanced communication services. The paper gives an overview of new database technologies and relates them to requirements of advanced communication services. A sample service is used to demonstrate the application of new data modeling and transaction concepts for advanced communication services.

Keywords. Database Systems, Communication Services, Object-Orientation, Cooperative/Collaborative Transaction Management, Quality of Service, Distance Education

1. Introduction

Recent developments in telecommunications are dominated by advances in high-speed networking, distributed multimedia applications, and mobility aspects. Advanced telecommunication services have to be carefully designed in order to meet the various Quality of Service (QoS) requirements of (a probably very high number of) end-users, including reliability, performance, and security aspects. Particular attention has to be paid to the software elements that are used to construct telecommunication services, because they represent a potential performance bottleneck. Gallersdörfer et al. [12] point out that due to the enormous data management problems in telecommunications the use of database technology seems a natural solution for several software elements, but the "ideal" database management system (DBMS) for telecommunication data management seems out of reach.

Conventional DBMSs such as relational, hierarchical, or network DBMSs have been successfully used for more than twenty years in business and administrative applications. These classical database applications retrieve mostly only small amount of data that contain numeric and short symbolic data. In the last years, many other (so-called non-standard) application domains like CAD/CAM, CIM, or office automation started to demand DBMS support. Besides the numeric and short symbolic data that classical database applications deal with, non-standard database applications store and retrieve complex nested data (e.g., VLSI chip design), compound data (e.g., sets, arrays, structures), and multimedia data (e.g., images, audio, text). The DBMS requirements for non-standard database application domains are very different from those of the classical database application domains and usage of conventional DBMSs in non-standard database application domains led thus to unsatisfactory results ([14], chapter 1 & 13). So-called next generation database technologies have been developed to solve these problems. The application of these next generation database technologies for advanced communication services is the main topic of this paper.

Generally, research activities concerned with DBMS aspects for telecommunications can be divided into two groups depending on their background. Research starting out from the database community is mainly concerned with next generation database technologies for todays telecommunication services [1], [9]. The telecommunication community in turn often applies conventional database technologies for advanced communication services [2]. However, it is our belief that only the application of next generation database technologies will enable the construction of advanced communication services that satisfy the high end-user requirements. The usage of a multimedia database system for a multimedia mail service [21], the implementation of an object-oriented multimedia server (MOSS) [13], and the INDIA [12] project are examples for a step in this direction. The Service and Network Database Research Activity (SANDRA) at the Center for Technology at Kjeller (UNIK), University of Oslo, focuses on the integrated research in the areas of telecommunication services and database systems. The aim of this paper is to present the first results of SANDRA by showing where and how next generation database technologies could be used to implement and maintain advanced communication services.

The remainder of this paper has two major parts: a database technology overview and an outline of a sample service used to describe the application of data modeling and transaction management aspects in more detail. Due to the space limitations for the paper this is no tutorial on new database technology, we can only roughly sketch the new functionality and indicate its potential usage for the realization of advanced communication services.

2. Database Technologies and Communication Services

In this chapter, we give an overview of next generation database technologies, summarize the requirements of advanced communication services, and indicate how next generation database technologies could be used to fulfill these requirements.

2.1 Next Generation Database Technologies - An Overview

DBMSs provide concepts for the reliable, long-term management of integrated data to fulfill requirements which are determined by the application domains and the computing environment. Examples for such concepts are: data structures and operations, views, authorization, transactions, and integrity constraints. In the last years new application domains demand for DBMS support. This results in a number of requirements to be fulfilled by next generation database technologies. In particular, next generation database technologies address non-standard application domains by improvement and integration of new concepts for the major DBMS aspects [6], [20], [14]:

- *DBMS architecture*: should be designed in such a way that it can easily accommodate various functional extensions. From an economic point of view, it is not practicable to construct several DBMSs with different capabilities from scratch over and over again. Extensible DBMSs try to overcome this deficiency by providing the possibility to integrate extensions, e.g., new data models, other transaction mechanisms, or alternative storage structures, depending on the characteristics of different application domains. The construction of a completely new DBMS for different requirements of application domains can be avoided by providing a flexible base DBMS part that can be extended in different ways [20]. The development effort, i.e., the costs for a specially tailored DBMS can thus be drastically reduced. There have been some prototypes implemented for academic use, but there is no commercial system available today.

- *Data modeling*: comprises object-oriented, deductive, temporal (history and versioning), spatial, and multimedia data modeling concepts, as well as imprecise data, and rules. The data model has to provide suitable modeling tools (e.g., object structures and operations) that support specific characteristics of application domains. At least the following data model concepts should be supported ([14], chapter 2 & 6), [18]: complex objects, object identity, types and classes, inheritance hierarchies, encapsulation, integrity management, abstract/user-defined data types, large unstructured data elements, and computational completeness of language. Most of the data model concepts listed above are covered by object-oriented data models.

- *Integrity maintenance*: is needed to support more powerful integrity concepts and active mechanisms ([14], chapter 21). For many applications it is important to monitor situations of interest and to trigger response when the situations occur. Active mechanisms should be provided to check automatically all required integrity constraints. Active mechanisms can be found today in some commercial systems.

- *Change management*: is a generic term which summarizes different concepts like version, configuration and transformation management, and schema evolution. It documents the evolution of objects for purposes of validation, traceability, and reuse. Change management may be provided as an integral part of the data model or as a distinct component decoupled from the data model [20]. It may be defined as a consistent set of techniques that aid in evolving the design and implementation of objects. These techniques may be applied at many levels to object history: explore alternatives (i.e., versions), manage a layered design (i.e., configurations), maintain consistency during evolution and across multiple representation (i.e., transformations), and manage changes to the set of type/class definitions (i.e., schema evolution). Some approaches for change management are already integrated in next generation DBMSs.

- *Query processing*: one of the basic functionalities of DBMSs is to be able to process declarative user queries. Information is accessed with data manipulation language statements based on its value, rather than its identity, like in SQL. The targets and results of queries are sets of objects. Queries to object-oriented DBMS differ from relational queries in three main respects ([14], chapter 3 & 8): possible predicates in queries and response sets of queries; semantics of relationships between objects and inheritance; and query optimization techniques. The first generation of object-oriented DBMSs did not provide declarative query capabilities. However, the last decade has seen significant research in defining query models and in techniques for processing and optimizing them, like query optimization and logical access path structures. Many of the current commercial object-oriented DBMSs provide at least rudimentary query capabilities.

- *Physical object and storage management*: DBMSs with new data models need extended or new storage and addressing structures, and have to support new media, e.g., CD-ROMs, or Jukeboxes, in order to provide an efficient system behavior [8]. A few physical object management techniques exist that are rooted in the relational context but were thoroughly adapted to the needs and requirements of object-oriented data modeling. Storage management comprises new addressing and indexing techniques, clustering mechanisms, physical access path structures and storage techniques for objects and versions.

- *Security*: applications must share information among users with different needs and authorizations by providing object-oriented authentication and authorization concepts ([14], chapter 7). The specific access rules vary from application to application, thus a flexible approach to discretionary access control is needed which led to the development of multilevel security mechanisms for object-oriented DBMS (per object: single uniform security classification).

- *Transaction and workflow management*: new transaction models have been proposed to address long-running operations on DBMSs, real-time processing and cooperation among multiple transactions. Advanced transaction models comprise long-duration transactions (lasting weeks or even months), multilevel transaction mechanisms, cooperative transactions, and sagas [11]. Today only few of them have been implemented, possibly because of their lack of flexibility required by todays complex applications. Workflow management systems bear a strong resemblance to the advanced transaction models both in their goals and modeling approach, yet they are quite different in that they address a much richer set of requirements than conventional and advanced transaction models [17]. Workflows merged with advanced transaction concepts can be used to enhance process correctness and performance.

- *Programming language coupling*: object-oriented programming languages (mainly C++) are used for database modeling and application programs solving the impedance mismatch problem. That means the type system of the DBMS should be as powerful as the type system of the programming language in which the application programs are implemented ([14], chapter 4, 5, 11).

After several years of research and prototype development object-oriented DBMS products, like ObjectStore, O_2, GemStone, or OpenODB, have been offered in the marketplace for some time. There are several approaches and companies that compete. Standardization efforts for object-oriented DBMSs have started but not yet led to settled results. For programming systems the CORBA architecture has brought first success for standardization and it seems that for DBMSs this will also provide a good basis ([14], chapter 2 & 15). However, available object-oriented DBMSs are less mature today than relational DBMSs but the number of installations increases remarkably, and market prognoses look promising. The list of object-oriented DBMS issues still to be improved includes: query language and optimization, handling of complex objects, openness of language, method handling, meta data management, management of dynamic class definitions and class extensions, consistency and integrity constraints, view mechanisms, security mechanisms (access control), integration with existing object-oriented programming systems, and especially interoperability with non-object-oriented DBMSs.

Next-generation DBMSs so far have not replaced conventional DBMSs. This has led to the coexistence of DBMSs of all generations. Therefore, a challenge is to bring about interoperability of DBMSs and file systems [3]. DBMS vendors have begun offering gateways that allow relational DBMSs to retrieve data from non-relational DBMSs or file systems (so-called heterogeneous database systems). Heterogeneous database systems provide a single database view over various independently developed database and file systems, thus allowing users to access various databases and files in one, uniform database language without being aware of the heterogeneity of the underlying systems ([14], part II). However, it will still take some time until there are commercially viable heterogeneous database systems available.

2.2   Requirements of Advanced Communication Services

Telecommunication services can be divided in bearer services and tele-services. The ATM Adaptation layer services or frame relay services are examples for bearer services. The data management needs of such bearer services are mainly concerned with handling of network management and routing information. Tele-services have to manage additional application data, e.g., storage and retrieval of multimedia documents in multimedia information services, or handling of group information in conference services. Leopold [15] gives an overview on tomorrow's tele-services. Generally, advanced communication services have stronger data management needs than traditional communication services, because they additionally include (at least) one of the following aspects:

- *Multimedia information* comprises basic data types like text, graphics, audio, and video and combines them to complex information units (e.g., multimedia documents or movies). The different structure and size of the basic data types results in different requirements for their storage, retrieval, and transmission. Timing relations have to be maintained within a single data type, like delay jitter of subsequent video frames, and between multiple data types, like the synchronization of audio and video.

- *Mobility and personalization* support for personal mobility, like the Universal Personal Telecommunication (UPT) service, and personalization of distributed computing environments. So-called user profiles contain user-specific location information and (optional) user-specific service control parameters have to be stored in the network and maintained [10]. For example, location changes of users require immediate updates of the user profiles.

- *Group communication* includes 1-to-1, 1-to-N, and M-to-N relationships between end-users. In addition to the passive participation of end-users in pure distribution services like video broadcasting, advanced communication services should support collaboration among end-users. Traditional collaboration among two parties is known by truly simultaneous access to a common data pool, like distributed whiteboard systems without system-enforced integrity. Advanced collaboration like cooperative group work should be supported by application independent data model and consistency, and system-enforced integrity to enable joint modification of a common data pool [17]. The cooperative group work concept is based on explicit synchronization of data access. Furthermore, the administration of groups requires operations like group creation, group deletion, join/leave members, edit membership roles, and group features [16].

In general, advanced communication services require software components for data management during service creation, service execution, service management, and service maintenance, comprising the following aspects:

- *Design and implementation support*: design methodologies, like object-oriented concepts or function-oriented approaches (e.g., Entity-Relationship concepts) represent a structured approach to develop a precise service specification out of some informal description [4]. In order to ease the implementation of conforming services, appropriate tools and data models are needed to map the design step directly onto an implementation, e.g., object-oriented design methodologies and object-oriented database systems.

- *Scalability*: the number of service users, as well as the amount of data that has to be handled by a service cannot always be foreseen. Therefore, communication services should be independent of data management aspects. Storage units, like disk arrays, should be easily extended to be scaleable in terms of data amounts, and processing units should support parallel data access to be scaleable in terms of number of service users.

- *Performance*: in particular multimedia data requires high performance because of its generally huge size. Several performance metrics are of importance, e.g., read throughput, write throughput, average seek time, transactions per second, response time, and number of parallel accesses. QoS parameters should be specified at a higher level of abstraction to control the preferred transactional and workflow process, and to control what transactional and workflow domain it conforms to.

- *Availability*: is defined as the probability that the system is up and running continuously throughout a specified period. The quality of a service is obviously depending on its availability respectively the availability of the corresponding data. The availability of user profiles, e.g., is vital for mobility and personalization related services. Main mechanism to increase the availability of data is data replication.

- *Reliability*: is defined as the probability that the system is up and running at any given moment (also for concurrent access, reliable transport and storage of data). Reliability mechanisms comprise, e.g., transactions, error codes, shadowing, or replication.

- *Data modeling and retrieval*: for advanced communication services has hardly been discussed in the literature, although it is an accepted fact that there are special requirements to support complex data structures, views, long-running processes, and history management [9], ([14], chapter 2 & 8).

- *Data integrity and consistency*: integrity refers to the accuracy or validity of data. Integrity involves ensuring that the things users are trying to do are correct. The operations have to preserve data(base) consistency. An operation transforms a consistent data state (a consistent state of the database) into another consistent state, without necessarily preserving consistency at all intermediate points.

- *Collaboration and cooperation support*: multiple service users might access concurrently the same data. Traditional transaction concepts have been shown to be too restrictive for advanced communication services [11], leading to insufficient performance for services, such as distributed design environments, cooperative services, collaborative services or distributed multimedia services.

- *Security*: refers to the protection of data against unauthorized disclosure, alteration, or destruction. Security involves ensuring that users are allowed to do the things they are trying to do, this includes authentication, access control, and auditing. Authentication means to verify that a request for an operation was originated by the principal and not modified by another process. Access control mechanisms allow to define and enforce which process or principal is allowed to perform which operation, like read, write, delete, and change, on which data element. For auditing all operations can be logged in order to have a history.

2.3 Using Database Technologies for Advanced Communication Services

Based on the described service requirements (section 2.2), we outline the usage of new database technologies for their realization as follows (structure similar to section 2.1):

- *Data modeling*: an efficient modeling solution is to use an object-oriented approach in which each information unit is treated as an object with a topological structural description (object hierarchy), or graph model, attached to it [18]. Currently, there is no underlying theory comparable to the relational model for DBMSs. The object-oriented data representation problem has been studied extensively ([14], chapter 2 & 15). They provide a rich set of functionality to manage diverse data types including text, images, audio, and video. Most of the current commercial DBMSs suffer from an inability to manage arbitrary types of data, arbitrarily large data, and data stored on devices other than magnetic disks. They understand a relatively limited set of data types, such as integer, real, date, monetary unit, short strings, and BLOBs (binary large objects). Furthermore, they are not designed to manage data stored on such increasingly important storage devices such as CD-ROMs and video disks.

- *View management*: is needed to support presentation according to user needs, e.g., in distance education. This comprises the consistency problem (update problem) when handling redundant data in the database caused, e.g., by materialized views [1], [9].

- *Query processing*: is needed for all services to retrieve and manipulate data efficiently. Due to the diverse indexing techniques required for multimedia information, the conventional query processing models are inadequate. An object-oriented query processing strategy adjusted to the object-oriented data modeling approach has to fulfill the requirements to filter out the required data for a special session according to specified user needs [18].

- *Transaction and workflow management*: to synchronize browsing and manipulation activities from multiple users, appropriate transaction management concepts tailored to the needs of a multimedia server, including cooperating processes, long-duration of activities, and complex process structures are needed [11]. The fact that multimedia data management requires dealing with large unstructured data, besides the conventional small data, implies changes in the unit of concurrency control and authorization. The unit of concurrency control in conventional DBMSs is often a page or even a table rather than a single record. Workflow and advanced transaction mechanisms can support services like joint editing and strengthen to formalize design work, and be open for a better coexistence of design and execution environments.

- *Storage management*: is important for each service that has to handle large amounts of data. In particular, multimedia data can be very large occupying multiple blocks in secondary storage. Objects can have complex logical structure and can be shared among other objects at various levels. Therefore, the design of a physical storage structure for multimedia data is more demanding than for conventional data. In order to improve I/O performance, the use of parallel disk systems [8] with suitable storage models (access path structures, caching, replication, and data placement strategies) has to be incorporated. There are inadequate means for dealing with the complexity and scale of large object systems [7], [18].

- Security: should be supported by all services according to application requirements. The unit of authorization in conventional database systems is usually an entire table or a column of a table rather than a single record. The objective towards addressing security in distributed, group-oriented systems is to provide an organizing framework within which the special security needs of group communications can be addressed.

3   DBMS Support for a Distance Education Service

In this chapter, we briefly describe an advanced communication service that is currently designed in the DEPEND (Distance Education for PEople with different NeeDs) project [19] at the Center for Technology at Kjeller (UNIK), University of Oslo. We will use this service to exemplify the usage of next generation database technologies for communication services and to study in particular the data modeling and transaction aspects of this sample service.

3.1   Sample Service

Besides the distribution of lectures at real-time within so-called electronic classrooms [5], the DEPEND project develops a World-Wide-Web (WWW) like information server for lecture and exercise hypermedia documents. The document structure is based on an extended version of SGML (Standard Generalized Markup Language) and comprises shared workspaces in addition to standard elements, like text, graphics, or video. In the first prototype, the hypermedia documents will be stored and maintained in a centralized DBMS, which is fully accessible from the network. Three application scenarios are supported by this service:

- *Design and development*: of hypermedia documents for lectures and exercises can be done by collaboration/cooperation of multiple authors. A dedicated person (team leader) has to define the type of collaboration/cooperation, and their integrity and consistency requirements.

- *Browsing*: corresponds to the traditional application of the WWW. Students can browse the hypermedia lectures and exercises. According to the particular user needs, elements of the hypermedia documents are automatically retrieved from the database, transferred over the network (with corresponding QoS), and presented to the end-user.

- *Solving exercise tasks*: could be done in groups or by individuals. Workspaces in the hypermedia document enable students to edit dedicated text, program source code, or graphics. For each individual and group, a version of the workspace (including its data) will be generated. For a workspace version owned by a group, the group leader has to define the type of collaboration/cooperation, as well as their integrity and consistency requirements.

3.2 Data Modeling

We have chosen a computer science lecture to illustrate the data modeling concepts with an example. The hypermedia document contains text with figures, workspaces, and video clips from the lecture. The concrete objects represent parts of a lecture. The class hierarchy shown in Figure 1-a sketches only roughly one possible modeling for a lecture. Both *Image* and *Video* can also have subclasses to represent different image and video formats. A video object has typically a complex internal structure to model scenes and frames. The subclass *VideoScene* is included to model links to specific parts of videos. The class *TextLink* is used to model hyperlinks to any other object in the database. The textual contents of the document is stored in objects of the classes *TextLink* (*tLink*) and *RunningText* (*rT*). Objects of the class *Chapter*, *Section* or *Subsection* can contain an optional combination of *TextLink* and *RunningText* to form a header, followed by any number of paragraphs (Figure 1-b). A paragraph can contain any combination of objects of the classes *TextLink* and *RunningText*.

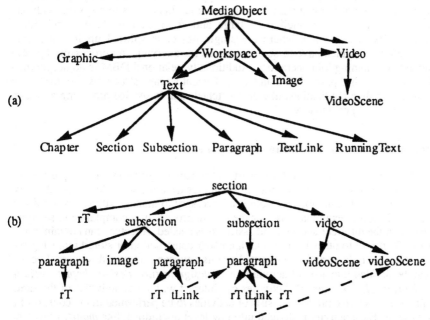

Figure 1. Class hierarchy (a) and instance (b) for lecture example

To view the lecture, the complex object is traversed from left to right. This very simple solution has no support for information about presentation of the document, like fonts or placement of the figures. We assume that the presentation facility uses the logical structure of the document to provide this, like in SGML. As an example of a complex object, we look at a subsection object. Such an object contains paragraphs, images, graphics, and video. The paragraphs are usually included in the subsection object as a subobject, while the images and video clips are only referenced. To handle and display images, we use the overloading mechanism for methods. It is possible to define a *display* method in the superclass *Image* and later redefine it for every different type of image. In this way, the use of image compression is completely transparent to users of the *Image* classes. Only the designer of a class needs to worry about implementing methods to compress and decompress images. Another advantage is that introducing new image types does not interfere with any existing applications, as long as the appropriate methods are implemented for the new class.

The use of object-oriented DBMSs - rather than relational DBMSs - allows to model services in terms of units that more closely resemble the units found in the real-world context to be modeled ([14], chapter 13 & 16), [18]. Object-orientation supports evolutionary system design in a better way by using inheritance mechanisms and polymorphism. The subclass mechanism results in less code to modify and less code to understand for the user of an existing application. The gap between the conceptual model and the database schema is much smaller than for traditional data models, and generally the task of mapping between these two levels is easier. The possibility of expressing more semantics of the modeled context in the database by using methods, reduces the size of application programs. The increased "semantic power" provides better support for user-defined data types. Every created class is an integrated part of the data types of the DBMS, and by implementing the necessary operations (methods), objects of any class are operated upon in exactly the same way as system-defined data types. However, the mapping of object-oriented design onto a relational realization is possible, this results in a system with bad runtime performance. The modeling of large and complex objects with relational structures results in structures where the single object is splitted up into multiple fragments (relations) which have to be reassembled for querying and processing by very large and time consuming join operations. Additionally, relational DBMSs do not provide extensible data models, appropriate data types, efficient storage structures, and accessing mechanisms to realize applications like our sketched sample service providing a system with required runtime performance.

3.3 Transaction and Workflow Management

We describe in this section the activity chart and processes given in our sample service (Figure 2). Circles describe service processes and arcs describe workflow dependencies. Courses are held every term and therefore it is necessary to manage adjusted versions of lectures, exercises, and examinations. For the design process of a complete course cooperation between the different teachers and assistants is needed. A lecture can contain cross-references (*TextLink*) to other lectures that are under construction or stored requiring cooperation to reveal only partially committed results between different designers. Lecture design can be divided into a set of activities, e.g., *design lecture, design chapter, include video, link external, link internal*. Activities can interact with other activities of the same service process. A service process consists of activities to be performed in the context of a specific process. Each activity is divided into low-level operations, like *modify object*, or *delete object*. The workflow dependencies indicate relationships between the different processes, to enforce global coordination providing better computational quality. The workflow forms a directed graph which is combined with one or more transaction models, e.g., nested and cooperative transaction models, and versioning techniques [17]. The service processes have to conform to the workflow policy and the applied transaction model(s). Consider for example the following scenario: two long-running transactions (called A and

B) are started in parallel to design lecturers. Transaction A is executing a *design chapter* activity. Transaction B is issuing a *link external* to some data designed by transaction A. To handle this situation cooperative messages are shipped between the collaborating parties.

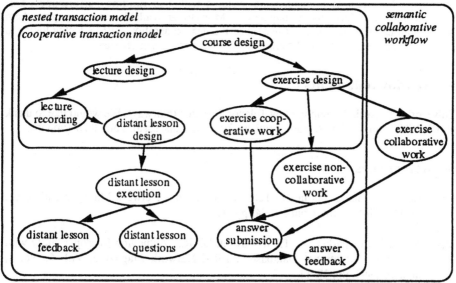

Figure 2. Transaction and workflow model for sample service

Transactions and workflows must always be handled but they are not always explicitly modeled like in the DBMS approach. Transactions and workflows are normally implicitly handled by the application programs. The advantages of the DBMS transaction concept are guaranteed rollback, recovery, and security [11]. Conventional DBMS transaction models guarantee concurrency and failure atomicity for consistent concurrent data access and persistence. These atomicity properties are too restrictive for applications like our sample service, even though we still need consistent concurrent access and persistence. This led us to the combined transaction and workflow approach we have taken to realize our sample service.

4. Conclusions and Outlook

Since some years many research groups deal world-wide with the development of next generation database technologies resulting in numerous prototypes. Today also commercially viable products begin to be available on the international marketplace. Thus, it is time for the users to see what the new database technologies offer in order to realize advanced communication services in a more efficient and effective way. This paper gives an overview of motives and backgrounds of this trends. We sketch important concepts and try to outline a perspective of the future DBMS market. The main focus lies on object-oriented DBMSs of different types extended with various of the other mentioned new database technologies like active or deductive mechanisms, and new transaction models.

In our opinion next generation database technologies should be used to realize advanced communication services because they provide specific software functionality which is needed for advanced communication services. This functionality is realized by DBMSs (especially next generation DBMSs) since a long time in a more appropriate and efficient manner. Conventional DBMSs are not suitable because especially their data modeling and transaction management concepts are not appropriate for advanced communication services, which results in inadequate systems with bad runtime performance. Our recommendation is thus either to buy new DBMSs (so far available) and to integrate them in communication systems, or at least, use the next generation DBMS technologies to design and implement advanced communication services in a better way.

5 Acknowledgments

This work was partially supported by Telenor Research (Norway). We would like to thank all DEPEND and SANDRA team members for fruitful discussions and their support.

6 References

[1] Ahn, I., "Database Issues in Telecommunications Network Management", Proc. ACM SIGMOD Int. Conf., 1994, pp. 37-43

[2] Aiken, J., A., Parker, S., T., Woodwell, D., R., "Achieving Interoperability with Distributed Relational Databases". IEEE Network Magazine, Vol. 5, No. 1, January 1991, pp. 38-44

[3] Berra, P., B., Chen, C., Y., R., Ghafoor, A., Lin, C., C., Little, T., D., C., Shin, D., "Architecture for Distributed Multimedia Database Systems", Computer Communications, Guilford, Vol. 13., No. 4, May 1990, pp. 217-231

[4] Bochmann, G., V., Poirier, S., Mondain-Monval, P., "Object-Oriented Design for Distributed Systems and OSI Standards", in: Neufeld, G., Plattner, B. (Eds.), "Upper Layer Protocols, Architectures and Applications", Elsvier Science Publishers (North-Holland), May 1992, pp. 265 - 280

[5] Bringsrud, K., A., Pedersen, G., "Distributed Electronic Class Rooms with Large Electronic White Boards", Proc. 4th Joint European Networking Conference (JENC 4), Trondheim (Norway), May 1993, pp. 132-144

[6] Brodie, M., L., Bancilhon, F., Harris, C., Kifer, M., Masunaga, Y., Sacerdoti, E., D., Tanaka, K., "Next Generation Database Management Systems Technology", in: Kim, W., Nicolas, J.-M., Nishio (Eds.), "Deductive and Object-Oriented Databases", Elsevere Science Publishers, Amsterdam, Netherlands, 1990, pp. 1-13

[7] Buddhikot, M., M., Parulkar, G., M., Cox, J., R., "Design of a Large Scale Multimedia Storage Server", Computer Networks and ISDN Systems, Elsevier, North Holland, Vol. 27, 1994, pp. 503-517

[8]   Chen, P., M., Lee, E., K., Gibson, G., A., Katz, R., H., Patterson, D., A., "RAID: High-Performance, Reliable Secondary Storage", ACM Computing Surveys, June 1994, pp. 145-185

[9]   Datta, A., "Research Issues in Databases for ARCS: Active Rapidly Changing Data Systems", ACM SIGMOD Record, Vol. 23, No. 3, September 1994, pp. 8-13

[10]   Eckhardt, T., Magedanz, T., "On the Personal Communication Impacts on Multimedia Teleservices", in: Steinmetz, R. (Ed.), "Multimedia: Advanced Teleservices and High-Speed Communication Architectures", 2nd Int. Workshop, IWACA'94, Heidelberg, Germany, Springer, September 1994, pp. 435-450

[11]   Elmagarmid, A., K. (Ed.), "Database Transaction Models for Advanced Applications", Morgan Kaufmann Publisher, 1992

[12]   Gallersdörfer, R., Jarke, M., Klabunde, K., "Intelligent Networks as a Data Intensive Application (INDIA)", Technical Report, RWTH Aachen, Germany, 1994

[13]   Käckenhoff, R., Merten, D., Meyer-Wegener, K., "MOSS as a Multimedia-Object Server, in: Steinmetz, R. (Ed.), "Multimedia: Advanced Teleservices and High-Speed Communication Architectures", 2nd Int. Workshop, IWACA'94, Heidelberg, Germany, Springer, September 1994, pp. 413-425

[14]   Kim, W. (Ed.), "Modern Database Systems - The Object Model, Interoperability, and Beyond", ACM Press, Addison-Wesley, 1995

[15]   Leopold, H., Frimpong-Ansah, Singer, N., "From Broadband Network Services to a Distributed Multimedia Support-Environment", in: Hutchinson, D. Danthine, A., Leopold, H., Coulson, G. (Eds.), "Multimedia Transport and Teleservices", Int. COST 237 Workshop, Springer, 1994, pp. 47-68

[16]   Mauthe, A., Hutchison, D., Coulson, G., Namuye, S., "From Requirements to Services: Group Communication Support for Distributed Multimedia Systems", in: Steinmetz, R. (Ed.), "Multimedia: Advanced Teleservices and High-Speed Communication Architectures", 2nd Int. Workshop, IWACA'94, Heidelberg, Germany, Springer, September 1994, pp. 266- 277

[17]   Mohan, C., Alonso, G., Günthör, R., Kamath, M., "Exotica: A Research Perspective on Workflow Management Systems", IEEE Data Engineering (Bulletin), Vol. 18, No. 1, March 1995, pp. 19-26

[18]   Özsu, M., T., Szafron, D., El-Medani, G., Vittal, C., "An Object-Oriented Multimedia Database System for a News-on-Demand Application", to appear in: Multimedia Systems, 1995

[19]   Plagemann, T., Goebel, V., Tollefsen, M., "DEPEND - Distance Education for People With Different Needs", to be published: 2nd IASTED/ISMM Int. Conf. on "Distributed Multimedia Systems and Applications", Stanford, California, August 1995

[20]   Stonebraker, M., Agrawal, R., Dayal, U., Neuhold, E., J., Reuter, A., "DBMS Research at a Crossroads: the Vienna Update", Proc. 19th IEEE VLDB Int. Conf., Dublin, Ireland, 1993, pp. 688-692

[21]   Thimm, H., Roehr, K., Rakow, T., C., "A Mail-Based Teleservice Architecture for Archiving and Retrieving Dynamically Composable Multimedia Documents", in: Hutchinson, D. Danthine, A., Leopold, H., Coulson, G. (Eds.), "Multimedia Transport and Teleservices", Int. COST 237 Workshop, Springer, 1994, pp. 14-34

Using SDL for Targeting Services to CORBA*

Anders Olsen and Bo Bichel Nørbæk

Tele Danmark Research, Lyngsø Allé 2, DK-2970 Hørsholm, Denmark
phone +45 4576 6444, fax +45 4576 6336, e-mail {bbn,anders}@tdr.dk

Abstract. The work presented in the paper has been performed in the
RACE project SCORE. The paper addresses the step from designing a
service to implementing it and executing it on a target platform. It is
shown how SDL can be used as a design language for service descriptions and how SDL service descriptions automatically can be translated
into C++ service implementations complying with the Common Object
Request Broker Architecture (CORBA).

1 Introduction

The paper addresses the step from designing a service to implementing it and
executing it on a target platform.

The specification technique selected for service design is the ITU standardized
language SDL, which is widely used within the telecommunication industry.
When using SDL, the service developer can utilize the abstraction level of a
formal description technique while creating the service, without having to deal
with implementation details. The use of SDL also makes it possible to use various
tools that offer a wide variety of capabilities.

The target for the implementation is products conforming to the Common
Object Request Broker Architecture (CORBA [3]). This architecture is an emerg-
ing ODP-like one for distributed computing, and is being defined by the Object
Management Group (OMG). To validate the work, a specific platform [5] has
been used, but the principles can, with minor modifications, be adapted to any
CORBA-conforming platform.

This paper describes the various aspects of combined use of CORBA and
SDL. It consist of two parts:

1. Section 1 describes the service developers view, i.e. how SDL can be used
   when the target is a CORBA platform.
2. Section 2 describes the SCE developers view in terms of a specific implementation, i.e. the SCORE SDL to CORBA compiler.

---

* This work was supported by the RACE SCORE project; however, it represents the
view of the authors

2   Service Developer's View

This section, describes how SDL can be used for service creation on a CORBA platform. CORBA is described in [3]. The description covers the service developer's view, but this view has strong impact on the SCE provider's view. The SCE providers view is described in Sect. 3.

2.1   Role of CORBA and SDL

With combined use of a CORBA platform and SDL, there are two rather different approaches:

1. *The CORBA-oriented approach*, where SDL is supported as implementation language for the definition of behaviour, treated in the same way as the already supported implementation language C (or C++).
2. *The SDL-oriented approach*, where a CORBA platform is used as execution system for SDL processes.

In the following, the two approaches will be described, but it should be noted that the CORBA-oriented approach is only included for completeness, since SCORE focus on the SDL-oriented approach.

SDL as Implementation Language for CORBA   In the CORBA-oriented approach, SDL is the *means* for defining behaviour of CORBA services. This is shown in Fig. 1.

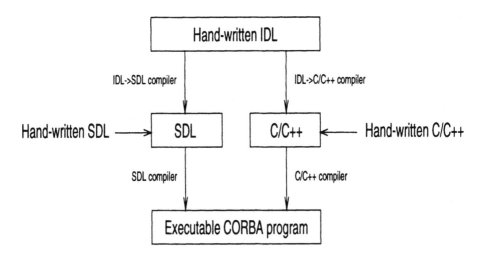

Fig. 1. The CORBA-oriented approach

1. The starting point for implementation is an IDL specification of the interfaces of *services* to be implemented. Note that SDL is primarily a specification language and most often it will be used also in activities prior to implementation.
2. The IDL specification is compiled into SDL packages containing signals, remote procedure specifications and process types.
3. The generated packages are used in the SDL system(s) which implement(s) the behaviour of the servers (and clients, see Sect. 2.2).

As SDL is a high-level specification and implementation language in its own right which is rather different from CORBA IDL, SDL may have to be modified to accomplish that all of the features of CORBA can be expressed in SDL. Likewise, those features of SDL which have no suitable CORBA counterpart or other runtime support, must be avoided (such as the SDL view concept).

To support properly the CORBA concepts of interfaces, objects, clients and servers, such modifications would probably include the introduction of gate types/instances, a close relation between gates and PId values and typed PId values (in SDL-92, gates have no representation during execution).

Other new SDL features, which also would be very useful for the SDL-oriented approach are system composition, external procedures and remote procedures on signal lists. Those features are described in Sect. 2.5.

CORBA as Execution Architecture for SDL Systems In the SDL-oriented approach, the CORBA platform is used as the underlying execution system for SDL processes in order to allow an SDL system to be distributed on a number of computers.

In this approach, the service developer needs only in special cases be concerned with the various CORBA concepts, such as IDL. This is illustrated in Fig. 2.

The exceptional cases include:

1. When some part of the system is developed using another programming language such as the CORBA supported language C. In this case, the relation between SDL concepts (signals, channels etc.) and the corresponding IDL must be known by the service developer such that he/she can properly interface the various heterogeneous sub-systems. C is for example better suited than SDL for low-level programming such as driver programming and terminal I/O. C/C++ can be used the same way in the two approaches.
2. When special programming principles are enforced by the nature of the underlying CORBA platform. For example, using SDL processes as a general modelling concept for objects (due to the lack of a data-oriented object concept in SDL), result in more processes during execution than strictly necessary for implementing the distributed system. The results will be a (usually significant) decrease in performance.

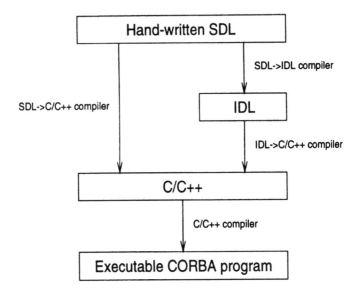

Fig. 2. The SDL-oriented approach

2.2 Reflections on the Client-Server Paradigm

As described in Sect. 3, CORBA is based on the server/client principle, i.e. a CORBA system consists of *client* processes requesting use of resources maintained by *server* processes.

In SDL, a system consist of a number of asynchronously communicating processes. This means that each SDL process will normally be both a server and a client in the CORBA sense. There cannot be guidelines for how "pure" client processes are defined (e.g. processes without states) since in SDL, a process issuing a remote procedure call acts both as a server and as a client. This is because a remote procedure call corresponds to exchange of a request signal followed by a reply signal issued by the server when the procedure call is completed. Furthermore, requests are queued in the input buffer of the process and the process can control when and in which order requests are handled (by using the save concept).

For remote variables, the situation is different: They do not trigger execution of a transition in the serving process and the serving process controls the access to the exported variables via the **export** construct rather than via the input buffer. This means that handling a remote variable request is completely transparent to the state machine and to the other variables of the server. Remote variables can therefore be efficiently implemented by handling remote variable requests directly rather than through the input buffer.

In the SDL standard, the semantics of remote variables is *modelled* using two signals in a similar way as for remote procedures, so the above strategy for remote variables can be considered as relaxing the SDL semantics. However, the relaxation is significant only if a client process (the importing process) relies on that a certain signal is consumed before a variable is imported. For example, consider sending a signal `updateremotevariable` to trigger updating of an exported variable in a server process:

```
OUTPUT updateremotevariable(newvalue) to p;
...
TASK x := IMPORT(remotevariable,p);
...
```

The SDL semantics imply that when the exported variable later is accessed, the new value will be returned provided that the serving process actually has consumed the signal and thereby updated the variable. However, it is actually a bad programming practice to rely on whether the receiving process is in a state where the signal will be consumed (rather than saved). Instead, a remote procedure should be used for updating the variable:

```
CALL updateremotevariable(newvalue) to p;
...
TASK x := IMPORT(remotevariable,p);
...
```

2.3 The Open Versus the Closed System Approach

Often, it is feasible to regard some part of the system environment as being part of the system. For example,

1. when a resource *logically* is part of the system, but *physically* is part of the network resources,
2. when an interface is of minor importance and therefore should be hidden,
3. when formalizing informal text results in access to external resources, or
4. when it allows for additional tool support. For example, for simulation or verification purposes, it might be preferred to include as much as possible of the behaviour in the system.

To regard some part of the system environment as being part of the system is in this document called *the closed approach*. This is as opposed to the usual *the open approach* where the access to an external resource is visible (as signals and remote procedures) on the channels leading to the system environment.

Consider, for example, access to a phone book which logically can be regarded as part of the system, but which during execution will result in accessing some external database. By including this entity of "secondary importance" as a process in the system, the interfaces can be significantly simplified.

Another example is a system which models a network and therefore includes a number of connected computers. Starting the network system does not necessarily mean starting the computers since they are systems of their own and possibly running whether or not their network connections are working.

Note that for composing a system of other systems in this way, the SDL extension proposed in Sect. 2.5 can conveniently be used.

Usually, a combination of the open and the closed approach is used. For each entity in the system on which the closed approach is used, the entity must be bound to an existing entity before the system is started.

2.4 Guidelines for Use of CORBA Platforms as Underlying SDL Systems

Below, the stepwise approach for production of an SDL specification as described in the *SDL methodology guidelines* ([1]) is extended to cover the additional considerations implied by using CORBA as underlying execution architecture for SDL. Note that just the additional guidelines are described; for more details, the reader is referred to [1].

Step 1 - System boundary *This step is concerned with delimiting the system from its environment and defining the interfaces to the system.*

1. Define the boundary between the system and the environment in an optimal way taking into account readability and intuition. The boundary need *not* necessarily reflect the physical boundaries since any combination of the *open approach* and the *closed approach* can be used (see Sect. 2.3).

2. Define every *signal* and *remote procedure* leading to/from the environment in a package such that the same definition can be used by other systems. Note that interfacing the environment with remote procedures is an extension of SDL. This is described in Sect. 2.5. Note also that with respect to identifying the PId values of destination processes, the same considerations must be taken for inter-system communication as for communication within a system, i.e. either a unique communication path must be used or the PId value of the destination process must be obtained before the communication takes place. The PId value may for instance be obtained through a global database process (a *trader*).

3. General rule for all steps: Use remote procedure identifiers in interfaces the same way as signals in accordance with the change proposed in Sect. 2.5. When using a commercial SDL-92 conforming tool, the remote procedure identifiers must be enclosed in comments, but the SDL/CORBA tool (e.g. see Sect. 3) must be able to handle the remote procedure identifiers.

Step 2- System structure *This step is concerned with defining the blocks of the system*

1. A block may also denote an autonomous unit (a sub system) which may execute independently and possibly exist before the system is started. This could for instance provide access to a system (e.g. a database) which for reason of simplicity or abstraction is considered part of the system (using the closed approach as described in Sect. 2.3) rather than part of the environment (using the open approach).

Step 3- Block partitioning *This step is concerned with partitioning the blocks into subblocks*

1. If a block denotes a (possibly already executing) sub-system, use the specification of that system as a block substructure, according to the proposed mechanism described in Sect. 2.5. If the SDL tool used does not support the mechanism, some preprocessing of the PR form or the CIF (Common Interchange Format) is required (replacing **system** by **substructure** and connecting the subchannels to the enclosing block's channels.) The binding of the block substructure to existing processes will take place during distribution configuration.

Step 4- Block constituents *This step is concerned with defining the processes for each block*

1. For convenience, some process instances which are part of the system, may exist beforehand and be shared by several systems (e.g. the user terminal). This is similar to the mechanism of composing systems of already existing sub-systems. But a process is a smaller unit than a system and a process corresponds more directly to a client/server process in CORBA. If the source specification of the process is not available (e.g. if the process is implemented using another programming language, like C++), a dummy process is inserted. The binding of the process instance to the existing process will take place during distribution configuration.

Step 5- Skeleton process specifications *This step is concerned with defining the skeleton of the process behaviour*

1. Identify the parts of the process behaviour which cannot conveniently be implemented using SDL. Define external procedures (the mechanism to do this is described in Sect. 2.5). During Steps 6 and 7, the external procedures are formalized using another language (e.g. C++).

The remaining steps (6-8) are not extended.

2.5 Adapting SDL to CORBA

In this Sect. three extensions to SDL are proposed; they are very important for harmonizing SDL and CORBA. The extensions are of general use and some of them have in fact been discussed earlier in ITU. The extensions are: harmonizing signals and remote procedures, harmonizing systems and block substructures and introducing external procedures. The first proposal is not real extension to SDL since the proposal harmonize existing concepts and thereby make SDL easier to learn.

In the following, the extensions will be described briefly.

Harmonizing Signals and Remote Procedures Remote procedure and variable specifications serve the same purpose as signal definitions: to define a communication primitive. However, when the interface between SDL entities are defined (using gates, channels and signal routes), only signals are mentioned. Another difference is that remote procedures/variables cannot be used to communicate with the system environment. Treating signals and remote procedures/variables equally will make the two concepts more consistent and thereby simplifying SDL. Another consequence will be the elimination of the remote procedure input and remote procedure save concepts. A detailed proposal can be found in [4].

Harmonizing Systems and Block Substructures The SDL view of an SDL system is that it is a self-contained entity which communicates with the its environment by means of signals. However, often more than one view exist. Consider, for example, a computer. It can be regarded as a well-defined system on its own, but it can also be a component of a larger system (for example a computer network). It would therefore be useful to allow composition of systems in SDL. This could be achieved by allowing a system definition to be used as a block substructure as systems and block substructures have almost the same structure (they both consist of a number of blocks connected to each other and to its environment with channels).

Specifically, it would imply that

1. A system type can have gates (this would also mean harmonizing with block types, process types and service types)
2. A block substructure can consist of a system type instantiation or it can have a system definition as **referenced** definition.
3. The system concept can be merged with substructures.

External Procedures As SDL is a specification and high-level implementation language, there will be situations where the service cannot or should not be described in SDL. Such situations include:

1. when it is convenient to reuse some existing code, written in another programming language,
2. when interfacing to an external program (e.g. Windows) and the external program does not define a language binding to SDL (typically, an external program demands C++ or Pascal to be used, if an external interface is supported at all),
3. when low-level code is needed, such as in development of device drivers, or
4. when accessing pointers.

In some cases, this can most convenient be done by defining a separate system (the open system approach, see Sect. 2.3) which is developed using another programming language, like the CORBA supported language C, or by defining one of the processes in the system using another programming language (the closed system approach).

To access external code locally in a process, it is more suitable to call a procedure which is developed using another programming language. This can be indicated by allowing the keyword **external** in referenced procedures and operators contained in packages. This keyword is to be used instead of **referenced**.

SDL has for example no built-in features for ordering of character strings, so a C library function could be added somewhere:

```
PROCEDURE strcmp FPAR op1,op2 String; RETURNS Integer; EXTERNAL;
```

For use with a CORBA platform, the alternative programming language must be one for which a language binding exists (e.g. C or C++) and any argument data type should be defined in a package such that the data type can be transformed from SDL, via IDL, into the alternative programming language for access when the external procedure is developed.

3 SCE Implementor's View

This section describes an implementation of an SDL compiler for the SDL-oriented approach for combining SDL with CORBA. (see Sect. 2.1).

Although the compiler has been specialized to support a specific platform (Orbix, see ([6]), it is expected that with minor modifications, the compiler can be generalized to support most of those CORBA platforms which support a language binding to C++ (rather than C)[2]

In the following, the mapping will be described with focus on the CORBA specific parts. The realization of the SDL state machine will not be described at all, since this part is not specific to CORBA.

3.1 Mapping Principles

The mapping from SDL to C++/C and IDL is shown in Fig. 3 for an SDL package and in Fig. 4 for an SDL system:

1. One IDL file is generated for each SDL package and one IDL file is generated for the SDL system. Each IDL file contains a oneway interface definition for each signal definition, two (oneway) interface definitions for each remote procedure (one for each direction, see Sect. 2.2) and one for each remote variable. For each data type definition and synonym definition, a corresponding IDL data type definition or constant definition is generated.
2. One C++ file is generated for the SDL *system process*. This process is responsible for:
   (a) finding a suitable process PId value whenever an SDL process sends a signal or calls a remote procedure without mentioning a destination process,
   (b) verifying that processes only communicate with the processes to which there exists a communication path (gates, channels and signal routes), and

---

[2] only the language binding to C is part of CORBA, but many CORBA platforms also supports a language binding to C++.

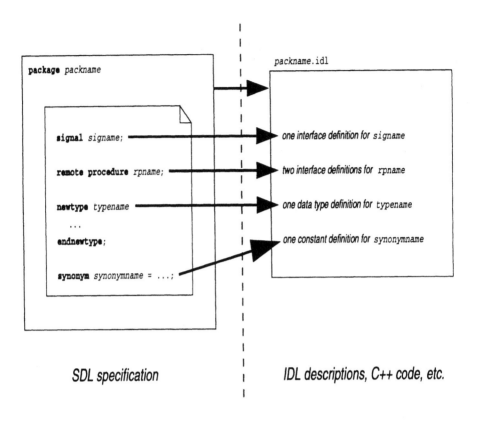

Fig. 3. Files generated from an SDL package

(c) Starting new processes. The SDL system is started by starting the executable file for the system process. The system process will then start all the processes which must exist initially. The system process uses the *configuration file* to determine on which machines the individual processes should run. An initial configuration file is generated by the compiler (to be edited manually).

The system process, in conjunction with the Orbix *locator* eliminates the need for a *trader* (see [4]).

3. For each SDL process instance set, a C++ file is generated. A new process instance in the set is started by starting the corresponding executable file for the C++ file (done by the system process).

4. The makefile takes care of compiling the IDL code into C++ code (the generated header files are included in the C++ files for the processes) and then compiling and linking the system process and the SDL processes.

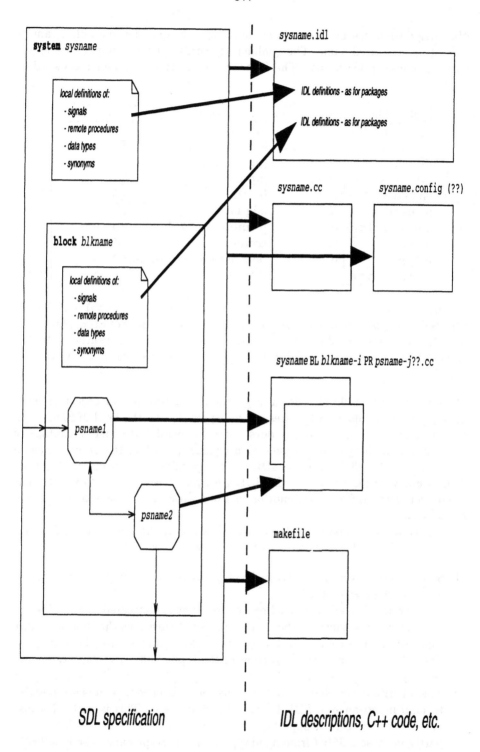

Fig. 4. Files generated from an SDL system

Naming Conventions To avoid name clashes, the qualifier of an SDL name is part of its generated name. The qualifier keywords appear in upper case letters and the names in lower case. The following translation of qualifier keywords is used:

Entity kind	Abbreviation	Entity kind	Abbreviation
system		block type	BT
package		service type	ST
block	BL	process type	PT
process	PR	operator	OP
service	SE	path	PA
procedure	PD	signal	SL
substructure	SU	remote variable	RM
type	TP	generator	GE
signal	SI	variable & synonym	VA
system type	ST	literal	

For example, a process instance set pset defined in block psetblock which in turn is defined in system psetsys has the name psetsysBLpsetblockPRpset. In most cases, the user need not be concerned with the naming conventions. But the names of the generated files and the configuration file follows the conventions.

Observations SDL-92 includes a large number of features for object oriented programming. While such features are indispensable on the level of SDL, both for structuring and for reuse, it does not seem possible to take advantage of the features in the generated code. Consequently, most of the object oriented features must be "flattened" to SDL-88-like specifications during compilation. This is not very important for the SDL-user, but it is rather unfortunate that the object oriented features cannot be retained without considerable decrease in performance.

This conclusion is based on the observation that an SDL process type cannot be compiled into a C++ class. This has several reasons:

1. SDL processes are active (they have behaviour of their own) while C++ objects are passive objects.

2. Input transitions and exported procedures cannot be treated as methods of a C++ class since there can be nested states in transitions through procedure calls. Another reason is that they cannot be executed directly on request since they must keep their position in the input buffer (which may involve saving).

3. The inheritance mechanism in SDL must be "flattened", i.e. it is not possible to map inheritance in SDL to inheritance in C++. This is partly because an asterisk state in a supertype also covers the states of a subtype and partly because a JOIN from a subtype to a supertype cannot be handled in a reasonable way.

3.2 Conformance to SDL

The compiler has primarily been developed with the purpose of demonstrating in practice how SDL can be combined with CORBA. Therefore, only a subset of SDL is supported and the compiler does not support full static semantic checking. However, a commercial graphical SDL tool capable of generating the textual SDL representation and having full static semantic checking, can be used as front-end for development such that the (more user-friendly) graphical representation can be used.

The SDL extensions described in Sect. 2.5 are supported by the compiler. For the extension described in Sect. 2.5 a special SDL comment can be used to keep the SDL description conforming to SDL-92 (to allow processing by SDL-92 tools), for example:

```
SIGNALLIST sl = s1,s2 /*ORB!! ,procedure rp !!END*/;
```

4 Conclusion

From the work with combined use of SDL and CORBA, the following conclusions can be drawn:

1. There are no real limitations on the use of SDL-92 for implementation of CORBA programs.
2. If SDL was enhanced with features for specifying remote procedures in signal lists (also to the environment), system composition and external procedures, the service provider could more easily take advantage of the features of CORBA.
3. The mapping from SDL to IDL/C++ is straightforward although some concepts might be mapped in a different way than expected. For example, SDL gates and CORBA interfaces are conceptually closely related, but in the generated CORBA code, there is no correspondence at all.

References

1. ITU. SDL Methodology Guidelines. Appendix 1 to Recommendation Z.100, June 1992.
2. P. Christensen, B. Nørbæk, O. Færgemand, A. Olsen. Exception Handling in SPECS-SDL. Contribution from SPECS to the CCITT meeting in Malmø, April 1991.
3. OMG. The Common Object Request Broker: Architecture and Specification. Object Management Group, Inc., Framingham, MA, U.S.A., December 1991.
4. SCORE-METHODS AND TOOLS. Report on Methods and Tools for Service Creation (Final Version). Deliverable D204—R2017/SCO/WP2/DS/P/028/b2, RACE project 2017 (SCORE), December 1994.
5. ORBIX distributed object technology - Programmer's Guide - Release 1.3.1, IONA Technologies Ltd., February 1995.
6. ORBIX distributed object technology - Advanced Programmer's Guide - Release 1.3.1, IONA Technologies Ltd., February 1995.

An Engineering Approach for Open Multimedia Services Management

Munir Tag, Telesystems
Paris, France, mt@syd.synergie.fr
Arek Lesch, Alcatel SEL
Stuttgart, Germany, alesch@rcs.sel.de
Phil Fisher, MARI Group Ltd.
Ashington, United Kingdom, phil.fisher@mari.co.uk
Alex Galis, UCL - Electrical Engineering Dept.
London, United Kingdom, a.galis@eleceng.ucl.ac.uk
Soren Sorensen, UCL - Computer Science Dept.
London, United Kingdom, s.sorensen@cs.ucl.ac.uk
Spyros Batistatos, Intracom SA
Athens, Greece, sbat@intranet.gr

Abstract : The development of systems according to the principles of Integrated Service Engineering (ISE) encompasses advanced multimedia services and usage scenarios, services management and configuration, platforms supporting service provision and management operation, and broadband network infrastructures. Important developments have been accomplished, concentrated on a specific aspect, in various RACE projects and demonstrated in appropriate established field trials. The main objective of the DRAGON project is to demonstrate the practical application of Service Management within the framework of the Open Services Architecture (OSA). The project is following an integrated service engineering approach, based on the concepts of OSA System, embracing multimedia service components being available to the project, a platform responsible for the interoperation of these components in a distributed environment, and wide-area ATM communication services representing the IBC network infrastructure provided to the potential users of the multimedia services. This paper is aiming to describe the service engineering approach followed in DRAGON, that can be generalized for the development of open systems for service deployment and provisioning over various infrastructures.

Keywords: Service Architectures, Service Machine, Services Management, Service Control Elements.

1. Service Design Architectures

The development and provision of services in an open environment, encompasses all activities from requirements analysis to service delivery supported by Service Design and Provision architectures (TINA-C, OSA). An essential constituent of the architectures is a basic structuring concept, used by service designers to capture the properties of the telecommunication services under design. The proposed structure for modelling the DRAGON services and management components has been derived after investigation of the OSA and TINA-C component models. Furthermore, concepts related to OSA System have been introduced to design the environment for service deployment and provisioning. Such environments provide abstractions of distributed processing, storage, telecommunications resources and their functional cooperation.

The **OSA component** is defined [1] as an abstract or concrete entity providing a set of services. It can be characterized by:

- something useful to be done, called Mission, representing the core service functionality.

- a set of constraints on the intended usage and life support, on the supporting environment and the used resources, representing the Ancillary facets.

Under the mission and ancillary facets, service primitives are grouped. In the perspective of a service designer, an OSA component model looks mostly appropriate to capture the semantics of the problem domain. During the development process, the component behaviour can be observed and refined onto an object model.

The **TINA-C component** is a classification scheme [2] for the components of a TINA compliant service. All services are analysed as consisting of :

- a Core, describing the nature of the service and its primary value to a user.

- an Access layer, ensuring that the core component is independent of a specific environment.

Within the Access layer, the Usage sector includes the service interfaces, the Substance sector includes the interactions and dependences upon external resources and other services. Management aspects are also included in this layer. Originally, the services are identified under the assumption of an idealized service environment (Idealized Level). In the following stages, they are expanded to cope with a realistic and physical environment (Pragmatic and Realized level).

At the time of demonstrator design, neither of the approaches had explicitly worked out component refinement, through all the stages of the service life-cycle they target. For instance, the OSA-component deployability is defined but not applied in the context of real systems, while the transformation of TINA services from the Idealized to the Pragmatic and the Realized levels had not been reported. For those reasons, DRAGON followed a practical modelling structure for the specification of the service and management components, with the following aspects :

- Access Interfaces : the interfaces must be identified with their attributes.

- Behaviour : the effect of each interface must be specified, referring to attributes involved, conditions and restrictions affecting each operation.

- Interaction : the interaction with the required resources of the distribution platform, the network and other components, must also be identified.

These can be thought as relevant to the TINA Usage, Core and Substance aspects, but also with the OSA Mission and Ancillary facets.

For the computational viewpoint of an OSA System [1], the **Service Machine** is a key concept used to describe a set of already deployed services or services of the underlying resources, that can be taken as available functionality to enable the deployment and provisioning of new or enhanced services. To fulfil its role, the Service Machine needs to involve functionality that caters for the following:

- to enable access and adaptation to, management and control of the resources of the underlying infrastructure (Nucleus).

- to represent services of the underlying resources (Resource support).

- to support distributed processing (Distributed Processing).

- to consider the already deployed services, which may provide a supporting role to the deployment of new services (Service support).

An OSA System is engineered [1] as a network which is a set of interconnected **Service Nodes**, defined as a logical entity representing a non-distributed configuration of information processing and storage resources, residing in an identifiable physical component. The functionality hosted by a Service Node should necessarily include Nucleus, Distributed Processing and Service Support functionality, as reflected in figure 1. A deployable service is introduced in a node, in order to get actually deployed in the OSA system. Although functionality hosted by other Service Nodes may possibly be employed, the end-user perceives the service as offered by the node, from which it was requested.

At the time of demonstrator design [4], the knowledge about the Service Machine had not yet been presented in a solid and consistent form. There was not a complete set of rules and guidelines regarding OSA and the definition of deployability was not clear. For those reasons, the DRAGON demonstrator implemented based on requirements extracted from the available applications, following the OSA system concepts.

2 Multimedia Services, Models And Management

The demonstration of the ISE principles can be achieved with the realisation of service applications which can exploit the high bandwidth and improved quality of service provided by the IBC environment. Those services must be capable of exercising the configuration and management features of the DRAGON demonstrator. Multimedia services offered in the project [3] provide the users with extensive access to hypermedia documents (joint navigation, joint composition, telepointing) in conjunction with conference services (audio conference, video conference, shared white-board). This combination provides a multimedia environment for advanced teleworking sessions integrated on a hypermedia terminal.

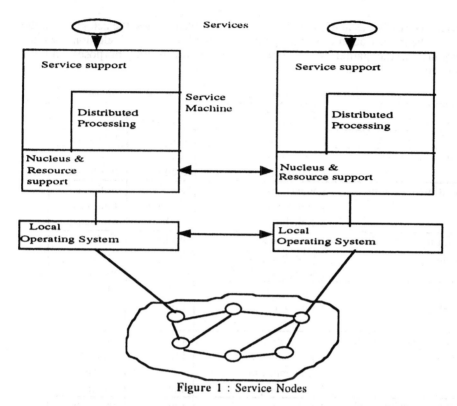

Figure 1 : Service Nodes

Hypermedia services : Joint navigation facilitates the behaviour that each request for presentation of multimedia documents is performed on the screens of all participants of the session in parallel, assuring consistent view on the set of hypermedia documents and co-operative work with them. Joint composition implies that new hypermedia documents can be defined interactively from monomedia objects, creating new links and storing. Telepointing provides a virtual pointing device which may be used by all participants to gain temporal control of the pointer, visualised by colour changes of the arrow displaying the telepointer.

Conference services : The video facilities adopted from the MICE project, providing multicast services to the participants of the conference. The specific tools support a range of options for codec types, compression and multiplexing techniques. The audio facilities are more sensitive to quality aspects, providing multicast services from the originator to the receivers. The shared whiteboard provides a shared surface for diagrams sketching and composition, that can be placed in multimedia documents. The telepointer services can be utilised for positioning and drawing. Table 1 summarizes the service descriptions available during the initial demonstrations of the project :

The first step, in the demonstration engineering approach, is to derive a scenario which in a later stage will enable to identify the requested service management functionality and realise the supporting environment, represented by a Service Machine. In general, the service user is faced with several activities, including the selection of the required service according to constraints, the invocation of the service, the session establishment between the involved parties, the actual use of the service, the modification of service parameters, the invocation of additional service components, the gathering of statistics mainly for billing purposes and the termination of the service.

Service name	Components		
	Name	Quality	Mode
audio conference	audio tool	high	multicast
video conference	video tool	low	multicast
	audio tool	medium	multicast
multimedia seminar	video tool	low	multicast
	audio tool	medium	multicast
	whiteboard	high	multicast
joint reading with audio	audio tool	high	multicast
	joint navigation	high	unicast
	joint pointing		unicast
joint reading with video	video tool	low	multicast
	joint navigation	high	unicast
	joint pointing		unicast
joint authoring with audio	audio tool	high	multicast
	joint navigation	high	unicast
	joint pointing		unicast
	joint composition	high	unicast
joint authoring with video	video tool	low	multicast
	joint navigation	high	unicast
	joint pointing		unicast
	joint composition	high	unicast

Table 1 : Service descriptions for the DRAGON demonstration

Within the RACE Project 1044-CSF, service models have been defined to support session management for establishing and managing multimedia, multiconnection and multiparty service sessions [6]. According to the service model as prescribed in the CSF, parties are connected through tasks, each task being made up of components. **Service control elements** (SCEs) are used to control services based on this model and provide flexibility to service users. Four pairs of SCE provide full service control (Setup and Release, Activate and Deactivate, Allocate and Deallocate, Join and Leave) with others providing additional flexibility (Modify, Report, Status, Main). In DRAGON, this model, has been improved to handle the requirements of the demonstrator for billing and trading, as depicted at figure 2.

User agent : represents the terminal user in the scenario, providing facilities to define service requirements from the User perspective. This agent initiates a dialogue with the Trader, trying to find a match between these requirements and the available services. Interaction with other users, invocation and configuration of services is accomplished through this agent. From the corresponding interface, the required service components and the parties involved in the specific service, are selected.

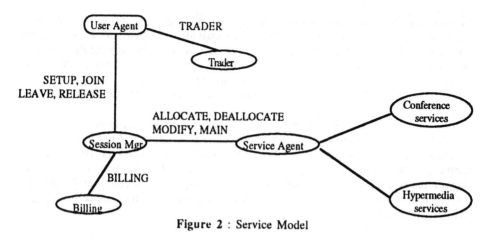

Figure 2 : Service Model

A message is sent to the Trader containing the required service components and quality of service for each component. Upon the receipt of the Trader reply, a message is forwarded to the Session Manager requesting the setup of the multimedia session, with the selected service components initiated between the participants. A new party can be involved in a an existing session, if a message is forwarded to the Session Manager requesting the new participant to join the multimedia session. For the session release, a relevant message is forwarded to the Session Manager.

Trader : the principal function of the trader is to match service offers with user requirements and provide the necessary interface reference for accessing the service in a distributed environment. It is a repository of information regarding available services, in which service providers register the services and the constraints applying to those services. An interface reference must contain sufficient information to allow the user to establish communication with one particular service instance.

The Trader is cooperating with the User Agent, in order to locate the required service components with specific quality of service. It searches the database and responds to the User Agent with a message, containing addressing information of the respective service components satisfying the user requirements.

Session Manager : it is the central part of the scenario, undertaking the establishment and management of multimedia service sessions, between the service participants. After performing user request validation, an instance of a session controller is created to handle the specific session and related information (session identifier, participants, components, startup time etc.) are forwarded to the Billing component for accounting purposes.

During the startup of the multimedia session, messages are being sent to the Service Agents to allocate the selected service components, containing the session identifier of the specific service. The proper Service Agent is located from the interface reference information extracted from the Trader. When a new party is wishing to join the multimedia session or the session release is requested, messages are being forwarded to the Service Agents. The added value of the Session Manager is to record the session, mainly for accounting and billing purposes.

Service Agent : the primary role is to provide a clear interface to the available multimedia service components. More specifically, to provide a conversion mechanism for the control messages of the service model, towards service specific actions carried out by the respective service component. In other words, this agent is calling the interface of the hypermedia and conference tools. At this point the propagation of data between the participants is started, modified or stopped.

Billing : it is the name given to the practice of recovering financial costs from end-users of the services deployed over the network. Subscription and usage charges are accumulated over a standard period of time and billed to the customer. Accounting is normally applied only to the settlement between different network operators, where they mutually cooperate to provide services. The purpose of Accounting and Billing Management system is to store all accounting data necessary for service billing and deliver appropriate reports.

3 Demonstrator Design

Distributed and object-oriented processing are key elements of the ISE technology approach. Therefore, distributed platforms are being used to facilitate the development of telecommunication services as object groups, distributed and executed at the various nodes of the underlying network. They are offering a uniform interface to the underlying network and operating system, enabling the transparent building of the telecommunication services from the actual resources. The current trends in the Information Technology world have been explored in DRAGON, taking into account the telecommunication problem domain. The best known products or standards in this category have been investigated (DCE, ANSA, CORBA).

DRAGON selected the **DOME platform**, which is based on the CORBA standard. This platform from the Object Oriented Technologies [7], provides a class hierarchy that prescribes distribution support and adaptation to the underlying resources. Client objects hold a repository of remote server instances to access, in the corresponding Object Request Brokers (ORBs). Furthermore, the RPCPeer class allow ORBs to be constructed and communicate over different networks. The lower layer of the toolkit provides the Connection class which support the transmission of information through Socket, among the other, network interfaces.

The distribution support of the DOME platform is achieved with the generation of multiple instances that invoke methods on a remote server. These multiple instances are derived from the class DObjectServer, that in turn is derived from the ORBcore class. For a specific invocation, a request is generated containing the method and remote server identifiers. This request is dispatched by calling the ORB method received-request, at the client side. The identifier is searched in the ORB data structure, at the server side. As a result, the dispatch-request method of the corresponding server class is called and the requested method is executed. The location of server instances with more complex properties, is undertaken by the Trader component.

The adaptation to the underlying network resources is achieved with the provision of the class Connection, that undertakes to dispatch requests between the local and remote RPCPeer. The class SocketConnection generated for a specific port-number and host system, is derived from the Connection class, that in turn is derived from the RPCPeer class. Methods exist to handle sockets (open, close, accept) and transmit requests (read, write) over the sockets.

Having derived the usage scenario for the multimedia services of the project, the next step in the service engineering approach is to specify the service support environment [5]. In other words to provide a computational view for the service model, covering the service selection, invocation and termination, the available service components and the distributed platform. For that purpose, the DRAGON Service Machine was designed including supporting components as implied by the service model. These components were specified in terms of the Access Interfaces, Behaviour and Interactions. At the development phase, they refined to objects derived from the ORBcore class, with object interfaces specified in CORBA Interface Definition Language (IDL). For each case, a DObjectServer was generated, as C++ class library, to access the remote instances at the operation phase.

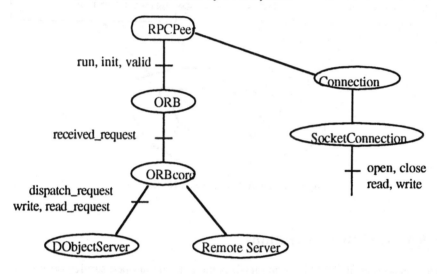

Figure 3 : DOME class hierarchy

The demonstration network, as depicted at figure 4, is comprised from nodes interconnected with ATM links. The Solaris 2.4 operating system including the TCP/IP protocols, is supporting the SocketConnection interface of the DOME platform. The IP protocol is configured to transmit datagrams over the ATM Adaptation Layer 5 layer of the network. The role of this layer is to map the packet data units into cells and vice versa. For the time frame of the project, the European ATM Pilot will offer virtual connections for the demonstration.

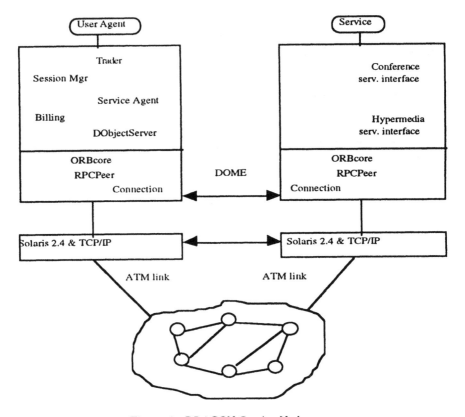

Figure 4 : DRAGON Service Nodes

4. DRAGON Demonstrations

The demonstration of DRAGON emphasizes the benefits in open service provision, emerging for the various stakeholders (service developers, service providers, network operators, end users). The User Agent and the multimedia services are the only visible part to the end-user of the demonstrator. At the end-user terminal a window presentation of the User Agent is provided, enabling the selection, configuration and activation of service components. After that, the user defines the parties for the session using also the drag-and-drop interacton principle of the RUS-editor (Reference for User Services).

The activities of the session manager and the trader are not directly visible to the end-user. During the session establishment, a complex control message flow between the various components is supervised by the session manager. The user takes advantage of the service provided over the IBC network. In order to visualize the operation of the Service Machine, monitoring of the control messages can be achieved with the DEMON tool of the BOOST service creation environment. In this way all the implemented components, their interactions and behaviour can be demonstrated.

5. Conclusions

The service engineering approach for the design and development of the DRAGON demonstrator, has closely following the ISE principles and especially the guidelines for the Design and Provision Architectures (OSA, TINA-C). The first step, in this approach is the derivation of a scenario from the User perspective, which in a later stage enabled to realize the supporting environment represented by a Service Machine. The DRAGON service scenario has been applied to six basic service components (video conference, audio conference, shared whiteboard, joint navigation, joint editing, joint composition). The Service Control Elements (SCE) provide a method for interchanging control messages supporting the necessary interactions between components. The message body, originally developed as part of the RACE CSF project, was enhanced to incorporate trading and billing requirements.

The next step in the engineering approach, is the computational view for the scenario, including the supporting components (User Agent, Trader, Service Agent, Session Manager, Billing), implied by the scenario, and the Distributed platform. User Agents are representing the user terminals for management and control in the whole scenario, while Service Agents provide the core service to the users. The latter are not involved in the actual propagation of data streams but only control them. Both agent components are crucial for a realistic demonstration and their definition must proceed towards standardization. The Trader is a repository of information regarding available services, in which service providers register the services and the constraints applying to those services. The supporting components specified with a practical modelling structure relevant to the OSA and TINA-C component models.

The selected distributed platform DOME, is CORBA compliant, allowing the DRAGON demonstrator to be relatively independent from the distribution mechanism and the underlying network. The same approach is being taken by other major developments including the TINA Concortium. The final step, deals with the implementation of the specified components, that subsume lower level functionality and mechanisms, over the particular operating system, transport protocols and network interconnections at the demonstration Service Nodes. Both the multimedia service components and the management components are based on the TCP/IP protocol and they have been tested at the operating systems Solaris 2.4 which is the target platform. Testing and demonstrations have shown the ability to provide multimedia applications and management, over an IBC network. The whole engineering approach resulted to the efficient and open design of the demonstrator.

6 References

[1] RACE R2049 Cassiopeia : *Open Service Architectural Framework for Integrated Service Engineering,* Draft deliverable, R2049/FUB/SAR/DS/I/23/b1, Nov. 1994.

[2] H. Berndt, P. Graubmann, M. Wakano : *Service Specification Concepts in TINA-C.* Proceedings of the IS&N Conference '94, Springer Verlag, Sept. 1994.

[3] RACE R2214 DRAGON : *Specification of DRAGON Demonstrator*. Deliverable 2, R2114/MARI/MCS/DS/P/002/b1, Sept. 1994.

[4] S. Batistatos, *"Use of OSA Principles for the DRAGON System Design"*. Presentation at the Joint RACE/TINA-C Workshop on Service Engineering, Brussels, Nov. 1994.

[5] M. Tag, *"Realising a minimal Service Machine - The DRAGON Experience"*. Presentation at the Joint RACE/TINA-C Workshop on Service Engineering, Brussels, Nov. 1994.

[6] RACE R1044, *Customer Service Functions (CSF)*. 44/RIC/CSF/DS/A/009/b1, Sept. 1991.

[7] Object Oriented Technologies Ltd., *The DOME Toolkit*. Issue 2.1D, 1993.

User Perspectives on Service Engineering

Session 6 Introduction

Chairman - Don Cochrane

Cray Communications
Cray Communications Ltd
Caxton Way, Watford Business Park
Watford, Herts WD1 8XH
Tel : +44 1923 258763
Fax : +44 1923 258890
email: don@case.co.uk

The progressive migration of telecommunications from bandwidth-limited simple telephony towards an IBC environment is exemplified by high speed islands interconnected by broadband links. The emphasis of development has moved from the simple provision of raw bandwidth towards consideration of exactly what the user is going to do with all this available resource.

This has led to the realisation that users do not want and should not have to be concerned with the exact mechanisms that are used to move their information around. Users have their own businesses to run and would rather not have to be telecommunications experts when their business is producing, for example, car designs. Users need telecommunications services to enable them to carry out their businesses which are in turn becoming, in some cases, totally dependant on the provision and delivery of high integrity, high availability services. The business of the telecommunications industry is to provide and deliver these services which range from the simple to extremely complex. To provide such services in a manner that users will buy, providers need to describe the services fully in order for them to be implemented in an economic and predictable manner.

Alongside the basic specification of a connection, or service usage instance, is the question of how the user achieves control over the connection and of the service features associated. The first paper of this session addresses this topic. It is from the ASCOT project, addressing the use of Integrated Service Engineering (ISE) as applied to In-Call control of services. A graphical approach has been taken and implemented as a toolset. This has, in turn been adopted by other RACE projects. The paper reports the implementation of a toolset and its success in controlling a security monitoring application involving remote video surveillance across B-ISDN channels.

At the same time as the expansion of the concepts of service has come an injection of ideas and concepts into telecommunications thinking from the computer science and information technology fields. These concepts address a wide range of aspects of telecommunications and reflect the pervasiveness of computers in the implementation of telecoms facilities.

The second paper of the session addresses the application of a particular technique, OMT (Object Modelling Technique). This was originated in General Electric in the US by James Rumbaugh et al and is well-established in general purpose programming development. It is being applied in several RACE projects to different aspects of telecommunications. RACE project SCORE has contributed this paper where the work was done while addressing Methods and Tools for Service Creation, especially for high-level analysis of services. This paper describes its use for telecommunications services, bringing together the traditional telephony requirements of IN Capability Set 1 with OMT. The discussion is detailed and concludes by addressing the problem of how object models can contribute to the detection of interactions between services and features (the Feature Interaction problem).

The third paper addresses another important consequence of the expansion into services; that users can and do now demand guarantees of a group of parameters such as availability, error rate, lost data, delay, etc. These parameters are collectively termed Quality of Service (QoS). This field of study has advanced considerably during the period of RACE. At the start, it was not even clear that users could perceive QoS and all QoS was believed to be entirely subjective. These seemingly archaic views are only five or six years old! The clear distinction between QoS and Network Performance and the description of techniques for QoS measurement and validation have dispelled these views for good. Another major advance is in the steady migration of subjective QoS to become objective QoS as work is done to apply instrumentation to hitherto unmeasurable parameters.

This paper reports the implementation of QoS Verification and Validation. A two level verification is described; firstly a simulation is performed to take the service specification and from it predict QoS levels. Secondly a measurement technique is applied to the service implementation to verify the QoS levels. To facilitate this, a method for generating relevant test scenarios is also described.

An Approach to User Management of Broadband Services

Paul Coates & Jeremy Ellman
MARI Computer Systems Ltd,
England

Abstract. This paper describes an approach to the In-Call control of services as developed by the ASCOT project. Emphasis is made on the application of usability principles to the practical problems presented by real service deployment within an Integrated Service Engineering environment. The ASCOT approach to the graphical control of services in-call will be described including details of practical experiments and any conclusions drawn.

1 Introduction

Broadband communications offer ordinary users the possibility of virtually limitless bandwidth. This in turn will potentially support a huge number of services. These services will have far greater variety than services in use today presenting systems designers with a significant challenge: Service architecture must be extensible to new service types, and use and control of new services must be "intuitive".

Whilst this may be an impossible goal, at the very least, services must be controlled in a common way. The situation has an exact parallel with the design of the control mechanisms in a car, a typewriter keyboard, or a computer operating system. Even if the skill is difficult to acquire it is transferable. As long as services have idiosyncratic controlling mechanisms their uptake by the general public will be limited.

The problem this paper addresses is consequently an approach to the design of a generic services control mechanism that reflects telecommunications engineering needs. A key aspect of this approach is usability. The general public must be able to learn the control mechanism, and be able to apply it to a variety of services.

In this paper we shall also review a generic view on Broadband Service Engineering within the RACE program known as Integrated Service Engineering, including the implied constraints this has for end user service control. We shall then go on to the user management of services that supports ISE. Finally we shall report actual experimental trials of the approach and conclude with a discussion of emerging ideas.

2 Integrated Service Engineering

Integrated Service Engineering (ISE) is the preferred RACE viewpoint on services. It is an architecture that is designed to allow the development and deployment of new services and to support the open market in telecommunications which is one of the policy goals of the EU. If there were no common approach for building and managing services then every service provider would have to provide its customers with different tools to use their services.

The Open Service Architecture (OSA) model of services uses an ISE approach, and was originally developed by ROSA (R1093) and further developed by CASSIOPEIA (R2049). It describes services in terms of parties which communicate using service tasks, each service task consisting of one or more service components, operating with a specific quality [CSF C110]. Figure 1 shows the relationship between these different parts of the service description.

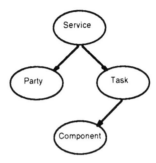

Fig. 1 The Service Model

The architecture is made up of several levels. At the lowest level is a point to point model of the telecommunications network that routes communications between network access points. For the sake of argument, we assume that any network access point can access any other network access point. Services are therefore universally accessible.

Services are defined in a object model that collects together service components. These components contain audio, video, etc. data types, and are further characterized by the service standard they support such as MPEG-2 for video, or JPEG for picture, etc. Depending on the data type, they will also contain other parameters such as a data stream is bursty, or components are required on demand, etc.

It is important to note that service users may include automatic devices (such as video sources, or computers) in addition to people. Therefore the concept of a service party, representing both people and machines, is used rather than end users.

Using a service implies the instantiation of a service object and making the connections in the network between service parties and their service access points. The service is then controlled using a generic signaling mechanism called service control elements [CFS C220].

Integrated Service Engineering presents a huge problem to the designer of a generic service user interface. In order to be flexible, the ISE approach explicitly includes huge quantities of information. At the very least it is unnecessary for users to see all this information, and at the worst it is would be extremely confusing for them. By taking an object oriented view of ISE, information can be selectively hidden.

Information Hiding is however a great advantage of the modern object oriented programming paradigm. This permits different views to be presented on the same information depending on who wishes to view it. Thus, an end user may only wish to see the parties in his call, whereas service engineers may wish to see the lowest level components, and their connection types.

3 The Ascot Approach

The ASCOT project has devised an intuitive view of services, that is easier to understand and control than a menu driven interface. To keep the interface simple non-essential information is hidden from the end-user and processed automatically. The approach also provides in-call service control that respects ISE, is generic, and is designed for usability. In ISE users control services by generating service control elements (SCEs). In ASCOT, the user has a distributed application that generates these signals.

The advantage of the ASCOT approach is that a distinction is enforced between the service and the end-users' service control interface. This entails that future broadband service users need only learn one service interface. Service provider's on the other habd will only need their services to interpret a standard set of control signals, SCEs. The actual interface to the service will be common between applications.

ASCOT represents services to end users using a graphical editor helping to provide a simple view of services by providing intuitive icons and symbols to represent the parties, tasks and components of a service. The graphical presentation allows an end-user to describe a service without the necessity of technical knowledge about a service. The end-users view of the service is also generic so a user only has to learn a single interface to be able to control all services. The graphical editor is capable of both creating service descriptions and configuring services while they are executing.

The ASCOT graphical editor was based on work by the RACE-1 usability project URM (R1077). This "User Reference Model" was designed to show what users consider important in a service. The view of the service is based on "Reference for User Services" (RUS) diagrams. RUS diagrams represent communication tasks between geographically separate entities. An alternate menu driven interface provides a more sophisticated configuration intended for network professionals. Figure 2 shows a RUS diagram of a simple service containing three parties connected by service tasks. The service components are shows as icons on the task lines.

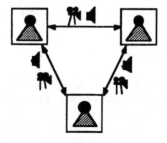

Fig.2 A RUS diagram

The RUS editor only requires a minimal amount of information from the user, and shows services in terms of parties, which are connected using service tasks. Each party is represented by a picture of the party and it's name. Each task is shown as a line between two parties. Service components are shown by small icons positioned on the task line. A service description can be built by adding and connecting parties, service tasks and service components.

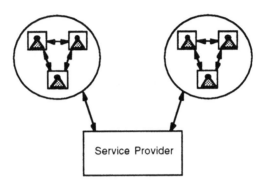

Fig.3 The ASCOT service architecture

The ASCOT toolset sends Service Control Element (SCE) messages to the service provider which runs the service. The SCEs are used to manipulate the service to suit the needs of the service parties. The service provider also send SCEs to the other parties in the service to maintain individual copies of the service description. Access control can be used to inform only those users that have permission to know about the change in the service. Figure 3 shows the service architecture used by the ASCOT project, the toolset for each party communicates with the service provider for the service being used. The service provider is being further developed by the DRAGON project (R2114).

The RUS editor was extended to provide in-call control of the service by manipulating the graphical service description. A service can be configured by selecting part of the diagram and either connecting or disconnect it from the service. Depending on the parts of the diagram selected the appropriate Service Control Element messages are sent to the service provider and the OSA service description is adjusted accordingly. Colour is used to indicate whether an item is connected or disconnected from the service. The quality of a service component is the only other property that can be changed using SCEs and is performed by selecting the appropriate component from the RUS diagram and using a quality slider.

4 Information Hiding

The ASCOT toolset provides a generic interface which can be used to control all services. An end-user only needs to learn how to use a single interface to be able to control all types of service. For some services it may be easier to use an interface specific to that service which removes any additional functionality required to handle other service types needed for a generic interface. A specific interface can also provide a simpler, more controllable interface to the user.

For a generic interface to become useful a degree of information hiding must be introduced, common interfaces should be as easy and powerful to use as possible, and information not relevant for a specific service should be hidden from the user. Information hiding is where all the technical details of a service are abstracted away from the user, e.g. an end-user should not need to know if a service component uses a bursty or continuous data stream, this information can be deduced by other means such as the type of service component being used, etc.

Many of these different approaches to information hiding are described here,

- The graphical view of a service provides a much clearer view of the entire service description presenting the majority of the service information in a single diagram which can be understood at a glance.

- Each party and service has its own icon used to represent it in the toolset. Icons are used to maintain the graphical view of services. Icons can be sent with service descriptions between toolsets and service providers to support a consistent graphical view of a service. Services can be stored with parties as part of the service description, so any icons used should be stored with the service description. Each user should be able to replace default icons with their own icons.

- Aliases are used so each user has shorter more familiar names for services and parties. Each user maintains their own set of aliases. Since only the icon and the alias are used to graphically identify each party or service, the icon and alias combination must be unique. An end-user should never have to see a numeric identifier for a party or service as that is what the alias is for.

- Parties can be collected together into groups to perform repetitive operations in a single step. An operation on a group will be applied to each member of the group in turn. Groups also have icons and aliases to graphically represent them, similar to parties. An alternative to groups is multiple selections. When multiple items are selected in the RUS editor any operations performed will be applied to all the selected items.

- For more complex services different views of the parties, tasks and components can be shown so that only the information needed by each party is presented. This information could be selected by personal preference or by using access controls for different types of information.

- When changing the quality of a service component the user should select a particular service quality that is appropriate to the type of service component (Such as 64KHz for audio, or 30 Frames Per Second for video etc.) In ASCOT the user chooses quality from a slider that goes from high to low, and toolset automatically selects the corresponding quality for that service component type.

- More detailed features such as Mandatory or Optional service components are hidden from the user completely. They can not be influenced by the user configuration of the service and instead are infered by the toolset based on the type of service, and pre-defined default settings in the service description.

- Nested operations such as a party leaving a service. For a party to leave a service all tasks that party is using must first be disconnected. For a task to be disconnected all components in that task must first be disconnected. This single operation will therefore generate a number of different SCEs. Figure 4 shows the result of a single party leaving a service. Each SCE is propagated to the other parties in the service.

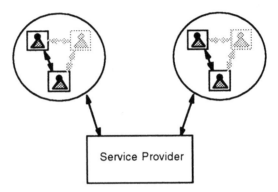

Fig.4 In-call control of services

- As more parties are added to the service the RUS diagram can become more complex and more difficult to read. There are two simple methods available to alleviate this problem, coordinates for parties can be stored with the service description, or automatic layout algorithms can be used to make best use of the available space. Automatic placement would work by applying an attraction between parties that are connected by a task, and applying a repulsion between parties that are not connected. After several iterations the diagram will settle down to it's best physical placement for the parties [GEM1]. This technique removes the need for users to worry over where to place party icons and how to arrange the parties to produce the clearest diagram.

- The toolset should be customizable. If you are using a generic interface you will be using it most of the time so it must be configured correctly for each individual user. Customizing colours, dialogue layout and different methods of entering data, do not provide information hiding, but other customizations do. Setting an experience level such as novice / expert can automatically adjust how much information is hidden from the user and how much control is assumed by the toolset. If certain facilities are used frequently with the same settings, for a more novice user this stage of configuration may be automatically performed for the user using this setting.

These approaches are practical applications of Information Hiding in an ISE context. They illustrate how the full power of ISE may be used to define services, but allow a much simpler view of the service. This has been conducive to the development of a generic, usable, service control interface.

For a useable system a fully featured help system should be provided. As well as providing the usual bubble and context sensitive help, the toolset help system could be oriented towards the particular service being used at any one time.

The generic interface described so far refers to the control of connections within the service, but do not describe the types of information sent down each connection. Components are described by type and sub-categorized to specify the specific type of data being transmitted, e.g. a video component in MPEG-2 format. Some component types can not be sub-categorized as they are service specific, so instead of sending information which the toolset can not interpret a user interface language should be used to describe a dialog which interfaces with the remote service. For example, a video retrieval service can be established using the RUS editor but the service requires its own interface to ask the user to select the required video signal, i.e. which piece of video do you want to watch. Several examples of a similar user interface language exist including HTML and Tcl/TK. Such a generic interface would be the next stage of development for the toolset.

5 Actual Implementation Trials

The ASCOT approach has been incrementally refined in actual implementation experiences in both the ASCOT and DRAGON projects.

The RUS editor and information hiding arose due to an early implementation of the toolset which used a dialog containing all information needed to describe a service. This interface was very complex and was only suitable for network / service providers. End-users found the information very confusing and unnecessary for their needs, and it was difficult to describe services in terms the toolset required. The RUS editor was then developed to provide a graphical interface which is simpler and more intuitive to use.

The first release of the ASCOT toolset could control a standalone security monitoring application involving a remote video scenario implemented over B-ISDN channels, implemented by Deutsche Telekom. The RUS editor could be used to construct simple service descriptions, but was deficient in a number of areas [DEL12]. In-call service control was performed using a separate dialogue with pairs of list boxes containing connected and unconnected, parties, tasks and components. This text based interface was difficult to use and it was decided to produce a graphical interface by incorporating it into the RUS editor.

The second release of the ASCOT toolset contains a number of improvements based on results from evaluation experiments, and requirements from the DRAGON project. The toolset is being used by the DRAGON project as a user agent to provide service control. Improvements include,

- Drag and drop control of the editor,
- Colour icons using scanned images,
- Improved integrity checks while a service description is being constructed and modified,
- Different views of the service,
- Group icons,
- Support for multicast tasks and components,
- In-call service configuration incorporating the latest version of the Service Control Element messages

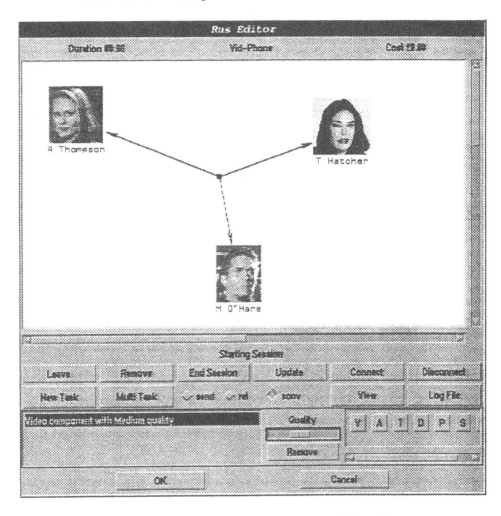

Fig.5 The RUS editor from the second release of the toolset

Service Control Elements were implemented as control message strings which were passed around the network using a CORBA based communications package.

6 Conclusions

ASCOT has illustrated how Integrated Service Engineering may be used without compromising either usability, or the ability of end-users to control services in-call. Furthermore, new SCE messages have been created and others modified to provide needed additional functionality.

There are very few projects that have used a Service Control Element approach. SCEs have influenced standards in CCITT and ETSI, as part of the RACE common functional specifications (CFSs). New SCE messages have been created and others modified to provide additional functionality, and a very useful graphical tool has been developed to configure a service while it is executing.

The toolset could become integrated into an end-users Personal Information Management (PIM) tool to reside along side diary and address book applications. An address book could record the Service Access Point (SAP) for a user or machine, and the diary could be used to reserve resources for services for scheduled meetings, etc.

A need for such a system is going to become more necessary when multimedia services become more widely available.

7 References

[CFS]   IBC Common Functional Specification, Issue D, December 1993.
    B210   "The Concept of IBC and it's Relationship to ISDN",
    B311   "Methods for the Characteristics of the Operational Requirements of IBC",
    C110   "Methods for the Specification of IBC Services",
    C220   "Service Control Elements".

[DEL12]   B.Hill, A.Whitefield, I.Denley, "Usability Evaluation". ASCOT Deliverable, R2089/UCL/ERG/DS/P/012/b1, May 1994.

[DEL17]   P.A.Coates, J.Middlemass, "Functional Specification of Improved ASCOT Toolset". ASCOT Deliverable, R2089/MAR/MCS/DS/P/017/b1, May 1995.

[GEM1]   Laszlo Szirmay-Kalos, "Dynamic Layout Algorithm to Display General Graphs". Graphic Gems IV, Edited by Paul S Heckbert, 1994, Academic Press, Inc.

[URM]   R1077 User Reference Model for IBC, 1992, Service Definition Methods

OMT Object Models of Telecommunications Services

Kathleen Milsted*

France Télécom CNET
38-40 rue du Général Leclerc
92131 Issy-les-Moulineaux
France
kathleen.milsted@issy.cnet.fr

1 Introduction

This paper describes an approach to modelling a telecommunications network and services using object-oriented concepts. The models are expressed in the object-oriented design notation, OMT (Object Modeling Technique) [4], and represent a high-level, user view of the network and services. The services targeted are Intelligent Network (IN) services [1, 2], in particular those of IN Capability Set 1 (CS-1) [3]. The models of three services are given here: basic call connection, terminating call screening, and call forwarding.

The paper is organised as follows. We first briefly describe certain concepts used in OMT object models, and introduce a new metaphor called a *storyboard*. This metaphor is not an extension of the OMT notation but rather a particular way of using the existing notation. We use storyboards to depict how execution of a service changes constraints and associations between network objects over time. Because of its visual nature, we propose that a storyboard can be used as a (partial) animation of a service. The paper ends with conclusions and directions for further work.

2 Object Modelling Concepts

OMT [4] stands for "Object Modeling Technique" and consists of a methodology and graphical notation for object-oriented analysis, specification and design of software. An object model abstractly represents a system in terms of its constituent elements (objects) and the relationships (associations) between them. In OMT, an object model consists of two kinds of diagrams: class diagrams and instance diagrams. Class diagrams define classes of objects and the possible associations between them at *any* point in time. Instance diagrams, on the other hand, describe the state of objects at a *particular* point in time, where the state

---

* This work was partially supported by the RACE SCORE (Service Creation in a Object-oriented Reuse Environment) project; however, it represents the view of the author.

of an object is given by the current values and constraints on its attributes, and the associations that the object currently participates in.

As objects in a system change state, clearly, the constraints on them and the associations between them may change. Thus, at one point in time, an object may be associated with a different set of objects than at another point in time. Indeed, an instance diagram is like a "snapshot" of a system, visually depicting a configuration of objects, associations, and constraints at a particular point in time.

Using this idea of instance diagrams as snapshots, then viewing a sequence of instance diagrams one after the other is like viewing consecutive snapshots of a system. This is the idea behind the *storyboard* metaphor, which is just an ordered sequence of instance diagrams visualising the changing states and configurations of objects in a system over time. In this sense, a storyboard can be regarded as a (partial) animation of a system. In particular, when used for a service, a storyboard can show how execution of the service changes constraints and associations between network objects over time.

Note that a storyboard is not an extension of the OMT notation; it is just a structured but still compatible way of using instance diagrams. Furthermore, a storyboard is not a new kind of model; it adds complementary information to an object model, by describing overall behaviour of the system as reflected in the changing states of objects. Note also that a storyboard cannot be generated automatically; it must be defined.

Our approach to constructing an object model of a system is summarised as follows:

- Define one or more OMT class diagrams defining the object classes and associations needed to model the system.
- Define one or more storyboards (sequence of OMT instance diagrams) showing how the associations and the constraints on objects change over time.

3 Object Models of Three Services

In this section, we use the approach described previously to model three IN services: basic call connection, terminating call screening, and call forwarding. The last two are CS-1 services [3], and are thus single-ended, single point-of-control services operating on two party calls only.

For reasons of space, we do not describe the meaning of the symbols used in the OMT diagrams that follow, assuming that the reader is familiar with OMT [4]. We do, however, provide a textual description of each diagram.

3.1 Basic Call Connection Service

Classes and Associations

A user view of a simple telephone network and basic call connection service can be modelled using the following object classes and associations:

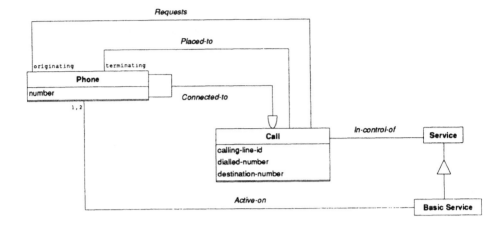

At this level of abstraction, four object classes are used. A *Phone* object is modelled with just one attribute: a number, while a *Call* instance (object) is modelled with three attributes: a calling line identity, a dialled number, and a destination number. A general *Service* object is modelled with no attributes, representing the fact that there are no attributes that all services have. Individual services are modelled as specialisations of *Service*, where each individual service may have its own attributes, if any. In particular, *Basic Service* is a specialisation of *Service*, but *Basic Service* has no attributes either.

Five associations are used. The association *Requests* associates a *Phone* with a *Call* (read as "a phone requests a call"). Its multiplicity is one-to-one: a phone may request only one call at a time, and a call can be requested by only one phone. *Requests* models the originating leg of a call.

The association *Placed-to* associates a *Call* with a *Phone* (read as "a call is placed to a phone"). This association models the situation where the terminating phone has been identified and the call is "routed" to that phone (without taking into account transmission paths, etc.). *Placed-to* models the terminating leg of a call. For two-party calling, its multiplicity is one-to-one: a call is placed to only one phone, and a phone has only one call placed to it. For multi-party calling, where a call could be placed to many phones, one-to-many multiplicity would be needed.

The association *Active-on* associates a *Service* with a *Phone*. The multiplicity of *Active-on* must be given separately for each individual service. For *Basic Service*, the multiplicity of *Active-on* is one-to-(one or two): a basic service instance may be active on one or two phones at a time but a phone has only one basic service instance active on it. This models two-party calling. For multi-party calling, the multiplicity would be one-to-many, whereby a basic service instance could be active on many phones at the same time. Note that, in contrast to CS-1 services, basic call connection is not single-ended since it may be active on more than one phone in a call.

The association *In-control-of* associates a *Service* with a *Call* (read as "a service is in control of a call"). Its multiplicity is one-to-one: a service instance is in control of only one call at a time, and a call is under the control of only one service instance at a time. This multiplicity models single point-of-control services, as defined for CS-1.

Finally, the association *Connected-to* associates a *Phone* with another *Phone*. Moreover, our interpretation is that, whenever two phones are connected, they are both taking part in a call. We model this by taking a *Call* instance to be an attribute of a pair of connected phones, i.e., as an attribute of a *Connected-to* link between two phone objects. The multiplicity of *Connected-to* is one-to-one, i.e., a phone may take part in a call with only one other phone at a time. This multiplicity models two-party calling; for multi-party calling, the multiplicity would be one-to-many.

Storyboard for Call Connection

We now give a storyboard for successful execution of the basic service. That is, a sequence of instance diagrams (snapshots) showing how associations and constraints between objects change during successful basic call connection.

1. **Access to the network is requested.** This first configuration of objects occurs when an off-hook signal is received from a phone. Then, a *Basic Service* instance becomes *Active-on* the *Phone*.

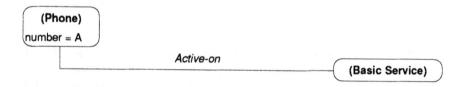

2. **A call is requested.** The next configuration of objects occurs when a number is dialled. Then, a *Call* instance is *Requested* by the originating phone, and the *Basic Service* instance is *In-control-of* that call. Equational constraints on attribute values assert that the values of the calling-line-id and the requesting phone's number are both equal to some value A. Likewise, the values of the destination-number and the dialled-number are constrained to both be equal to some value B.

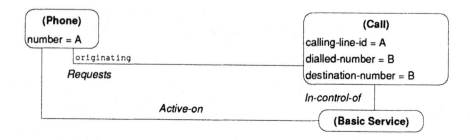

3. **The call is placed.** This configuration of objects includes the terminating *Phone*, with the call being *Placed-to* it. A further equational constraint asserts that the terminating phone's number is equal in value to **B**.

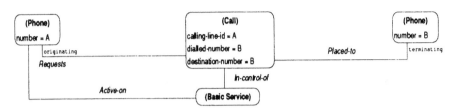

4. **The terminating phone is alerted.** In this configuration, the *Basic Service* instance also becomes *Active-on* the terminating phone. This can model the phase where the terminating phone rings.

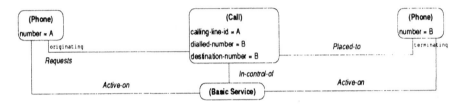

5. **The call is connected.** This final configuration occurs when the terminating phone is answered. Then, the two phones are *Connected* and participate in the call. The *Basic Service* remains *Active-on* the two phones and *In-control-of* the call. Note that the *Requests* and *Placed-to* links no longer exist, i.e., the objects no longer take part in these associations at this stage of the service.

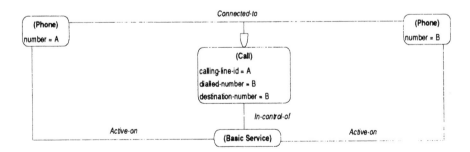

This storyboard models a successful basic call attempt and connection. To model error or other cases, it is a simple matter of specifying the configurations of objects (instance diagrams) that would or would not occur. For example, in the case that the terminating phone is busy, the basic service would only get as far as configuration 3; configurations 4 and 5 would not occur.

3.2 Terminating Call Screening Service

The Terminating Call Screening (TCS) service allows a subscriber to screen incoming calls. Here, we model the version of the service whereby an incoming call whose number appears on a list will not be connected. Note that the object model of TCS is just an extension of that of the basic call connection service.

Classes and Associations

A new class *TCS* is introduced as a specialisation of *Service*, and it has one attribute, a screening list of numbers. No new associations are needed, but we must specify that the association *Active-on* also associates *TCS* with a *Phone*, and that, in this case, the multiplicity of *Active-on* is one-to-one: an instance of *TCS* can only be active on one phone at a time, and only one instance of *TCS* can be active on a phone. This models the "single-ended" nature of TCS, as defined for CS-1.

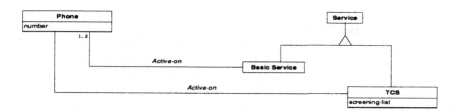

Storyboard for TCS

Successful execution of TCS is modelled by the following storyboard. Since a call connection attempt always occurs first, we will start this storyboard with the configuration of objects just before TCS gets control of the call.

1. **Network access requested, call requested, call placed.** This is the standard configuration of objects that occurs when an off-hook signal is received from a phone, a number is dialled, then the call is placed to the terminating phone. It corresponds to configuration 3 of the previous storyboard for the basic service alone, except with *TCS* active on the terminating phone.

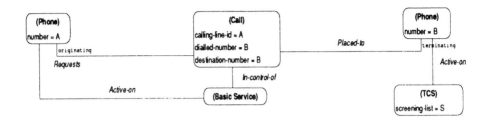

2. **TCS is invoked.** This configuration of objects represents the fact that control of the call passes from the basic service to TCS. Thus, by contrast with the previous configuration, the association *In-control-of* now links *TCS* and the *Call*. Note that this respects the one-to-one multiplicity defined for *In-control-of*, which implies the strong constraint that only one service at a time can be in control of a call.

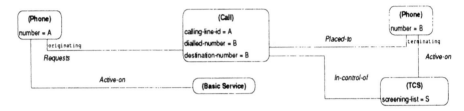

3. **The call is allowed.** This configuration occurs after *TCS* has verified that the call may be connected to the terminating phone. This verification is indicated by the extra constraint on the call's calling-line-id attribute, that it must not be on the *TCS* screening list S. Only then can this configuration of objects occur, in which control of the call has passed back to the basic service, which is active on the terminating phone. As before, this can model where the terminating phone rings.

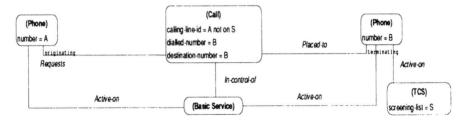

4. **The call is connected.** When the terminating phone is answered, the two phones are connected via the call. *TCS* remains active on the terminating phone but it is not associated with the call.

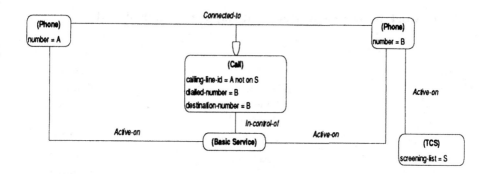

As in the case of the basic service, errors encountered during execution of a service are modelled by giving the configuration of objects that would or would not be obtained. For example, if an incoming call is from a number on the TCS screening list, then the service would only get as far as configuration 2; the configurations after that would not occur.

3.3 Call Forwarding Service

The Call Forwarding (CF) service allows a subscriber to redirect incoming calls to another number.

Classes and Associations

A new class *CF* is introduced as a specialisation of *Service*. It has one attribute, *target*, representing the number to which calls are forwarded. As with other CS-1 services, the association *Active-on* has one-to-one multiplicity between *CF* and a *Phone*.

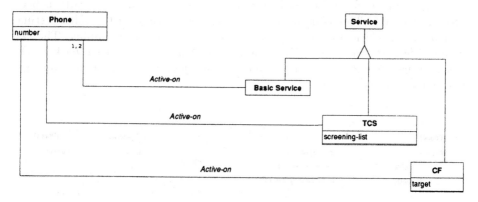

Storyboard for CF

The storyboard for CF starts with the usual configuration of objects: access to the network has been requested, a number dialled, and the call placed.

1. Network access requested, call requested, call placed.

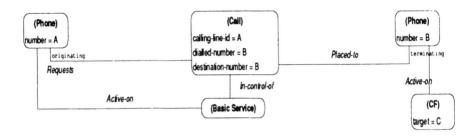

2. CF is invoked. Control of the call passes from the basic service to CF.

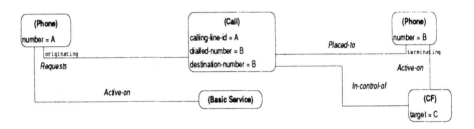

3. **The call is forwarded.** This configuration occurs after the call has been processed by CF. This is indicated by the change in value of the call's destination number, which now has the same value as CF's target number. The call is now *Placed-to* the target phone, and control of the call passes back to the basic service. Observe that the original terminating phone and CF are no longer associated with the other objects.

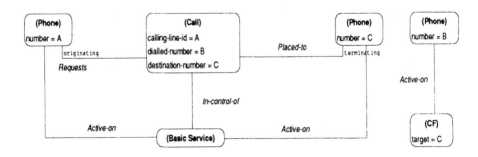

4. The call is connected.

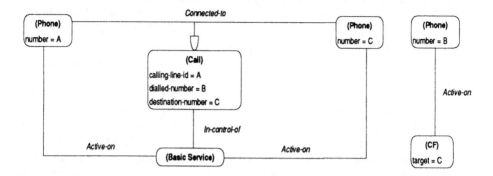

4 Conclusions and Directions for Further Work

In this work, we have described an approach to modelling telecommunications services using object-oriented concepts. Basically, it consists of two steps: defining the object classes and associations that are needed, then giving a storyboard depicting how execution of the service changes the states and associations of network objects over time.

Since this work concerns just object models, the first step (defining classes and associations) is exactly the same as in other object modelling approaches. It is just a question of which notions one chooses to model, and the amount of detail to include. For example, we have chosen not to model concepts from the lower IN planes (the two functional planes and the physical planes) but, obviously, this can be done, and the models here extended with much more detail.

What is new in this work is the *storyboard* metaphor. A storyboard is just a sequence of instance diagrams, where each instance diagram represents a snapshot of the system. A storyboard visually depicts how configurations of objects, their states and associations, change over time due to execution of the service.

Note that, even though we have used the OMT notation to express the object models and storyboards, a storyboard does not intrinsically depend on OMT, and it is certainly not an extension of OMT. Rather, a storyboard is a particular way of using the OMT notation (instance diagrams), exploiting, in an essential way, the graphical nature of this notation. Clearly, though, a storyboard would be less easily expressed and understood if a purely textual language was used!

We propose that a storyboard can be used as a (partial) animation of a service, indeed of any system, since it shows how the service executes by describing the changing states of objects and the changes in associations between them. In some sense, a storyboard describes part of the overall dynamic behaviour of a system. This is relevant as, currently, in OMT, there is no way to express behaviour of the entire system: OMT dynamic models describe the behaviour of individual classes of objects but do not describe how these individual behaviours are related in order to produce a model of the entire system's behaviour.

There are (at least) three directions for further work:

- Understanding how object models may contribute to the detection, and perhaps even resolution, of interactions between services/features. One idea we wish to pursue is that by combining storyboards of different services, one may arrive at configurations of objects (snapshots) that represent interactions. Identifying which configurations represent interactions would thus be a means of detecting interactions early on in the life-cycle of a service.
- Understanding how an object model of a service is related to a dynamic model of the same service where the latter is usually given by a finite state machine. Equivalently, this means understanding how an object-oriented view of a service relates to a process-oriented view. One step in this direction would be to relate storyboards to event traces since the change from one configuration of objects to another in a storyboard is typically brought about by some event. Indeed, a storyboard can be regarded as describing the overall finite state machine behaviour of a system, where each configuration in the storyboard represents a global state of the system.
- Tool support for storyboards. Currently, storyboards are not implemented in any OMT tool, so, for example, the order of instance diagrams in a storyboard can not be enforced. However, it should not be difficult to extend existing instance diagram editors to handle such restrictions.

References

1. Principles of Intelligent Network Architecture. ITU-T Recommendation Q.1201. October 1992.
2. Intelligent Network - Service Plane Architecture. ITU-T Recommendation Q.1202. October 1992.
3. Introduction to Intelligent Network Capability Set 1. ITU-T Recommendation Q.1211. March 1993.
4. J. Rumbaugh, M. Blaha, W. Premerlani, F. Eddy, and W. Lorensen. Object-Oriented Modeling and Design. Prentice-Hall, New Jersey, 1991.

QoS Modelling of Distributed Teleoperating Services[1]

J. de Meer, A. Hafid[#]

GMD-FOKUS,
Hardenbergplatz 2, D-10623 Berlin, Germany
phone: +49 30 25499-239, fax: +49 30 25499-286
email: jdm@fokus.berlin.gmd.d400.de

[#] Universite de Montreal, Dept. d'IRO,
Montreal, H3C 3J7, Canada
email: hafid@iro.umontreal.ca

Abstract:

The quality of a teleoperating service is mainly determined by the resilience of the embedding communication environment. The notion resilience represents constraints under which the communication environment must adaptive react to faulty or unreliable conditions in the system in order to maintain the requested quality of service (QoS). To evaluate the communication behaviour for maintaining an expected level of quality, modelling is an adequate tool for getting views on the effects of dynamic strategies which depend from the current configuration. A simulation technique is applied to the service specification to get an assessment of the behaviour of the service. However, these results generally, are to be applied to the service implementation to verify the models' conformity to the specification. Thus, a modelling approach which supports QoS validation for teleoperating services is proposed. The Joint viewing and teleoperating service is taken as a case study to apply our proposals.

Keywords: QoS Modelling, QoS Validation, Multi Casting Synchronisation and Reliability, Joint Viewing and Tele-Operation Service (*JVTOS*).

---

1. This work was partially supported by the RACE II Project R2088 TOPIC.

1.0 Introduction

The introduction of the new services and protocol features provides a new quality in communications. To study the quality of a service and particularly its performance, several techniques can be applied: simulation, analytical modelling and measurements techniques. The most suitable and accurate approach is the direct derivation of simulation models of a given specification before the implementation phase.

Our QoS verification approach is based on model simulation which we call also QoS validation and on QoS Measurement. QoS validation permits the design alternatives' evaluation, the identification of quality bottlenecks and the prediction of the quality of the service when implemented under certain conditions. QoS measurements at the service implementation validates the observations obtained from the model which should be conformant to a specification coping with performance properties. More detailed description of our approach, the techniques used, and the obtained results can be found in [1].

We decided for our case study of the telepointer service (joint viewing and tele-operation service (*JVTOS*)[6][7][8][9][10] to use *SIMTIS[4][5]* as the specification technique because of the availability of the *SES*/workbench[13], which allows to run a *SIMTIS* specification as an executable performance model. *TIS* is a type of process algebraic specification language that supports time and resources management in the formal model. A formal relation to the process algebra description technique *LOTOS* thus exists. This enables us to annotate our existing behavioural specifications with time and resource consumption constraints. *SIMTIS* is a dedicated software interface to the *SES*/workbench that supports animated execution and simulation runs along with collecting performance estimates. The *SES*/Workbench is an integrated collection of software tools for the design and the performance evaluation of complex systems and particularly the telecommunication protocols and services. It is an object oriented system since it supports object types, methods instances and references.To run models of time and resource consuming services, an appropriate simulation environment is required. The environment that we have chosen is based on the *SES*/workbench. On the top of the *SES*/workbench the *SIMTIS* library is used. Using this library, we are able to translate a *TIS* specification to a *SES*/workbench model.

The main components of the *SES*/Wokbench are the *SES/design*: a graphic interface tool for specifying a system design and the *SES/simulation* tool for converting the graphic design specification into an executable simulation model. To display an animated execution of the modelled system *SES/Scope* can be used. Hence an unexpected behaviour (functional and performance behaviours) of the model may be easily detected. The functional behaviour error (e.g. the receiver does not send the connect response when it receives the connect request) can be detected by the flow of transactions while the performance behaviour error (e.g. the delay to receive a connect request is zero) can be detected by the performance parameters values since at run time different statistical values can be displayed. The statistical output data may be processed by *SES/graph* to produce meaningful (simple) graphical representations. Hence one can display and plot statistical data produced during the model run.

The paper is structured as follows. Section 2 introduces a modelling approach that supports QoS validation. In Section 3 we apply our proposals to the *JVTOS* telepointer service and discuss in detail results on the two QoS constraints of synchronisation and reliability. Obviously, this is just a review of all the measurement results obtain from the TOPIC project and documented in our technical reports [12][1][3]. Finally Section 4 concludes the paper reviewing our contribution and the ongoing work.

2.0 QoS Modelling Method

The method that we propose for QoS validation consists of five iterative steps, that initially start with the original design decisions and eventually ends with a QoS analysis based on the model observations. In the first step the service or protocol using a language which supports QoS constraints is specified. To make use of an extended FDT which supports time, resource and probability concepts *LOTOS* together with a SES-related notation for specifying the quality constraints has been chosen. In the second step the service specifications consisting of a behavioural and quality constraint part are both integrated into one specification using the *SES/design* tool and the *SIMTIS* library, to an equivalent *SES/workbench* model (graphical representation). The specification of statistics, e.g. QoS parameters, to collect at identified *LOTOS* gates is made during this step as well. The graphical integrated specification of the model then is compiled, by using the *SES/simulation* tool to an executable service or protocol model. So, the resulting model contains the values of QoS parameters under specific underlying system conditions. We can run the model several times to measure the same QoS parameters under different underlying system constraints. Result analysis, using for example the *SES/graph* tool to eventually make design choices, to identify quality bottlenecks and produce QoS verification statements concludes the cycle of iteration. Either the model observations coincide exactly with QoS requirements specification which enables the final step of an implementation or the observed bottlenecks lead to a revision of the model and possibly of the constrains too.

The new thing is, that the QoS requirement specification is at first separately specified from a given behavioural or functional specification. After, the quality constraints are simply added to the design of service or protocol. Thus, a design must not be changed when specifying the qualities of communication. Or vice versa, if the qualities cannot be optimized, design changes are done by means of a separate specification. By the *SES/design* tool an integrated specification is produced, which then can be executed and analysed.

3.0 A Case Study with the *JVTOS* Telepointer Service

For a demonstration of the proposed method, we have chosen the *JVTOS*, i.e. the joint viewing and telepointer service. The *JVTOS* is an experimental multicasting application development environment. The users of *JVTOS* can jointly view multimedia documents and conduct tele-conferencing type of applications. For QoS result validation, an implementation of the telepointer service[7][9][10] at Technical University of Berlin has been used. In that realisation video data retrieval is used as the application; *ATM* is used as the networking technology and *XTP* as the underlying transportation mechanism.

To perform the QoS validation, we have developed a complete *LOTOS* telepointer service specification[3]. A simple version of the latter has been translated to an equivalent *TIS* specification. Using *SIMTIS* library the *TIS* specification has been translated, manually, to a corresponding *SES*/workbench model according to our modelling method proposed above. The performance statistics can be collected at any point in the model. In the following, the different steps to perform telepointer service QoS verification according to our approach will be presented.

3.1 The Telepointer Service Components

The telepointer service permits monitoring of the telepointers owned by a user and the mouse pointer of the floor holder and to display these movements locally and at remote user sites. The telepointer owner access this service through his *tep_user_agent*, which represents an interface to the user and permits him to create and control the appearance of telepointers. The telepointer service is provided by the telepointer service entity *tep_entity* which is composed by the *tep_source* and a collection of *tep_sinks* (figure 1). Each user, a *JVTOS* session participant, has his telepointer sink (*tep_sink*) which

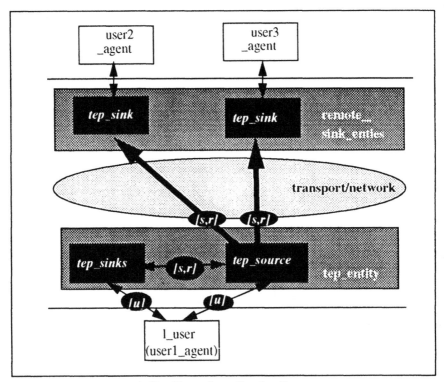

Figure 1: The Telepointer Service Components

handles the display of telepointers and the mouse of the floor holder. The telepointer source (*tep_source*) handles the telepointer input of its owner, furthermore it traces the mouse pointer of the floor holder.

A behavioural specification of the telepointer service components is given in figure 2

```
spec telepointer_service[u]:noexit
behaviour
let s_q,v_q,t_id, ...
in ( u?cr_treq[tep_id(cr_treq).eq.id];
     tep_sinks[s,r](cr_trq,s_q,v_q))
     |[s,r]|
     tep_source[s,r]
     |[s,r]|
     (remote_sink_entities[s,r](s_q,t_id)
     |||
     l_user[u](t_id)) where ...
```

Figure 2: Behavioural Specification

The components are linked with *Points of Control and Observation (PCO)* or simply called *gates*. In the formal behavioural specification gates occur in square brackets. So the local user agent interacts with his telepointer entity *tep_entity* at the [u]- gate. Internally, the tep_entity communicates with its telepointer sinks and with its *remote sink entities* via send and receive gates, i.e. *[s]* resp. *[r]*. (Gates, of course, may also be combined into one term comprising more than one gate, i.e. *[s,r]*, simply separated by commas. A behaviour expression is preceded by the underlined keyword *behaviour*, followed by the initialisation expression of the variables used in that expression, followed by the specification of the teleoperating service behaviour and the declaration of the processes referenced by that specification. Specification and declarations are embedded in the expression: *in specification where declarations*. The initialisation expression is bracketed by the keywords *let* and *in*.

The constituent parts of the behaviour specification are formed either by process instantiations or action expressions. Process instantiations are formed by a process name with actual gate names and value expressions, e.g. *tep_sinks[s,r](cr_trq,s_q,v_q)*, where *tep_sinks* is the process name and *[s,r]* is the set of gates and *cr_trq,s_q,v_q* are the actual values of the create telepointer request data unit and of two actualized queues used at the sink sites.

An action expression is a gate name followed by a sequence of an input or output indicators, i.e. ?, ! respectively with value expressions, e.g. *u?cr_treq*. Provided, the interaction can only be executed under certain conditions, the interaction expression will be annotated by the appropriate condition, e.g. *[tep_id(cr_treq).eq.id]* which must be a boolean expression, i.e. the equality between the two identifiers *tep_id(cr_treq)* and *id*. Notice, before comparison one of the identifier is read from the input of a *cr_treq* unit, i.e. the user-created telepointer request (data) unit.

The synchronized and asynchronous communication behaviours are represented by the operators *|[s,r]|* and *|||* respectively. In the first case, synchronisation takes place at the set of gates *[s,r]*.

Now the complete behaviour expression tells us the communicating components of the specified teleoperating service *JVTOS* consisting of three sets of processes: the dynamically created set of telepointer sinks, exactly one telepointer source and the set of remote sink entities.

Following the proposed modelling method, the behavioural specification is to be enhanced by the actualized quality parameters, which are the telepointer interval update *tep_interval*, a presentation offset *tep_pres_offset* to improve the data loss rate, and the presentation policy *tep_pres_policy* for far, medium and close sinks, the parameter impacting the loss rate at the sink considered. The QoS requirement specification, which is also used in the case study is presented in figure 3.

---

parameter int tep_interval "tep_interval:" = 20;
parameter int tep_pres_offset "tep_pres_offset:" = 40;
parameter int tep_pres_policy "tep_pres_policy:" = 1;

Figure 3: QoS Requirement Specification

The separate specifications of the behaviour and the QoS constraints are integrated by means of the *SES/workbench*. The resulting specification is a cyclic graph of transactions. A transaction is represented as a separate thread of execution in the specification, as one can find in figure 4. Transactions move along the arcs of the graph from and to nodes containing the actions just executed or to be executed next respectively. At each node the actualized QoS requirement parameters can be recognized literally. Parallelism in the model is supported by the capability of running more than one transaction at the same instance of time. Transactions are able to be forked and joined by particular nodes. In figure 4, the modelled transactions are bracketed by one entering and two return nodes. That means the transaction must have been split into two alternatives internally. This is represented by the follow-up splitting node labelled with *t_race*. Non-deterministicly one output path out of two is chosen; i.e. one returns with finishing the correct protocol transaction of having assembled some statistical data by the operation *data_ass* and issued a *net_send_data* protocol package. The QoS parameters of that transaction of interest are enumerated in the return operation as the sent or received packages. Notice that for both paths of that transaction the QoS parameter enumeration is identical. The split transaction path is the protocol error path, that eventually gets true.

Figure 5 indeed represents an instance of an animated path. The transaction just passes the *send_packet* node. The transaction is labelled by an identifier, e.g. T19 which moves along the arcs of the graph. When it is passing a node, the actions specified are immediately executed and statistics are collected.

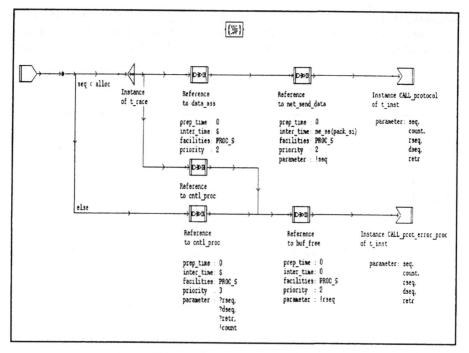

Figure 4: Integrated Behavioural and QoS Requirement Specification

3.2 Telepointer Service QoS Measurements

Because of the distributed architecture of the telepointer service various presentation times of the local and remote sinks are of special interest. Especially, there are the time to display the telepointer changes locally and at remote sites; the time of creation of a telepointer in response to a creation request of the local user.

Furthermore the loss ratio caused by temporal constraints violation and the buffer usage rate, i.e the buffer costs at each recipient are characteristics of special interests of the telepointer service.

In order to demonstrate the QoS validation method, we have developed a modelling scheme which is parameterized by a set of presentation strategies. The objective is to make a choice among different strategies. For demonstration purposes of the method, the strategies that we will consider are less complicated:

1) The shortest time of the three sinks is taken as a common strategy parameter for all recipients to display telepointer changes.

2) The average time is taken as a common strategy parameter to display telepointer changes.

3) The longest time is taken as a common strategy parameter to display telepointer changes.

Based on the QoS requirements as described in this section, two QoS constraints are identified: synchronisation, e.g. the response time to display the telepointer changes at remote sites or the response time of the telepointer creation, and reliability, e.g. loss rate, categories. To perform the QoS measurements, each QoS constraint requires specific observation techniques and service scenarios.

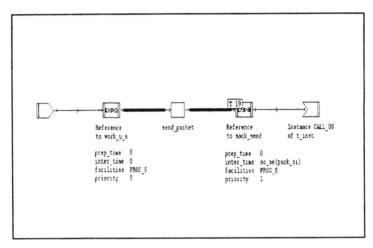

Figure 5: Animation of the Tep Service Model

3.2.1 Synchronisation Constraint

The two techniques that we propose to measure the telepointer QoS for the *synchronisation* constraint are:

(1) Each workstation (a *JVTOS* session participant) issues a timestamp for the moment of sending or/and receiving a specific primitive. To have accurate measurements, the clocks of the different workstations must be synchronised. It is difficult to have synchronised clocks in a highly distributed environment. Furthermore, in our model environment the workstations are running non-real-time operating systems.

(2) To use the round trip delay to derive one way delay. Hence a workstation issues time-stamps for the moment of sending a primitive and of receiving the corresponding response primitive. It is worth noting that in a paper[14] it was reported that round trip latencies are an insufficient and sometimes misleading method to determine unidirectional delays.

Since the two proposed techniques are not accurate enough to get objective measurements we decided to use them both simultaneously to compare the results obtained.

To perform such measurements, the *tep_source* and the *tep_sinks* must issue timestamps for the moments of sending and receiving a primitive. To be able to use the second technique the *tep_sinks* must issue a timestamped specific primitive (response_primitive) for each one received from the *tep_source*. The timestamped response_primitives will be stored in a specific database (*DBMR*: database measurement results) to be processed after the measurement campaign to produce meaningful results (curves, histograms, etc.).

3.2.2 Reliability Constraint

To measure a QoS parameter (e.g., loss rate) for reliability, we have only to measure the number n of data units sent by the sender and the number m of the corresponding data units received by the receiver for a predefined time interval t. Thus *loss rate = (n - m) / n for time interval t.*

The telepointer service has two operation types:

(1) Operations to send the telepointer repositioning information, e.g. *tep_put* or *tep_unput.*

(2) Operations to send the telepointer attribute changes, e.g. *set_tep_text* or *set_tep_color.*

Consequently the loss rate parameter must be measured for each type and must take the varying time of transmission into account. If the update interval is too short the service becomes unreliable. If the update distance becomes too long the WYSIWIS quality declines. A detailed demonstration sample is discussed in section 3.4.

3.2.3 QoS Constraints Impacting Parameters

The parameters which may have an impact on the QoS constraints to be measured are:
- tep_*interval* to update telepointer information,
- maximal acceptable difference value to present at each site WYSIWIS that we call *presentation offset*

$$tep\_pres\_offset = MAX\,(delay_{ki} - delay_{kj})_{1 \le \{i,j\} \le N}$$, where $delay_{ij}$ repre-

sents the transmission time to display telepointer, located at source i, changes at sink j and N is the number of active sinks or users respectively.

The number of active sinks, the number of active shared windows, the number of created telepointers, the local shared window characteristics (size, position, etc.), the geographic location of the active sinks, since it influences the delays encountered by a primitive issued by the telepointer source to reach the sinks, the underlying transport system, the machine types used (Sun, Mac, etc.) and their characteristics and the underlying system characteristics (load, errors, etc.) are all impacting parameters by which different scenarios can be specified.

Thus, by repetitive measurements under the same scenario an average of the measured QoS parameter value can be found. We then vary each of the above parameters (impact parameters) to obtain the measured QoS parameter value plotted against each of them. Three-dimensional plots involving two impact parameters are also possible if care is taken to avoid correlation.

The determination of the run length of a system, e.g. service, to obtain statistically representative results is not obvious. To overcome this problem the interval confidence interval notion has been used. Hence the specification simulation and the implementation measurements will be stopped when a desired accuracy is reached. Consequently our statistical declarations for telepointer service have consisted of the confidence interval specification (confidence level, confidence width, etc.) for each QoS parameter.

3.3 Synchronisation Constraint Modelling Results

For our case study we have considered three telepointer service participants geographically distributed through e.g. Europe. So, the response times to display the telepointer changes at the modelled three sites (user1, user2, user3) vary a lot. Consequently, the WYSIWIS (What You See Is What I See) concept is far to be satisfied. Hence, a multi-casting synchronisation scheme is required. The description of the synchronisation scheme that we have used and the complete telepointer service results can be found in more detail in[2,4].

In figure 6 you'll find a discussion of the model results for finding the optimum presentation offset between three active sinks. In figure 6 only the results measured at sink three are represented. A complete consideration however, must also include the results measured at the two remaining sinks. Measurements are taken for the given three presentation policies taking the assumed distances of the three sinks close, medium, far into account. Assuming further, that sink three is the farest (slowest), applying policy 1 for the closest (quickest) sink will result into a high rate of discarding unless presentation offset does not eliminate the distance to sink three. In the model this becomes the case for presentation offsets greater than 60 time units. In applying policy 2 for a medium distant sink, the discarding rate decreases already for a presentation offset greater than 40 time units. As expected, the minimum presentation offset will occur with policy 3 which is already suited for the farest distant sink three.

In the upper half of figure 6 an example of the minimum presentation offset applying policy 2 at sink three is illustrated. As can be concluded from the graphical representation of the measured results, with policy 2 almost all telepointer update information will be discarded by sink three if sink three will not be able to buffer at least 40 time units the received information. The curve shows that the discard rate is declining quite rapidly with presentation offsets higher than this limit. With a presentation offset of 80 time units the discard rate sinks under the limit of 10%. Thus, each 10th update information gets lost which might be an acceptable quality. The higher the presentation offset the higher are the costs of buffering the update information. The relation between buffering costs and the presentation quality must be negotiated with the user at the considered sink.

3.4 Reliability Constraint Modelling Results

In figure 7 the relation between buffer costs and the updating interval in dependence of some presentation policies is shown. The designer can vary the size of buffers and the distance between two telepointer updates. Obviously, the update interval depends of the source-sink distances. Big update intervals for all sinks means bad WYSIWIS quality, i.e. all remote sinks make their updates in big steps. So, the overall quality increases if some very distant sinks accept some losses, but the update interval

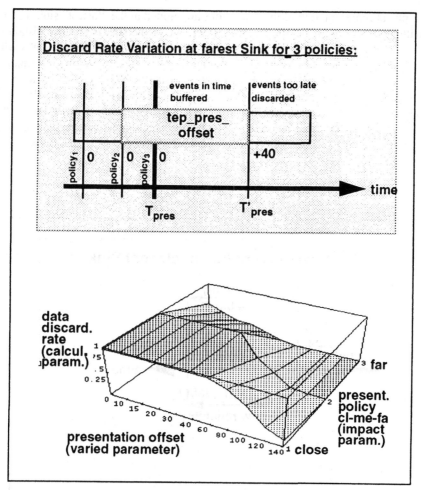

Figure 6: Synchronisation Constraint Modelling Results

becomes shortened. At those sinks the losses have the same effect as big intervals. If updating information is lost or arrives too late the pointers remain still. Thus, at different sinks, there exist different qualities. in the following table, the dependencies between update interval, WYSIWIS quality and reliability are outlined.

small buffer size + far distance sinks -> buffer cost increases

greatest update interval -> low WYSIWIS quality - high reliability

smallest update interval -> best WYSIWIS quality - low reliability

The graphical representation of those relations are plotted at the "fastest" sink, i.e. the sink with possibly has chosen the biggest buffer size or the sink is in close distant to the telepointer owner source. If you move towards slower or more distant sinks with the smallest update interval the more buffers must be used. Much updating information is broadcasted on a high frequency. However at far distant sinks the chance of information loss increases non-linear. If the updating interval copes with speed and distance, i.e. gets reliable, the overall quality becomes almost perfect. In the drawing above the telepointer interval variations are illustrated. The observed quality is reverse reciprocal to the service reliability. Low quality means high reliability and high quality induces a reduction of reliability.

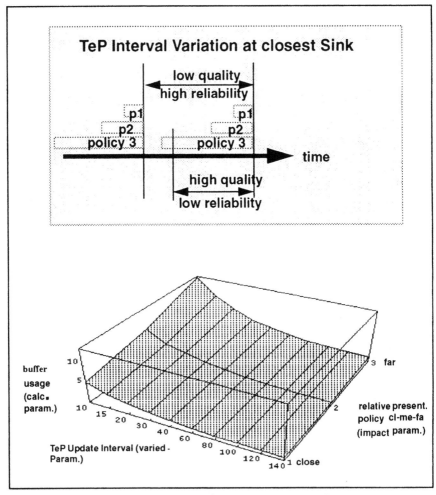

Figure 7: Reliability Constraint Modelling Results

4.0 Conclusion

A QoS specification and validation method has been presented. Part if which a simulation technique is applied to a composite behavioural and QoS requirement specification to predict the quality of service (QoS validation). Measurements obtained from an implementation will help to verify the conformity between model and the specification of the QoS requirements. The different steps of the proposed method have been defined. We have taken *JVTOS* telepointer service as a case study to apply our proposals: Different implementation strategies for a multi-casting synchronization mechanism have been analysed and presented during simulation experiments. For the interested reader results about the testing features and measurement analysis for QoS constraints can be found in the paper "Quality of Service Verification Experiments"[2]. It has been shown appropriate to handle the specification of the service behaviour separately from the specification of the QoS requirements, since this was supported by tools which did the integration into one representation. This approach has the benefit that existing behavioural specifications can easily be enhanced by additional QoS requirement specification without re-writing the existing ones.

5.0 References

[1]   A.Hafid, On JVTOS QoS Experiments, GMD-Studies Nr. 236, Berlin, Germany, July 1994.

[2]   A.Hafid, Jan de Meer, A.Rennoch, R.Roth: Quality Verification Experiments, IEEE MmNet95 Conference Proceedings, Aizu-Wakamatsu Fukishima Japan 1995

[3]   A.Hafid, Telepointer Service Simulation, Technical Report, GMD-FOKUS, Berlin, Germany, 1994.

[4]   A.Wolisz, Timed Interacting Systems for the Performance Enhanced Specification of Communication Protocols, SICON, september 1991.

[5]   J.Wolf-Günter and A.Wolisz, SIMTIS a Simulation Package for the Performance Evaluation of Communication Protocols Specified as Timed Interacting Systems, TUB.

[6]   T.Gutekunst and B.Plattner, Sharing Multimedia Among Heterogeneous Workstations, 2 Inter. Confer. on Broadband Islands, Greece 1993.

[7]   G.Dermler and K.Froitzheim, JVTOS A Reference Model for a New Multimedia Service, IFIP 1992.

[8]   T.Gutekunst et al., A Distributed Multimedia Joint Viewing and Tele-Operation Service for Heterogeneous Workstation Environments, Workshop on Distributed Multimedia Systems, Stuttgart 1993.

[9]   G.Dermler, et al., Constructing a Distributed Multimedia Joint Viewing and Tele-Operation Service for Heterogeneous Workstation Environments, IEEE Workshop on Future Trends of Distributed Computing Systems, Lisboa 1993.

[10]   G.Dermler, 1st Version of Telepointer Protocol Issues, CIO Internal report, 1992.

[11]   RACE TOPIC, Experience Report, EU project deliverable R2088/CLE/TEE/DS/P/020/b2, December 1994 (http://www.fokus.gmd.de/sem/topic/deliverables.html).

[12]   [Scientific and Engineering Software (SES) Inc., 1989, SES/Workbench users guide, SES 1301 west 25th, Austin, Texas 78705.

[13]   C.Klaffi, G.Polyzos and H.Braun, Measurement Considerations for Assessing Unidirectional Latencies, Internetworking: Research and Experience, Vol. 4, 1993.

Intelligent Distributed Management

Session 7 Introduction

Chairman- Patrick McLaughlin

Broadcom, Kestrel House,
Clanwilliam Place, Dublin 2
Phone: +353 1 6761531
Fax: +353 1 6761532
Email: pml@broadcom.ie

Standards bodies in the telecommunications services area have traditionally taken an engineering approach to defining architectures, reference models and reference configurations. They usually wind-up defining function blocks and associated interfaces, in sufficient detail to enable application/service interoperability. IN and TMN are two prime examples.

For both of these cases the standards do not as yet exploit some attractive characteristics of distributed technologies such as that found in ODP/TINA, nor do they rely on any Distributed Artificial Intelligence (DAI) approaches. The static nature of the interfaces in TMN and the absence of guidelines for implementing TMN conformant systems has been the topic of many R&D papers within RACE and elsewhere. The need for distributed control and localised intelligence under the control of higher layer *lean*-management systems, is a topic of active discussion within the research community.

A new paradigm called Intelligent Agents (IA), which is adapted from DAI, has received a lot of press lately. Paper 1 presents a broad overview of the various IA concepts and technologies that exist today. A tentative taxonomy for IAs is proposed. The paper further examines how fixed and mobile IAs can be used to enable new telecoms-based services, and to enable distributed intelligent management systems. The paper postulates how IN might evolve to incorporate agent technologies.

Two RACE II projects namely PREPARE and ICM have focused heavily on the implementation of TMN/OSI management based systems. This has enabled the developers to identify shortcomings in taking the 'obvious' approach to implementing managing (managers) and managed (agents) systems.

Papers 2 and 3 in this session are partly inspired by work carried out in the ICM project. Both papers explore the advantages of putting more intelligence into the managed systems by creating special 'Active' managed objects that implement management policies and directives by carrying out significant processing on their local managed objects before emitting notifications to the physically separated managers. It is argued, quite convincingly, that this management by delegation approach is efficient at reducing management traffic overhead and latency, and distributes the load effectively in the overall management environment.

Paper 2 describes such an idealised system and points out the need a management programming/scripting language to be used for dynamically changing the policies of Active Managed Objects at run time. This is supported by a description of a pilot implementation produced in UCL. The system consists of a combination of TCL and OSIMIS working in unison. Outstanding research issues are also presented, for example supporting guarantees for security and integrity in an environment where scripts can easily be changed within the managed system.

Paper 3 is similar to paper 2 in its aspiration to distribute intelligence as close to the resources being managed/monitored as possible. The usefulness of the recently defined monitor metric and summarisation OSI systems management functions for enabling intelligent remote monitoring is described in detail. Examples of how to use these for FDDI and ATM network management are provided. Both papers therefore try to break away from the statically compiled managed system capabilities, by enhancing managed system intelligence while still being TMN/OSI management system compliant.

On the Impacts of Intelligent Agent Concepts on Future Telecommunication Environments

Thomas Magedanz
Technical University of Berlin / Fokus / De·Te·Berkom
Department for Open Communication Systems
Hardenbergplatz 2, D-10623 Berlin, Germany
Phone: +49-30-25499229, Fax: +49-30-25499202
Email: magedanz@fokus.gmd.de

Abstract. The telecommunications environment is changing its face towards an open market of services where the vision is (access to) "information any time, any place, in any form". Mobility, service personalization and interoperability of services represent fundamental aspects of this environment. In this context a new paradigm is gaining momentum, referred to as "Intelligent Agents". Unfortunately the term "Intelligent Agent" has become a buzzword and is used in many different contexts. Therefore, this paper provides an overview of the emerging field of Intelligent Agents (IAs) by categorizing the existing types of IAs and identifying their fundamental properties. In addition, the paper identifies potential application areas of IA concepts in the telecommunication environment, focusing on intelligent communications and management, where also the relationships to existing key concepts, such as Intelligent Networks, Telecommunication Management Network and TINA will be addressed.

Keywords: Fixed Agents, Intelligent Agents, Mobile Agents, Network Management, Electronic Service Market, Personal Communications, User Agents

1 Introduction

In the light of global connectivity and increasing communication options for accessing and exchanging information the vision for future communications is *"information anytime, any place in any form""*. Based on the idea of an open "electronic" market of services, an unlimited spectrum of communication and information services will be offered ranging from simple communication services up to complex distributed applications. Therefore, besides the traditional aspect of information transport, the fundamental issue for new telecommunication services is information processing, i.e. the filtering, transformation and presentation of multimedia information in accordance with customer needs, available communication networks and end systems currently used by the customer. It will be the aspects of service personalization and service interworking that are of fundamental importance for future telecommunications, enabling customers to define when, where, for whom and how they will be reachable or not. Consequently, adequate means are required to support end users in configuring their communications environment in accord to individual preferences.

In this context the term *"Intelligent Agent (IA)"* has become a buzzword in the last years. Today agent technology is considered in a wide range of application areas, such as intelligent user interfaces, distributed problem solving, PDA connection, information retrieval & filtering, smart messaging, workflow systems, intelligent communications, network management, and the electronic market place [OVUM-94]. The reason for this increasing interest in emerging agent technology originates from the progressing convergence of computing and telecommunications, where in particular the recent advances in the field of mobile / portable computing and personal telecommunications are major drivers.

However, IA ideas and technologies have been influenced by a wide variety of disciplines and practices. The origin of IAs is in the field of Artificial Intelligence, in particular Distributed Artificial Intelligence (DAI) in the beginning of the 1980's, based on the idea of creating "objects that think", i.e. (intelligent) autonomous software entities that behave in accord with self-contained intelligence. Today *"Agent-Oriented Technology (AOT)"* can be considered as an interdisciplinary research field, influenced by object-oriented programming, distributed systems concepts, telecommunications, knowledge-based systems, decision theory, and organization theory.

Consequently agents substantially differ in the functionality provided. Therefore, it seems essential to clearly identify the basic characteristics and the functional scope of this new concept and to investigate potential application areas for that new paradigm in the context of telecommunications. Chapter 2 identifies the basic characteristics of agent-based computing and introduces a classification of today's agent technology. Chapter 3 investigates potential application areas for IA concepts in the field of telecommunications, focusing on personal communications and management, whereas chapter 4 concludes this paper.

2 A Taxonomy of Intelligent Agents

The term "Intelligent Agent" is associated with various expectations and is used in many different contexts. In general, the term "agent" ranges from adaptive user interfaces, communities of "intelligent" processes that cooperate with each other to achieve a common task, to mobile software entities which roam through networks in order to perform tasks at different network nodes in a coordinated way. Agent technology is applied in a wide range of applications and the methods that can be applied for agent development also differ substantially: agents may be developed on the basis of inference mechanism and knowledge bases, by means of script languages or even macros. Due to this fact, it is nearly impossible to give a sharp and at the same time comprehensive definition of the term "agent". Therefore, rather than giving an exact definition, it seems more valuable to identify the characteristics of agent-based computing and then introduce a classification of today's agent technology.

The following attributes are typical for agents: An intelligent agent is a self-contained software element responsible for performing some set of operations on behalf of a user or a another program with some degree of independence or autonomy. This means that an agent performs specific tasks delegated to it. Therefore, it employs some level of intelligence, i.e. knowledge or representation of the user's goals or desires, ranging from preference statements (i.e. predefined rules), some form of understanding and reasoning up to self-learning and adapting AI inference machines. Additionally, agents operate rather asynchronously (they are often event or time triggered) and may communicate with the user, system resources and other agents as required to perform their tasks. Moreover, more advanced agents may cooperate with other agents to carry out tasks beyond the capability of a single agent. Finally, as transportable or even active objects, they move from one system to another to access remote resources or even meet other agents and cooperate with them. Consequently, agents can be characterized by the following attributes in order to classify them:

- *Agent Intelligence:* This attribute indicates what method is used for developing the agent logic or "intelligence", and is closely related to agent languages, where two aspects are predominant. One aspect is the creation the programmatic content of the agent, the second is the knowledge representation, which provides the means to express goals, tasks, preferences, and vocabularies appropriate to various domains. Basically, three different approaches can be found in today's agent-based systems: macros, scripting languages, and AI techniques. Obviously the highest degree of "agent intelligence" can be achieved with AI techniques and the lowest with macros. However, scripting languages are currently in the focus of attention.

- *Asynchronous Operation:* An agent may execute its task(s) totally decoupled from its user or other agents. This means that agents may be triggered by a certain event occurring, or by the time of day. For example, an agent placed within the network can operate totally asynchronously to the user, it performs its task by talking to various system resources and potentially to other agents. In addition, an agent injected by an user into the network may perform a specific task and may return to the user when its task is finished or when some unforeseen problems occur.

- *Agent Communication:* During their operation, agents may communicate with various system resources and users. Agents that directly interact with the user are often called personal assistants or interface agents. From an agent's point of view, resources may be local or remote. There is a wide range of system resources agents may access, for example, application programs, data bases, information systems, instances of a management information base, sensor devices of a robot system, and so on.

- *Agent Cooperation:* This attribute indicates that the agent system allows for cooperation between agent entities. The complexity of cooperation may range from a client/server style of interaction to negotiations and cooperation based on AI methods, such as contract nets and protocols derived from speech act theory. This cooperation may necessitate the exchange of knowledge information and represents the prerequisite for multi-agent systems.

- *Agent Mobility:* In order to do their job, agents may be transported throughout a network to remote sites in order to perform specific tasks, usually in a specific run-time environment. Note that this approach departs fundamentally from the classical form of client/server computing and is also referred to as "Remote Programming". Generally, two levels of agent mobility can be identified:

a) *Remote Execution*: An agent (i.e. program code and data) is transferred to a remote system, where it is activated and entirely executed. The transport mechanims used for agent tranport could vary from TCP/IP up to electronic mail, such as used for emerging mail-enabled applications.

b) *Migration*: During its execution, an "active" agent may move from node to node in order to accomplish progressively its task. In other words, agents are capable of suspending their execution, transporting themselves (i.e. program code, data, and execution state) to another node in the network, and resuming execution from the point at which they were suspended. In addition, agents may launch new agents during their journey, e.g. for delivering acquired information to the client or perform specific subtasks.

After having introduced these basic characteristics, the various agent technologies existing today can be classified. In single-agent systems, an agent performs a task on behalf of a user or some process. While performing its task, the agent may communicate with the user as well as with local or remote system resources, but it will never communicate with other agents. In contrast, the agents in a multi-agent system may extensively cooperate with each other to achieve their individual goals. Here agents may be either fixed entities communicating via a network with other agents or they may be mobile entities, which move to remote sites to cooperate. Of course, in those systems, agents may also interact with users and system resources. Subsequently, we distinguish in the context of single-agent systems "*Interface Agents*" and "*Networked Agents*, while in the area of multi-agent systems "*DAI-based Agents*" and "*Mobile Agents*" can be identified.

2.1 Interface Agents

The major goal of this agent type is to collaborate with the user, and hence the main emphasis of investigations clearly lies in the field of user/agent interaction. They realize customized interfaces, which learn a user's individual habits and preferences. Therefore, this type of agent is called "*intelligent interface*" or "*interface agent*". Usually interface agents act as Advisory Agents (e.g. intelligent help systems) or Personal Assistants of users, who the agents assist during their daily work. Members of this class access local resources only.

Interface agents may assist users in a range of different ways: they perform task's of the user's behalf, they can hide the complexity of difficult tasks, or can even train and teach the user. The set of tasks an agent can assist in is virtually unlimited: local information retrieval and filtering, local mail management, meeting scheduling, etc. [ACM-94]. The idea of employing agents in the user interface was introduced many years ago [Kay-84]. Even though a great amount of research has gone into the construction of interface agents, currently available techniques are far from being able to produce high-level, human-like interactions [Maes-94].

For building PAs three different approaches are conceivable, namely end-user programming (e.g. in HP's New Wave Agent [HP-89]), the "knowledge-based approach" (e.g. in UCEgo [Chin-91]), and "machine learning techniques" (e.g. in Maxims [Lash-94]).

2.2 Networked Agents

In contrast to agents of the previous agent class, networked agents are able to access not only local but also remote resources, and thus have a more or less detailed knowledge about the network infrastructure and the services available in the network. The main difference between this agent class and multi-agent systems treated below is that networked agents are assumed not to cooperate with each other.

Many of the existing networked agents act as personal assistants. "Smart Mailboxes" and "search engines" are probably the two most popular examples for this agent type. They not only provide an intelligent interface to the user, but also make extensive use of the various services available in the network. In contrast to smart mailboxes which perform advanced mail filtering based on the user's preferences, search engines collect knowledge available in the network on the user's behalf. This type of agent has also been called "KnowBot" or "Softbot" [Etzi-94]. When collecting the information, the agent not only hides the complexity of the network infrastructure but also makes the heterogeneity of the network's wide range of information sources and access protocols invisible for the user. A Softbot usually provides an expressive and uniform interface, where the user can specify his/her information needs on a rather high-level of abstraction. The agent takes this specification and autonomously collects and combines the desired information potentially from multiple information sources in the network, filters out information relevant for the user, and presents the resulting information in a form tailored to the specific user needs. This type of agent is especially relevant for the Internet, where along with the explosive growth of the World-wide Web (WWW) and the corresponding demand for tools for managing the vast amount of available information has resulted in several agent implementations (variously known as WebCrawlers, Spiders, and Robots [Inde-95]).

So far, we have only considered agents acting on behalf of users. Networked agents, however, may also perform tasks on behalf of non-human clients, such as applications, system processes or management entities. Of course, in this case the intelligent interface is replaced by an appropriate communication protocol allowing the agent to interact with its client.

2.3 DAI-based Agents

In the context of Distributed Artificial Intelligence (DAI), multi-agent systems are concerned with coordinating intelligent behaviour among a collection of autonomous intelligent agents, how they coordinate their knowledge, goals, skills, and plans jointly to take actions or solve problems [Bond-88]. DAI-based agents are used in a wide range of applications, such as distributed vehicle monitoring, computer integrated manufacturing, natural language parsing, transportation planning, and in particular telecommunications management [Goya-91].

Agents are developed by means of AI techniques, such as rule-based systems, case-based reasoning, and example-based reasoning. The knowledge representations and inference mechanisms applied are basically the same as those used in (non-distributed) knowledge-based systems. DAI-based agents may communicate with the user, system resources and other agents.

These agents communicate and cooperate with each other either by "message passing" or the so-called "blackboard" approach. In the latter, agents communicate by writing to and reading from the blackboard. If the message-passing paradigm is adopted, agents interact with each other by exchanging messages. Many of the proposed inter-agent negotiation and cooperation protocols are based on the concept of contract nets, and on speech act theory, have been developed. The *contract net* protocol is a bidding protocol that allows for the negotiation between agents. The *speech act theory* is used to define the semantics of messages, where for example *Knowledge Query and Manipulation Language (KQML)* [Gene-94, Fini-94] and *Knowledge Interchange Format (KIF)* [Gins-91] are representing languages especially designed for knowledge transfer between various types knowledge-based systems and the modelling of rules and contents.

In DAI, agent mobility was no issue so far. Agents are usually fixed, i.e. they cannot travel while performing their job. Consequently, once an agent has been created at a node, it will stay there until it will be destroyed.

2.4 Mobile Agents

While the previous agent classes represent *fixed* agents, which remain in a single location throughout their execution, mobile agents are dispatched from a source computer and roam among a set of network servers in large computer networks until they are able to accomplish their tasks. Mobile agents are used for offering a huge number of sophisticated services. Examples are ranging from smart messaging [Bors-92], mobile computing [Ches-95] [Wayn-94], intelligent communication [Rein-94], (network) management, and the electronic market place [Whit-94].

Agent mobility represents probably the most challenging property in emerging IA technologies, enabling the concept of *"remote programming"*. Remote programming can be seen as an alternative to the traditional client/server computing paradigm. Rather than "shout" requests across the network, the agent transports itself from the client to the remote (server) computer, where the work must be done. When the agent arrives, the remote site executes the agent within an *"agent execution environment"*, which provides access to local data and resources. Consequently, with the remote programming approach, functions are shipped to the data. However, the prerequisite for this approach is the existence of an appropriate agent execution environment at the server(s).

It has to be stressed that the idea of performing client/server computing by the transmission of executable programs between clients and servers, i.e. dispatching a program for execution on a remote computer is quite old and has been called "remote (batch) job processing" in the 70's and "function shipping" or "remote evaluation" [Stam-90] in the 80's. The basic motivation for this has been that local computers did not have the capacity to execute the programs locally or just for resource sharing and load balancing in distributed systems. By contrast to these concepts designed for rather specific or closed environments, the new challenge in this concept is to enable this idea in open environments (e.g. within the Internet), enabling for example the spontaneous provision of new services.

Therefore, agents of this class are typically developed by means of scripting languages, since they are considered to be highly portable and provide a high degree of security. The most prominent languages are *Save-Tcl* [Bors-94], which is an "extended subset" of the Tool Command Language (Tcl) [Oust-94], SUN Microsystem's Java Language [Java-95] and General Magic's *Telescript* [Whit-94].
Whereas Safe-Tcl and Java are primarily used to enable asynchronous operation and remote execution of "mail-enabled" and "WWW-enabled" applications" within the Internet without any agent cooperation, the metaphor used for motivating the Telescript technology, which supports agent migration, is the electronic market place, which provides a dynamically changing set of services to potentially mobile users. This anticipated market can be seen as an electronic community, where people meet, work, learn, shop and may obtain entertainment. Within this market, agents asynchronously perform tasks on behalf of users. They may communicate with the user, the services available in the network and other agents. In particular, dynamic intinerary would allow agents to achieve their goals even when the capabilities of specific remote servers are not known ahead of time.

Sophisticated cooperation mechanisms, such as those present in DAI-based agents, are still out of the scope of current mobile agent technology. In Telescript, for example, agents cooperate by mutually calling methods of their object-oriented interfaces.

2.5 Summary

Agents may be classified in many ways and that the presented taxonomy of IA classes represents only one possible approach. It should be regarded only as a means to categorize existing agent concepts and to help readers in categorizing agent concepts. Table 1 summarizes the properties of the identified agent classes.

Attribute \ Class	Local Agents	Networked Agents	DAI-based Agents	Mobile Agents
Agent Intelligence	✕	✕	✕	✕
Asynchronous Operation	✕	✕	✕	✕
Agent Communication		✕	✕	✕
Agent Cooperation			✕	✕
Agent Mobility				✕

Tab. 1: Intelligent Agent Classes and their Characteristics

It has to be stressed that with the ongoing research activities in agent-oriented technologies the borderlines between the above given agent classes is becoming fuzzy. For example DAI-based agents will be getting mobile in the future, whereas cooperation between mobile agents will be supported by KQML in the near future. In the following our main interest will be devoted to the impacts of the agent properties asynchronous operation, cooperation and in particular mobility on the telecommunications environment.

3 Application Fields of IA Concepts in Telecommunications

The area of telecommunications represents probably the shining edge of agent-based services, where IAs could be considered as enabling technology. In particular, the fields of intelligent/personal communications and management are in the focus of ongoing research. The driving force for this view is the assumption that IAs will provide better support for distribution of control and management in telecommunication systems. This is due to the fact that the IA concept can be seen as an evolution step beyond the object-oriented paradigm, since IAs enhance objects by distributing control and goals to them. This means that IAs, i.e. their identified characteristics, such as intelligence, co-operation, asynchronous operation and agent mobility, could enable control and management tasks at the most appropriate locations within the telecommunications environment in contrast to existing architectures. This may have significant impacts on the architecture and the related (signalling) protocols of existing telecommunication systems.

In the following section we will look briefly at the existing key concepts in the existing telecommunications environment, where in section 3.2 we will extract the basic findings of the presented IA concepts of chapter two, and look at their impacts on the evolution of today's telecommunications concepts.

3.1 Todays Telecommunications Environment

The current telecommunications environment is based primarily on international standards, such as the *Intelligent Network (IN)* [Q.12xx] and *Telecommunication Management Network (TMN)* [M.3010], representing the foundations for the uniform and efficient creation, provision and management of advanced telecommunication services, such as *Universal Personal Telecommunications (UPT)*. In this context the integration of IN and TMN represents a fundamental issue, due to the convergence of service control and management (i.e. customer profile management) in future telecommunications services (e.g. UPT) [Mage-95]. In addition, the evolution of this environment increasingly takes into account the standards in the field of *Open Distributed Processing (ODP)*, due to the progressing convergence of distributed computing and telecommunications. This process results in the definition of new so called "information networking architectures", such as the *TINA-C* architecture [Depu-95], aiming for the provision of generic platforms for both telecommunications and management applications based on the object-oriented paradigm.

However, current IN and TMN architectures are based on highly centralized approaches for the location of service control and network (and service) management "intelligence", where the related protocols (i.e. INAP and CMIP) rely on the client/server paradigm. This means that centralized nodes are used for hosting service and management programs, respectively.

For example, in the IN the "intelligence", i.e. the telecommunications service logic and data resides statically in a centralized computer, referred to as Service Control Point (SCP), which will be interrogated by the distributed switches, referred to as Service Switching Points (SSPs) during call set-up, i.e. for service execution. The fundamental basis for this approach is the definition of an appropriate call model (in the switches) and a corresponding signaling protocol (i.e. INAP [INAP-93]), enabling the remote control of the switches by the SCP. The central part of the service data forms the "Subscriber Profile", which allows for customization of the service. Although customers are allowed to perform limited profile data modifications (known as customer profile management), the service logic is rather static, i.e. once deployed in the network no modification of service logic is possible. This means that the SCP could be regarded as a Fixed Agent, which acts on behalf of the calling or called party, depending on the type of service. Interactions between SCPs are not yet defined, due to the focus on single point-of-control services within the first set of IN standards. However, this will change within future IN standards, which requires modifications of the call model and the signaling protocol.

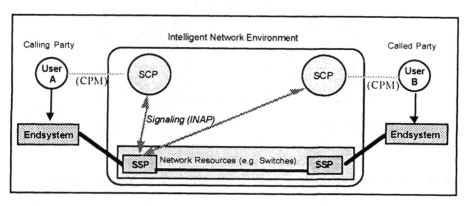

Fig. 1: Centralized Control within the IN Environment

Similarly, today's network management is based on centralized intelligence in corresponding management systems, e.g. management services residing on rather centralized operations systems (acting as "managers"). These "managing" systems query the management information located in numerous distributed "managed" systems (e.g. network elements), referred to as "(management) agents" by means of an appropriate management protocol as depicted in Figure 2. It has to be stated that the level of functionality offered by existing standard management protocols differs tremendously. In contrast to the Internet's Simple Network Management Protocol (SNMP) [RFC1157], OSI's Common Management Information Protocol (CMIP) [X.711], which is used in TMN, allows for more complex management interactions. However, the distribution of management activities, e.g. interactions between operations system (and between agents), management delegation and the construction of operations systems hierarchies (e.g. the separation of network and service management), as well as placement of intelligence into the (management) agents is still in its infancy. One reason for this is probably the lack of adequate concepts for the modeling of management services and their placement within the management architecture.

In contrast to IN and TMN, TINA is based on an object-oriented approach, following the ODP standards (and the ODP viewpoints) for the modeling of distributed service and management applications. Focusing on the computational modeling, there is the explicit notion of "User Agents" which represent the end users within the system and represent also the contact points for users in order to gain access to services. Although user agents are fixed, they could interact with other user agents (via session managers) for service session (i.e. communication) establishment. In addition, there is also the notion of terminal agents, which should adapt communication services to the used terminal capabilities of the users [Bern-95].

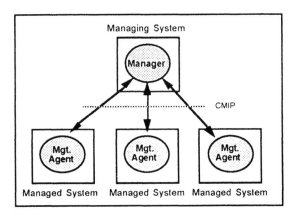

Fig. 2: Centralized Manager controlling Distributed Management Agents

3.2 Towards an Agent-based Telecommunications Environment

In accord to the previous considerations two basic approaches of agent-based service architectures can be identified:

1. "Smart network" approach where the agents are stationary entities in the network and provide the necessary intelligence, and are able to perform specific predefined tasks autonomously (on behalf of a user or an application). The fundamental aspects of this type of agent are their ability to act asynchronously, to communicate, to cooperate with other agents, and to be dynamically configurable. In order to enable this dynamic reconfiguration of the user agent, the dynamic downloading and / or exchange of corresponding control scripts seems a promising issue. This means that this kind of agent could be considered more likely as a specialized *agent execution environment*, which executes scripts (i.e. remote execution type agents). The communication and cooperation between user agents could be realized by using traditional client/server communication mechanisms.

2. "Smart message" approach, where agents are mobile entities, which travel between different computers/systems and perform specific tasks at remote locations. In respect to the task to be performed either remote execution agents or migration agents may be deployed. This means that a mobile agent contains all necessary control information (i.e. the service logic of a particular telecommunications or management service), instead of the corresponding end systems and /or nodes within the network. This means that corresponding agent execution environments have to be provided by the potential end user systems and within the network, in order to perform the execution of agents and thus the realization of their intended services.

Obviously the second approach is more complex. However, it has to be stressed, that these approaches could also be combined in order to gain maximum benefits. In the following we elaborate more on these two approaches in the fields of intelligent communications and management.

Agent-based Intelligent Communications

Taking the above classification into account, the following approaches for future communications could be identified:

A. Fixed Agent based Communications

In case of a smart system approach *fixed agents* will be deployed in the system/network. Here an agent is sort of a personal assistant, mostly referred to as "User Agent" or *"Personal Agent"*, which knows the communication preferences in respect to time, space, medium, cost, security, quality, accessibility and privacy of its user, and controls on behalf of that user all incoming and outgoing communications in respect to intelligent routing, information filtering, and service interworking (e.g. information conversion). The respective user agents of the involved communication partners, i.e. calling and called parties (note that this could be more than two in case of multi-party calls) have to negotiate and cooperate for establishing the desired communication session between the users. Note that this approach could be used for both asynchronous (e.g. mail) and synchronous (e.g. telephony) services. Additional Terminal Agents could provide independence of access devices and provide for service adaptions. In addition it has to be stressed, that the user agents act totally autonomous and asynchronously, this means that the users do not recognize any agent negotiations. This is illustrated in Figure 3.

The prerequisite for this scenario is the definition of an appropriate user profile for each user, representing the agent's knowledge base for a specific user. Correspondingly, appropriate customer profile management capabilities are required. Exactly this could be achieved with mobile scripts, i.e. the user profile could be realized by a corresonding "agent script", where the user agent could be regarded as the corresponding user specific "agent execution environment"! This means that traditional IN service logic programs could be replaced by agent-based scripts, where the user agent (i.e. its execution environment) will invoke the corresponding service script in case of a related service call.

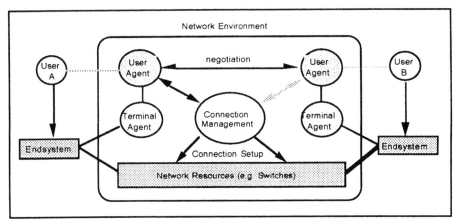

Fig. 3: Fixed Agent-based Telecommunications Architecture

However, in the light of the emerging integration of IN and B-ISDN it seems likely, that the current IN architecture evolves towards the fixed agent approach, where the SCP could be used to realize the user agents [Laue-95]. Additionally, it has to be noted that the TINA-C architecture, globally considered as the long-term IN architecture, defines already a similar architecture. In addition, it has to be stressed that a variety of research activities are currently investigating the area of intelligent/personal communications, which all address the issues of service integration (e.g. telephone, mail, fax, paging, multimedia, etc.), mobility of users, and information conversion. An extensive description of these approaches is not possible within this paper. However interested readers are referred to [Ecka-95], [Iida-95], [Moha-95], and [Rizz-95].

B. Mobile Agent based Communications

Adopting the smart message approach *mobile agents* could be used in two domains, namely asynchronous information exchange, such as mail services and pre-establishment of real-time information exchange, such as services on demand (video on demand) and multimedia communication services.

In the first case, two options exist: first the agent is contained in a mail allowing new types of information exchange (‡ la Active Mail), second the agent transfers the information (e.g. the agent uses a specific agent-transport service ‡ la PersonaLink or uses any kind of network, i.e. it will find the best way through the network(s) to the recipient).

In the second case a mobile agent will be used for primarily for signaling and system configuration, i.e. the establishment of a following real-time communication service (e.g. a video conference). This means that an agent will be generated by the calling user's end system prior the real call (ranging from reservation of up to immediate service invocations) and sent to the destination user's end system in order to set up the required communication path. This task could include reservation of corresponding network resources (e.g. conference bridges), configuration of the involved end systems in accord to the planned communication session (e.g. a multimedia conference), and service adaptation in accord to calling party and called party communication preferences. In addition, both remote execution as well as migration agent types can be used in this scenario, where a remote execution type agent will once travel to the target system and then communicate in some RPC style remotely with other systems and network elements, whereas a migration type agent will travel to the different systems and perform the operations locally. However, the prerequisite for this approach are corresponding agent engines on the involved systems.

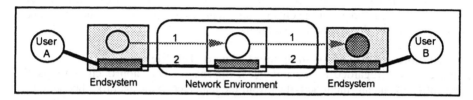

Fig. 4: Mobile Agent-based Signaling for Synchronous Service Sessions

Figure 4 depicts the agent usage for signaling (step 1) in order to setup a following multimedia connection (step 2). On its way to user B the agent could take care for the reservation of necessary network resources. In addition, the agent could take care for the endsystem configuration of both end users.

In summary, it can be stated that the IN will probably evolve towards a fixed agent-based network architecture (which would make the IN evolution towards TINA much easier). In particular the incorporation of mobile agent concepts within the IN for the dynamic downloading of customized service scripts, seems to be the most attractive scenario for the medium term time frame. The integration of mobile agents for signalling purposes does not seem sensible for todays IN, since the IN architecture has been designed for enabling realtime exchange of service control (i.e. signaling) information, based on centralized service logic. Hence the smart message approach could be regarded as an alternative to the IN approach.

In this context new telecommunications architectures, such as the TINA-C architecture seem to be more appropriate candidates for this incorporation of mobile agents. In this context it has to be stressed, that OMG studies already the incorporation of static and mobile agents within its Common Facilities Architecture [OMG-95]. This is important, since the TINA DPE can be considered as an extension of the OMG CORBA platform. Other architectures promoting mobile agent-based services are AT&T's PersonaLink and IBM's Intelligent Communications [Rein-94].

Agent-based Telecommunications Management
The distribution of management tasks in (network) management represents a major research issue since many years. This field of research, also referred to as "Management by Delegation (MBD)" (see [Yemi-91, Meye-95]), adopts a decentralized management paradigm that takes advantage of the increased computational power in network nodes (i.e. management agents) and decreases pressure on centralized network management systems and network bandwidth. MBD enables both temporal distribution (i.e. distribution over time) and spatial distribution (i.e. distribution over different network nodes). The basic idea is to increase the management autonomy of management agents by running socalled "elastic procceses" on the "elastic" agents, which could absorb dynamically (i.e. by

delegation) new functions. This approach is based on a "Management Scripting Language" [Yemi-91] and the idea of "remote evaluation" [Stam-90], where programs could be delegated from managers to management agents.

The basic goal should be to bring the management intelligence, i.e. the management services, as close as possible to the managed resource, i.e. its logical representation in form of managed objects. In accord to our initial separation of static (system) agents and mobile (message) agents we are facing two application scenarios of agent technology in the domain of management.

A. Enhancements of Management Agent Intelligence

This could be regarded as a *fixed agent* approach. This approach features the delegation of management activities from the manager to the agent(s), becoming (a) proxy manager(s), and hence reducing the amount of communication between manager and agent(s). This means that the management agent will be regarded as a specific agent execution environment, supporting the remote execution of management scripts. This could be achieved by downloading management scripts containing sequences of management operations. Hence the management agents are able to act autonomously of the managers for specific tasks. These scripts could be activated based on time, management actions or occurrence of events in the agent. This provides also the basis for cooperation between management agents (and managers), i.e. between corresponding management scripts located at different management agents.

Note that migration of management scripts will not be supported in this scenario. This means that a management script once installed at a specific management agent will not migrate to another agent. Figure 5 illustrates this approach where a manager downloads the same management script to multiple management agents. It has tobe stresses that this downloading could be achieved in many different ways, e.g. file transfer, or specific management protocol extension.

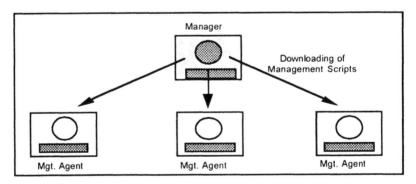

Fig. 5: Delegation of Management Intelligence to Management Agents

In this context currently some activities in the OSI management standards bodies can be recognized. In addition to the already defined log control and event report management functions, OSI management standardization is currently defining a new "command sequencer" management function [N2139], which allows the delegation of management activities from the manager to proxy managers (i.e. agents). This will be realized by defining corresponding scripts of management operations (i.e. command records), which will be executed by the command sequencer, where in particular, pre-scheduled or delayed execution of systems management operations will be supported, with capabilities for initiating, suspending, and resuming management operations based on time, management actions or occurrence of events in the agent. This includes also capabilities for reporting the outcome of pre-scheduled or delayed execution of these "launch scripts". This approach will significantly reduce the amount of communication between managers and agents, and make the agents more independent from the managers, e.g. in case of connection problems.

B. Mobile Management Applications

This scenario represents an extension of the previous one, where management applications will be realized by means of *mobile agents* supporting also migration. Here a manager could generate a mobile agent (i.e. management script) performing specific management tasks automonously and purposefully at specific / multiple appropriate management agents in order to collect, download or modify specific distributed or replicated management information or management scripts on multiple agents in a coordinated way (see Figure 6). Furthermore, also the downloading of scripts between management agents seems sensible, e.g. for load balancing purposes. This means that software management in general can be decentralized and automated by means of mobile management IAs.

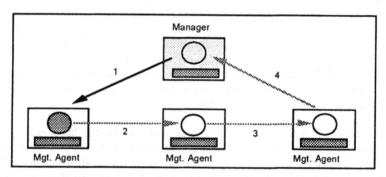

Fig. 6: Mobile Agent-based Management Services

Additionally, a mobile management script could act as a front end to a customer and travel automatically to the corresponding management system or agent, in order to perform the desired management operations (see Figure 7). For example, a management service could be provided by sending an appropriate mobile agent from a management user to the corresponding operations system (representing the managing system), which subsequently performs the requested management operations locally at the opeations system. Note that the agent could also travel to several operations systems.

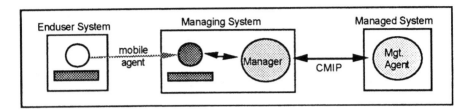

Fig. 7: Mobile Agent-based Management Service Customization (e.g. for CPM)

Note that the aspect of mobile agents, i.e. mobile management scipts, will have fundamental impacts on existing management architectures and the related management protocols, since management relies from the early beginning on the client/server model. However first management platforms consider already the possibility of management scripts [Perr-95]. This issue has to be studied in more detail in the future.

4 Conclusion

Based on a taxonomy of Intelligent Agents, existing agent technologies have been classified within this paper. With the focus on script-based multi-agent systems, the impacts of the concept of remote programing based on the property of agent mobility in the field of telecommunications have been investigated. Basically, two agent approaches relevant in this context have been identified: the smart network approach based on fixed agents, i.e. agent execution environments supporting the remote execution of mobile scripts, and the smart message approach based on migrating agent scripts. These approaches have been projected onto both existing personal communications and network management architectures, such as IN and TMN.

It can be assumed that in particular the aspect of agent mobility has a great potential to influence the development of future telecommunications and management applications, since it represents an alternative to the traditional client/server interaction model [Harr-95]. However, the delegation and distribution of service control and management by means of mobile agent technology is not necessarily the ultimate solution for any type of problem. Whether mobile agents provide for better performance than traditional client/server computing strongly depends on the interaction patterns and the "size" (code plus state information) of the agent to be transfered. Therefore, agent technology will probably not remove the traditional way of client/server computing. Rather mobile agents should be regarded as an "add on" to existing computing platforms, providing to basis for future telecommunications and management platforms. This means that it seems most likely, that state-of-the-art computing platforms, such as OMG's CORBA, will be enhanced with the capability to support mobile agents. This will allow the application developer to decide himself, which technology to use for solving a particular (distributed) computing problem.

References

[ACM-94]   Communications of the ACM Journal "Intelligent Agents",
              Vol.37, No.7, July 1994
[Bern-95]   H. Berndt, et.al.: "Service and Management Architecture in TINA-
              C", Proceedings of the 5th TINA Workshop, Melbourne, Australia,
              February 1995
[Bond-88]   A. Bond, L. Gassner: "Readings in Distributed Artificial
              Intelligence", Morgan Kaufmann, Los Angeles, 1988
[Bors-92]   N.S. Borenstein: "Computational Mail as Network Infrastructure
              for Computer-Supported Cooperative Work", CSCW'92
              Proceedings, Toronto, November 1992
[Bors-94]   N.S.Borenstein: "EMail with a Mind of its own: The Safe-Tcl
              Language for Enabled Mail", ULPAA, Barcelona, 1994
[Ches-95]   D. Chess et.al.: "Itinerant Agents for Mobile Computing", Internal
              IBM paper submitted for publication in IEEE Personal
              Communications Magazine, 1995
[Depu-95]   F. Depuy et.al.: "The TINA Consortium: Towards Networking
              Telecommunications Information Services", XV. Int. Switching
              Symposium, Berlin, Germany, April 1995
[Ecka-95]   T. Eckardt, T. Magedanz, R. Popescu-Zeletin: "Application of
              X.500 and X.700 Standards for Supporting Personal
              Communications in Distributed Computing Environments", 5th
              Workshop on Future Trends of Distributed Computing Systems
              (FTDCS), South Korea, August 1995
[Etzi-94]   E. Etzioni, D. Weld: "A Softbot-Based Interface to the Internet",
              pp.72-76, Communications of the ACM, Vol.37, No.7, July 1994
[Fini-94]   T. Finin et.al.: "KQML as an Agent Communication Language",
              3rd International Conference on Information and Knowledge
              Management (CIKM'94), ACM Press, December 1994
[Gene-94]   Geneserth, M.R., Ketchpel, S.P.: "Software Agents",
              Communications of the ACM, Vol.37, No.7, July 1994
[Gins-91]   M. Ginsberg: "Knowledge Interchange Format: the kif of death",
              AI Magazine, Vol.5, No.63, 1991
[Goya-91]   S.K. Goyal: "Knowledge Technologies for Evolving Networks",
              Integrated Network Management II, Krishnan & Zimmer (Eds.),
              Elsevier Science Publishers, 1991
[Harr-94]   C.G. Harrison et.al.: "Mobile Agents: Are they a good Idea", IBM
              Research Report, RC 19887, October 1994
[HP-89]   Hewlett Packard. "New Wave Agent Guide", Santa Clara, 1989
[Iida-95]   I. Iida et.al.: " DUET: Agend-based Personal Communications
              Network", XV. Int. Switching Symposium, Berlin, Germany,
              April 1995
[INAP-93]   ETSI DE/SPS-3015: "Signalling Protocols and Switching - IN
              CS-1 Core Intelligent Network Apllication Protocol (INAP)",
              Version 08, Mai 93
[Inde-95]   K. Indermaur: "Baby Steps", BYTE Magazine, pp. 97-104, March
              1995

[Java-95]	Sun Microsystems: "The Java Language: A White Paper", WWW: "http//java.sun.com", 1995
[Kay-84]	A. Kay: "Computer Software", Scientific American, Vol.251, No.3, pp.191-207, 1984
[Laue-95]	G. Lauer et.al.: "Broadband Intelligent Network Architecture", Intelligent Network Workshop (IN'95), Ottawa, Canada, May 1995
[Lash-94]	Y. Lashkari et.al.: "Collaborative Interface Agents", Int. Conference on Artificial Intelligence, MIT Press, 1994
[Maes-94]	P. Maes: "Agents that Reduce Work and Information Overload", pp. 31-40, Communications of the ACM, Vol.37, No.7, July 1994
[Mage-95]	T. Magedanz: "On the Integration of IN and TMN - Modeling IN-based Service Control Capabilities as Part of TMN-based Service Management", ISINM'95, Santa Barbara, USA, May 1-5, 1995
[Meye-95]	K. Meyer: "Decentralizing Control and Intelligence in Network Management", Integrated Network Management IV, Ed. Sethi et.al., ISBN 0412715708, Chapman&Hall, 1995
[Moha-95]	S. Mohan et.al.: "A Personal Messenger Application Based on TINA-C", XV. Int. Switching Symposium, Berlin, Germany, April 1995
[M.3010]	ITU-T Recommendation M.3010: "Principles for a Telecommunications Management Network", Geneva, November 1991
[N2139]	ISO/IEC JTC1/SC21/WG4 Working Document N2139: "Command Sequencer", March 1995
[OMG-95]	OMG Document 95-1-2: "Common Facilities Architecture", Revision 4.0, January 3, 1994
[Oust-94]	John K. Ousterhout: "Tcl and the Tk Toolkit", Addison-Wesley, 1994
[OVUM-94]	C. Guilfoyle, E. Warner: "Intelligent Agents: the New Revolution in Software", Technical Report, OVUM Limited, 1994
[Perr-95]	G.S. Perrow et.al.: "The Abstraction and Modelling of Management Agents", Integrated Network Management IV, Ed. Sethi et.al., Chapman&Hall, 1995
[Q.12xx]	ITU-T Recommendations Q.12xx Series on Intelligent Networks, Geneva, March 1992
[Rein-94]	A. Reinhardt: "The Network with Smarts", BYTE Magazine, pp. 51-64, October 1994
[RFC1157]	J. Case et.al.: "A Simple Network Management Protocol (SNMP)", May 1990
[Rizz-95]	M. Rizzo, I.A. Utting: "An Agent-based Model for the Provision of Advanced Telecommunication Services", pp 205-218, Proceedings of the 5th Telecommunications Information Networking Architecture (TINA) Workshop, Melbourne, Australia, February 1995
[Stam-90]	J.W. Stamos, D.K. Grifford: "Implementing Remote Evaluation", IEEE Transactions on Software Engineering, Vol.16, No.7, pp. 710-22, July 1990
[Wayn-94]	P. Wayner: "Agents Away", BYTE Magazine, pp. 113-118, May 1994

[Wayn-95]   P. Wayner: "Free Agents", BYTE Magazine, pp. 105-114, March
             1995
[Whit-94]   J.E. White: "Telescript Technology: The Foundation for the
             Electronic Marketplace", General Magic White Paper, 1994
[X.711]   ITU-T Recommendation X.711: Information Technology - Open
             Systems Interconnection - Common Management Information
             Protocol Definition, 1992
[Yemi-91]   Y. Yemini et.al.: "Network Management by Delegation",
             Integrated Network Management II, Krishnan & Zimmer (Eds.),
             Elsevier Science Publishers, 1991

Introducing Active Managed Objects for Effective and Autonomous Distributed Management

Anastasia Vassila and Graham J. Knight

University College London, UK

Abstract. In this paper, we criticise the manager - agent model as it is used in current systems, and propose a powerful solution towards flexible management by introducing *Active Managed Objects (AMOs)* which are remotely programmable entities that carry out management tasks locally; these tasks are defined using a scripting language. We examine the notion of an AMO system, describe its architecture and present some early results. We finally attempt to place the concept within the TMN[1] infrastructure.

1 Introduction

Network Management has become a significant field in the past few years. Standardisation bodies, network equipment vendors, as well as the research community, put a lot of effort in visualising requirements as well as producing solutions for the effective management of large networks. The established approach to the representation of management information comes mainly from the object oriented and distributed database fields. Management data is stored in databases kept at the managed elements and managing applications monitor the data, which must be transferred to the managing application environment before being transformed to management information. Management in this way, tends to require bandwidth, especially at times when the system or the link is problematic and requires urgent management intervention.

The focus of our study is to extend the functionality of existing architectures by shifting some of the management intelligence close to where management data resides. Most decisions will then be made locally, while data is transformed into useful management information before leaving the network element, thus, the link and the managing application will not be overloaded with unstructured information.

The remainder of this paper is structured as follows: In the next section, we provide some general information on network management models and terminology and we then discuss the present approaches and try to identify the existing problems. Following that, in section 3 we suggest and discuss a solution towards more flexible and effective management in a hierarchical environment. This allows us to introduce our mechanism of *Active Managed Objects* and explain its structure. To support this, in section 5 we describe a prototype implementation of such a system, using OSIMIS[2] and the Tcl[3] scripting language. Finally, we give a view of our future work on the subject and a brief description of related work in the field.

2 The Current Approach to Network Management

One of the basic concepts on which most of the existing management architectures are based is the Manager – Agent model. This model is a special case of the client – server model. In the general case, numerous clients request services from few specific servers. This fact justifies placing the bulk of the load on the clients, where there is the most computing power, instead of overloading servers which have to execute requests. In network management, the situation is reversed; entities are grouped into management domains, and the model becomes a one - to - many relationship; a client (manager) collects information concerning numerous servers (agents) in the system.

Today's dominant management architectures are all based on this model and implement it in different fashions[4],[5]. In the OSI management framework, agent processes can perform complex selection operations and calculations on MIB objects.

Nevertheless, they act as database interfaces, and as such they cannot take any decisions concerning management activities. As OSI management is significantly dependent on statistical measurements on aspects of the system, one can imagine the amount of data that may reach a managing process after the invocation of a single management operation, and the resulting overhead of such a procedure both on the link and on the manager. Moreover, management functionality is lost in the event of communications failure between manager and agents, as agents are not autonomous entities and do not have any knowledge of management objectives.

To illustrate the problems that the static manner in which management operations take place at present, we will consider an example from the MIB - II of SNMP [6]. We can evaluate the percentage of erroneous packets received on an interface, by calculating :

$$A = ifInErrors/(ifInUcastPkts + ifInNUcastPkts)$$

Each time there is a considerable increase in the error rate on a particular interface, we would like to decrease the value of the $ifMtu$ attribute. This attribute represents the size of the largest packet which the specified interface can handle (send or receive) and by decreasing it, we hope to reduce the possibility of encountering errors[1]. The new value of the $ifMtu$ attribute is provided by executing an algorithm $LowerMtu(A, ifMtu)$ in the manager. This algorithm is devised by the local administration while its description is out of the scope of this paper. The conditions used to decide when $LowerMtu()$ should be invoked are determined by another algorithm $error(A, A^\sim, t)$ (where $A^\sim$ represents the

---

[1] It is a common technique in packet - switched networks to reduce the size of the packets that can be handled by a specific interface in order to diminish the probability of packets being affected by errors and lower the cost of retransmission.

previous value of A), which calculates the increase of A in the time elapsed between two consecutive calculations of A. The outcome of *error*() is compared against a threshold b. Therefore, the procedure realising the above management task will be:

```
for every interface from 1 to ifNumber
    get the values for ifInErrors,
        ifInUcastPkts, ifInNUcastPkts
    calculate A
    if error(A, A˜, t) > b
        get the value for ifMtu
        execute LowerMtu(A, ifMtu)
sleep(t)
```

When using the mechanisms provided by current architectures, we can see that the link is heavily used, as this task must be performed for every interface of the managed entity. The selection mechanisms of CMIP[7] can be used in order to retrieve the required attribute values from the managed system effectively, but the algorithms are executed in the manager node, while threshold and event reporting mechanisms cannot be used as b is a threshold devised by the local administration and is not specified in the standard MIB.

The use of passive agents in the manager - agent model, leads to ineffective distribution of management responsibilities as can be seen from the above example and its illustration in Fig. 1. In particular, network traffic increases as management operations become more demanding and sophisticated. Managing processes carry the load of processing and evaluating data coming from the system in a static manner; after receiving some notification of an event, the manager requests some additional information concerning the object that generated the notification and/or other related objects. The network elements retrieve and send MIB values which become useful information only after some operation is performed on them (e.g. comparison, transformation and presentation) upon their receipt by the managing application. Only after performing these operations can a manager issue control actions back on NEs.

The provision of more powerful managers does not offer an adequate solution, especially in a hierarchical architecture. Even very powerful managing applications will be overwhelmed by these tasks when dealing with a continuously increasing number of network elements and other management entities, as requirements build up in an enterprise network.

3   Adding Flexibility to Managed Systems

In a hierarchical environment, lower layers provide information that is summoned by the layers above. NEs provide raw data, as mentioned before, to the network management layer. Manager applications there, collect and transform this data to specific information concerning their domain (performance statistics, status

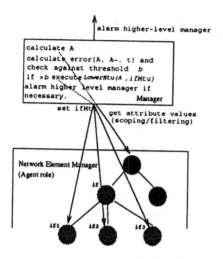

Fig. 1. Realisation of a management task in a conventional system.

of important devices in the network etc.), which is presented to the layer above in the form of an MO.

This model incorporates the idea of disseminating management control to several layers of activity, maintaining though the traditional assignment of roles to management entities, if we examine each layer separately; agent processes remain passive whether in a high or low level of hierarchy. All data is manipulated within the managing processes.

When we consider the way current networks are structured, it pays to concentrate on the managed system side; we can see that NEs become more and more powerful and in most cases, it is now expected that significant computing power should be available for management. Consequently, it is a reasonable consideration to migrate more of the management intelligence to managed systems in order to tackle the problems noted above.

This is not an entirely novel approach; in the SNMP world we see RMON[8] boxes gathering low level network status variables and OSI have defined monitor metric objects[9] that are bound to attributes or whole managed objects and perform statistic calculations to help analyse the performance of the system, and emit notifications when certain thresholds are exceeded. However, RMON boxes make calculations on a per packet basis and cannot be used for high - level management calculations. Furthermore, the RMON MIB is static and cannot be extended as requirements build up. On the other hand, OSI monitor metric objects can only be bound to counter or gauge attributes so their functionality is limited by what can be expressed as an arithmetic value. Furthermore, they can only monitor single attributes from a specific managed object and the algorithm that is executed on them is static and cannot be modified. In this case, the example from the previous section could not have been implemented using metric objects, as the parameters required for algorithm $LowerMtu()$ are acquired from more than one attributes of the interface managed object.

These two efforts both try to reduce network traffic as well as to migrate some intelligence close to management data. We attempt to go further and extend the functionality of managed systems in a way that could qualify them to take decisions and initiate management actions dynamically. In particular, it would be ideal if managers could give directions to the management agents of what the management requirements are, how they can be reached, and then leave the managed system to operate autonomously. What is more, we would like managers to be able to empower agent processes to perform some operations when an event occurs, or at intervals automatically, or even when they receive a special command to do so. They should be provided with the capability of controlling the management behaviour[2]of the managed system dynamically, depending on its state and on management needs. Management overload by "notification storms" at times when the system is problematic must be diminished, as well as that of MIB data gathering and evaluation from numerous sources. That way, the distribution of responsibility would be better divided and the overall performance of the management network would be increased, as the shifting of data manipulation closer to the data repository instead of a central platform (a manager) enables managed systems to make calculations on the requested data and transmit back useful management information. Network traffic relevant to management would be minimised and managed systems would be able to operate independently from managers even if the network (or the manager process) is dead.

In order to fulfill the objectives set out above and to provide autonomous management of systems, we propose a new framework. We will realise management activity as a set of management programs that are downloaded on managed systems, are interpreted and executed locally. We define a management program (MP) as a set of management and generic commands that can access and manipulate local MIB data and perform calculations on it. The set of these commands specifies the expected behaviour of the system[3]and the methods that should be followed when events affecting that behaviour occur. This mechanism allows managed systems to monitor and control local managed objects effectively, with the least possible involvement by managing applications. The managing applications, or the human manager in the simple case, is responsible for providing this program, taking into account the state of the system as well as the current requirements concerning particular systems. The MP is written in a high – level language, that encapsulates management operations and provides mechanisms

---

[2] In the current management systems, a human manager provides an informal description for the behaviour of the management system which is fixed when the MIB is specified and cannot be altered unless the system is shut down. The management behaviour in this case, includes the circumstances in which notifications are issued, how statistical attributes should be calculated etc. It does not involve any notion of control actions which should be triggered when particular conditions arise.

[3] This term is a combination of management behaviour as well as the overall system behaviour, how the entity operates.

for control and access of the system and is downloaded at run - time, enabling the dynamic reconfiguration of the system in respect to its demands. The engine that hosts MPs is located in managed systems and must be able to provide an appropriate execution environment. In this scope, we furnish network entities with a programmable agent component that interprets and executes downloaded management programs. The term *programmable agent component* denotes an execution environment which accepts requests for the execution of MPs and has access to the internals of managed systems in order to carry out the objective of the MPs. These agents do not require continuous communication with the manager; once the program is loaded, they can operate independently from any manager node, so their functionality is not lost on the occasion of a network or manager failure. They nevertheless cooperate with managing applications, as the latter are the providers of MPs, and with agent applications as they form the interface to management information.

4 Active Managed Objects

In order to be able to assess the utility of our ideas as quickly as possible, we have decided to approach them within an existing management framework. We have chosen the TMN/OSI framework, which is especially well - suited, given its well - structured MIB and the richness of its management operations. Moreover, the hierarchical architecture of the TMN provides a fruitful field of utilisation for the Active Managed Object scheme.

Downloaded MPs should be able to access other managed objects and perform operations affecting managed resources. Furthermore, the agent component that hosts and executes MPs, must itself be visible to management in order for a manager to control its operation. Therefore, the most straightforward way to design such an agent component, is in terms of a special managed object which is defined in the MIB of managed systems and can be accessed by managing applications like any other object, according to existing access control rules. We call it an **Active Managed Object (AMO)**.

This approach provides all the necessary conditions for operation outlined above, as:

▷ Managed objects (MOs) in OSI are structured in a way that by inheriting functions from a single superclass, they contain the same set of routines to access information and can therefore communicate with each other for the purpose of management. By representing the programmable agent as a managed object in the MIB, data can be acquired from other MOs in the system without violating encapsulation[10]. AMOs can also exchange information with each other at their MO boundary.

▷ The managing application can control the execution and operation of the AMO using normal management operations, without the need for the development of a specialised interface.

Additionally, AMOs are provided with full management functionality as they have access to event reporting mechanisms. This enables them to report their results to the managing application via notifications, for example, and not to require any regular monitoring as they can execute using local scheduling mechanisms as described in section 5.1.

AMOs, once defined, can be created, deleted, MPs can be changed at run - time, so, AMOs provide an engine for dynamic system behaviour specification. Changes in management policy goals can easily be reflected by the replacement of MPs. The management behaviour of the AMOs themselves is predictable when MPs are carefully written, in a carefully implemented language. Hence, an AMO seen on the outside, is a normal MO with predefined attributes and behaviour. It nevertheless includes a powerful mechanism that activates the managed system it is hosted on, to an autonomous management entity.

The management script from the example from section 2 could be downloaded as an MP on managed systems and executed locally as can be seen in Fig. 2. The AMO would poll the interfaces object regularly and apply *LowerMtu()* when needed, without the need for communication with a manager. This model requires sufficient computing power by managed systems but this is balanced by the gain in bandwidth and manager load.

The internal layout of an AMO is described in the next section.

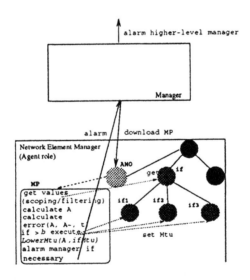

Fig. 2. Realisation of a management task using AMOs.

4.1 The Function of AMOs

The internal structure of the AMO enables it to encapsulate the information necessary to interpret and execute MPs. The active part of the AMO is its ability to communicate with a language interpreter/compiler as well as a mechanism empowering it to be aware of events happening in the system and to take actions according to the MP. Specifically, the AMO has access to mechanisms enabling it to:

▷ *Apply strict access control rules to entities attempting to contact or modify it.*

The managing application that invoked the execution of an MP must be the only entity, or must belong to a limited group of entities that can manipulate it. Moreover and most importantly, not all manager entities should be empowered to bind MPs to AMOs. Actually, this will come as a side - effect of the general access control mechanisms applied to the system. AMOs act on behalf of a manager and as such, they should inherit the manager's access rights.

▷ *Allow and schedule repeated execution of MPs.*

The need for monitoring AMOs must be limited. Instead, the AMO must be event driven and should communicate with the manager asynchronously.

▷ *Check the integrity and safety of MPs.*

There should be mechanisms that reject MPs that include commands and functions that are dangerous to the system. These include unlimited file access, tight loops, memory intensive operations and access to the system kernel. The provision for this kind of security mechanisms is very difficult and requires further investigation. However, the architecture of the MPL (see next section) could include such safety features.

▷ *Present results to the managing application.*

As MPs have a specific function, a managing application will expect results from their execution. Therefore, AMOs must include presentation mechanisms for this purpose.

▷ *Support storage facilities for MPs.*

MPs should be "persistent" once downloaded on MSs. They should survive any kind of system failure and they should be accessible for further use without the need for repeated downloading.

When considering the MP as a whole, we can see that it defines a high - level management goal, as seen in the example for controlling erroneous packets on interfaces. The operations specified in MPs however, are procedures indicating the methods for achieving these goals. These operations are written in a predefined language that the managed system can understand. The issues concerning the language are described in the following section.

4.2 The Management Programming Language (MPL)

The choice for an appropriate management language is very important. We would like to use an existing language and not invent one for the particular application even if in the end we could consider this fact. The MPL provides an interface to management operations performed on managed objects. It therefore must respect the structuring rules of MOs. Resources can only be accessed via operations at the managed object boundary so, the MPL must not contain tools that could be used for accessing them in any other way. It must model protocol operations and express them as a remote managing application would. There are several other restrictions which we attempt to identify in the following sections.

Options. The existing choices for an appropriate MPL are numerous and can variate in structure, functionality and pertinence. We can choose between compiled and interpreted languages, general purpose and special purpose ones (management languages as in [11]).

In [12] there is a detailed analysis of existing programming styles in respect to the requirements of an active database. Our requirements are not that strict, so we can allow our system to adopt an *imperative* programming style as opposed to more expensive and complex ones.

We will therefore select and extend a general purpose procedural language and we will incorporate management procedures via auxiliary routines. Various external events and actions will be handled by the local agent process.

In order to avoid compiling and linking overhead within the agent environment as well as issues concerning the underlying system architecture, we will select an interpreted language by sacrificing execution performance which is nonetheless tolerable in our initial approach to the matter.

Requirements. The programs must be independent from the underlying architecture of the machine they are executed on. We can identify the following requirements regarding the structure of the language:

▷ It must be small in size, so as to minimise the impact that AMOs will have on a NE.

▷ It must be high - level and simple at the same time, so that it can express management operations in a straightforward way and that it is easily learned. It must be oriented towards the requirements of the human manager and support trivial and sophisticated management schemes. Nevertheless, it must be able to form an interface for existing standards and be platform independent.

▷ It must not have unsafe points, in the aspect of harmful operations (e.g. access to the operating system or to other local applications), and must have execution stability. The possibility of an error at run - time must be diminished.

▷ Finally, it must be equipped with mechanisms for concurrent execution of management programs.

The MPL must provide an adequate level of abstraction in order for management operations as well as control/security/access mechanisms to be encapsulated in commands and functions, and model real management requirements. Therefore,

▷ The language must provide access to commands recognizable at the MO boundary.

▷ It must be equipped with notification generating mechanisms for exceptional communication with the managing application.

▷ It must understand the status of the system. This can be accomplished by providing notification receipt and evaluation mechanisms.

5   A Pilot Implementation

We have implemented a simple programmable agent component based on AMOs in order to identify the requirements of such a system by using it, and to experiment with the concept of distributed autonomous management through an existing platform.

Our system, as mentioned before, is based on the OSI management model. This implementation is actually an application of OSIMIS.

5.1   OSIMIS

OSIMIS is an object oriented management platform based on the OSI management model developed at UCL. It implements a generic management system (GMS) which is an object oriented OSI agent that offers the opportunity of implementing new managed object classes by using its high level API. It provides management facilities using CMIP. The application developer has to implement only the intelligent part of managed objects, that is their communication with the real resource. OSIMIS includes a scheduling mechanism that provides the capability of polling the real resource at regular time intervals by using *wake up* mechanisms on the relevant MO routines. This scheduler is implemented as a special managed object class and it is called the Coordinator object.

AMOs can be easily implemented in order to cooperate with the Coordinator and MP execution can be repeated in regular time intervals which can be altered as wished. The Coordinator can also be used to support event - driven communication between the agent and the real resource wherever possible (supported by the real resource kernel). Therefore, this object will be used to provide the necessary support for the repeated execution of MPs in the system as well as their knowledge of asynchronous events happening in specific resources. OSIMIS supports only single threads of execution, therefore there is no notion of concurrent execution of MPs. However, there is work underway, in order to explore the possibility of extending the function of the Coordinator to suspend and resume the execution of MPs according to specific events on real resources and to enable it to keep track of their execution time in order to detect MPs that seem to block. Despite its limitations though, OSIMIS was the most appropriate platform to use, as it is a fully implemented OSI management platform and we also have access to its source code.

5.2   The AMO in OSIMIS

We have extended the OSIMIS agent, enabling it to support AMOs. This has been done in the following way: An Active Managed Object class has been defined in GDMO[13] and the MP is included as an attribute. This attribute can contain the script itself or a filename pointing to it. The AMO also contains attributes that contribute to the execution of the program; we have defined an attribute which keeps the results returned after each execution of the MP in order to enable the manager to consult this attribute when required instead of results being sent after each invocation of the MP. Results are structured as name - value pairs and they represent the outcome of the MP execution. An additional attribute has been defined which declares the interval of time in seconds that elapses between each invocation of the MP. An interval of 0 seconds "turns off" the AMO. The downloading of programs, their execution and the retrieval of results can be achieved through the use of CMIP operations on AMOs. For example, simple MPs could be downloaded using an M_SET operation, activated by an M_ACTION operation, or eventually by a local event, and results can be obtained using M_GET or asynchronously using M_EVENT-REPORT.

In addition, a notification is tied to an AMO, in order to report exceptional states detected in the system as defined in the MP. It reports the managed object that caused the exception and additional information on the nature of the exception (and may include the contents of the result attribute if required). We have also extended the OSIMIS agent enabling it to start the language interpreter execution, and support its communication with the AMO code.

Realisation. As we stressed earlier, the dominant idea was to provide functional extensions to the existing management applications instead of providing another infrastructure for management of large networks. As such, we had to accept the various design limitations such systems impose. AMOs are not tasks with their own execution threads and control, instead they are controlled by a real - world management system. Therefore, our system is limited by the power of OSIMIS and the OSI management framework in general. The actual GDMO definition of the AMO, as well as the attributes it will finally include, is still under consideration.

5.3   Tcl as an AMO Programming Language (CMIS - Tcl)

In order to provide a language that would represent management procedures, we have extended the Tcl scripting language to enable it to perform operations at the managed object boundary. The extended Tcl interpreter, which we call CMIS–Tcl, runs in managed systems and executes the scripts that reside in the relevant attribute of the AMOs. As mentioned earlier (section 4.2), we have decided on a sub - optimal choice for an MPL in order to obtain early results for our system and benefit from the feedback from these results.

We have chosen Tcl as it is an interpreted, public domain, complete language. Tcl could be easily extended to encapsulate OSIMIS operations on managed objects and to provide an interface to CMIS operations within the managed system run – time environment, as it can be easily linked with C and C++ which were the languages used in developing OSIMIS. Apart from the traditional Tcl, a safe version of the language has been introduced[14] and is currently developed to provide a safe environment to scripts that wander round the network and require to execute on various hosts. The concepts of this version could be of use to our system, as our language must protect the system from unreasonable utilisation of computing power as well as from programs that try to access system files and commands of no relevance to management. Tcl is broadly used in the research and industrial community, and the Tcl interpreter is present on many platforms.

The Tcl extensions we developed form an interface to OSIMIS low - level routines that access other MOs in the system, and convert MO information to strings. The existence of these routines made OSIMIS an ideal platform for us to use, as we should otherwise have to implement the bulk of MO manipulation code in respect to Tcl from scratch. The Tcl script is downloaded to the managed system as an ordinary file and an attribute of the AMO points to that file and passes its filename to the Tcl interpreter in order to execute it; see Fig. 3 for a description of the layout of the system.

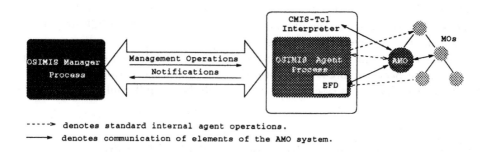

----> denotes standard internal agent operations.
——> denotes communication of elements of the AMO system.

Fig. 3. The execution model of the OSIMIS AMO system.

The OSIMIS/CMIS agent is created as a Tcl application. This was necessary because of the internal structure of Tcl, as the interpreter is given an identifier and if it were created from within the AMO, every AMO would create a new interpreter (except if AMOs could exchange information about the current interpreter in use, which we found unnecessary).

The event reporting mechanism is not very effective at present, as the script has to return from execution in order for a notification to be emitted. This accounts to OSIMIS being single threaded and blocking. The program being run detects a situation in which a notification has to be emitted and writes the appropriate information on a file (string). Upon return, the AMO code detects that it has to emit a notification and does so, after reading the file and including the relevant information in the report. Nevertheless, most of the requirements listed in section 4.2 were satisfied in our hybrid system, except the security and concurrency issues which are currently investigated.

The basic extensions to Tcl that we have implemented so far can be seen in Tab. 1.

Command	Parameters	Description
TclGet	*distingushed_name* *attribute_name*	Retrieves the value of the specified attribute from the given MO.
TclSet	*distinguished_name* *attribute_name* *operation*[4], *new_value*	Sets the value of the specified attribute to *new_value*
TclEvent	*distinguished_name* *string*	Triggers an event concerning the MO specified. The *string* is the information specific to the event.

Table 1. Basic CMIS - Tcl commands.

To summarise, a first version of our AMO system is operational and has been tested with the SNMP MIB-II, proxied by our extended OSIMIS agent. Our simple AMO includes the attributes and notifications discussed in section 5.2. The existence of the Tcl interpreter does not increase either the size of the agent executable, or the execution speed, thus providing a lightweight but powerful and effective agent process.

Nevertheless, not all the relevant issues and possibilities presented in this paper have been explored, as discussed in the following section, and as such, further investigation is essential in order to produce a fully autonomous AMO system.

6 Issues for Further Research and Conclusions

The embryonic state of our system gives rise to many questions and equally many prospective solutions. We will try to put forward some of the dominant issues that concern the development of autonomous managed systems using the AMO concept.

▷ The execution model of the AMO plays a significant role in the functionality of the system. At present, there is a separation between script invocation, which is controlled by the agent application, and execution, which is driven by the script. We could move to a multithreaded model which would treat each AMO as a separate task. In this approach there need to be mechanisms to resolve scheduling and synchronisation problems between tasks.

▷ The choice of the MPL is of primary importance. We have decided on Tcl as a first choice on the grounds of simplicity but we have to consider the performance trade - off of a compiled language versus the interpreted style of Tcl. Furthermore, we could incorporate MIB objects as objects of the language. At present, MOs are "outside" the language mechanisms and rather have to be accessed via function calls.

---

[4] The allowed operation types are replace, setToDefault, addValue, removeValue.

▷ AMOs located in the same or in remote managed systems should be able to communicate and exchange information using sophisticated mechanisms (at present AMOs can be accessed via operations at their MO boundary).

▷ The interpreter should execute regardless the state of the agent process in order for the system to have a chance of recovering when the agent is dead.

▷ The construction of a flexible engine on top of a rigidly built system (the OSI Framework) is bound to yield to compromise. The AMO concept could be realised to operate in a distributed object management architecture like a CORBA[15] platform, when such a system seems powerful enough to provide the mechanisms that an AMO requires (powerful selection methods and naming schemes).

The AMO system would be of much use in a TMN, as MPs could circulate from higher level managing applications to lower level managed systems and realise any management policy currently in effect. Such a hierarchical architecture currently suffers from manipulation of inevitably out of date information, as latency increases when layers of management responsibility have to be updated continuously in order to propagate information higher in the hierarchy. This limitation can be diminished by disseminating management policy to management tasks that can be performed as close to the information they affect as possible, in the form of MPs downloaded on to AMOs residing in managed systems.

Relevant work is being carried out by several research groups as it is identified that effective and autonomous management is an important and pressing issue. In particular, in the RACE ICM project, they have implemented intelligent monitoring objects[16] which have been specified by ISO/ITU. At Columbia University, work is being carried out towards a flexible management solution following a delegation model[17]. In INRS-Telecommunications Canada[18], another delegation framework is being developed for application management and for systems administration as well.

In this paper we argued that it is needed to migrate management intelligence closer to the managed resources in order to reduce traffic specific to management and make reaction to management activities more immediate. This is possible as network elements become powerful, with remarkable processing power. We suggested the concept of Active Managed Objects, and have furnished them with the capability of communicating with a language interpreter and executing programs that realise management operations as commands of a high - level language. We have extended the Tcl language to make it "management - aware" and implemented a prototype system using the OSIMIS management platform as a vehicle for testing the concept of AMOs. Early results suggest that network traffic between manager and managed systems can be remarkably reduced even by implementing simple routines as MPs. Even though our concept is still in an early stage and requires further research, we can see that AMO systems equipped with the mechanisms discussed in this paper will provide an effective tool for distributed, event based, autonomous management.

References

1. ITU/CCITT Recommendation M.3010 – Principles For A Telecommunications Management Network, 1992.

2. G. Pavlou, K. McCarthy, S. Bhatti, and G. Knight. The OSIMIS Platform: Making OSI Management Simple. In *In IFIP Fourth International Symposium on Integrated Network Manag*, May 1995.

3. John K. Ousterhout. *Tcl and the Tk Toolkit*. Addison–Wessley, 1994.

4. J. Case, M. Fedor, M. Schoffstall, and C. Davin. RFC 1157 Simple Network Management Protocol (SNMP), May 1990.

5. ITU Recommendation X.700 – ISO/IEC 7498-4 Information Technology - Open Systems Interconnection - Management Framework, 1992.

6. K. McCloghrie and M. Rose. RFC 1213 Management Information Base for Network Management of TCP/IP-based internets: MIB-II, March 1991.

7. ITU Recommendation X.711 – ISO/IEC 9596 Information Technology - Open Systems Interconnection - Common Management Information Protocol Specification, 1991.

8. S. Waldbusser. RFC 1271 Remote Network Monitoring Management Information Base, November 1991.

9. ITU Recommendation X.739 – ISO/IEC 1064-ii Information Technology - Open Systems Interconnection - Systems Management: Metric Objects and Attributes, 1993.

10. ITU Recommendation X.720 – ISO/IEC 10165-1 Information Technology - Open Systems Interconnection - Structure of Management Information: Management Information Model, 1992.

11. U. Warrier, P. Relan, O. Berry, and J. Bannister. A Network Management Language for OSI Networks. In *Proceedings of SIGCOMM 1988, ACM Press*, pages 98 – 105, 1988.

12. A. Gal and O. Etzion. Maintaining Data-Driven Rules in Databases. *IEEE Computer*, pages 28 – 38, January 1995.

13. ITU Recommendation X.722 – ISO/IEC 10165-4 Information Technology - Open Systems Interconnection - Structure of Management Information: Guidelines for the Definition of Managed Objects, 1992.

14. Nathaniel S. Borenstein. EMail With A Mind of Its Own: The Safe-Tcl Language for Enabled Mail. In *ULPAA*, pages 377–390, June 1994.

15. OMG Document Number 91.12.1. *The Common Object Request Broker: Architecture and Specification*.

16. G Pavlou, K. McCarthy, G. Mykoniatis, and J. Sanchez. Intelligent Remote Monitoring. In *this volume*.

17. G. Goldszmidt and Yemini Y. Evaluating Management Decisions via Delegation. In *IFIP International Symposium on Network Management*, April 1993.

18. Jean-Charles Gregoire. Management with Delegation. In *IFIP, AIPs Techniques for LAN and MAN Management, Paris, France*, 1993.

Intelligent Remote Monitoring

George Pavlou, Kevin McCarthy - University College London, UK
George Mykoniatis, Jorge Sanchez - National Technical University of Athens, Greece

Abstract. Intelligent monitoring facilities are of paramount importance in both service and network management as they provide the capability to monitor quality of service and utilisation parameters and notify degradations so that corrective action can be taken. By intelligent, we refer to the capability of performing the monitoring tasks in a way that has the smallest possible impact on the managed network, facilitates the observation and summarisation of information according to a number of criteria and, in its most advanced form, permits the specification of these criteria dynamically to suit the particular policy in hand. The OSI management metric monitoring and summarisation management functions provide models that only partially satisfy the above requirements. This paper describes our extensions to the proposed models to support further capabilities, with the intention to eventually lead to fully dynamically defined monitoring policies. The concept of distributing intelligence is also discussed, including the consideration of security issues and the applicability of the model in CORBA-based environments.

Keywords: Management, Monitoring, OSI, TMN, GDMO, CORBA

1. Introduction

Managing Quality of Service (QoS) is of paramount importance in both current and future service architectures operating over underlying broadband technologies (SDH, ATM etc.). Engineering considerations such as minimising the impact of management systems on the managed network and reacting in a timely fashion to performance degradations need to be supported by relevant computational and information models. In the TMN framework [M3010], OSI Management principles [X701] are used to structure and access management information in an object-oriented fashion. The need for generic information models supporting intelligent monitoring activities was recognised long ago, resulting in the metric monitoring [X738] and summarisation specifications [X739].

These move essential intelligence close to the managed resources, reducing the amount of management traffic and providing support for a sophisticated event-driven operation paradigm. Though of undoubted usefulness, it soon becomes apparent that the intelligent monitoring policies they support are predefined. A slightly different policy than the ones supported needs to be specified in object-oriented terms, realised and then fielded in systems supporting such facilities as pre-compiled logic i.e. it is necessary to go through the full development cycle. Ideally, fully flexible generic facilities are necessary, where new policies can be realised dynamically, without the need to alter anything at the managed end of the spectrum.

Having realised these limitations, in the context of the RACE-II ICM project we have carried out research leading in the provision of such advanced facilities. This research culminated in the specification of the Generic Support Monitoring Objects (GSMO) [Sanchez] and their object-oriented realisation as part of the OSIMIS TMN platform

[Pav95a]. In this paper, we explain the relevant issues and operational models, present two examples to show the applicability of these concepts and discuss issues related to dynamic intelligence, security and future mappings on CORBA.

2. The OSI Management and Directory Model

The Telecommunications Management Network (TMN) [M3010] provides the framework for managing heterogeneous networks and services. It proposes a hierarchical layered architecture with functional blocks communicating management information across reference points. These become interfaces when functional blocks are mapped onto distinct physical entities that exist at different network nodes.

The TMN reference points and interfaces are based on OSI application layer services. They are of a transaction-oriented nature, involving both protocols as well as associated information models. OSI Management [X701] is the main base technology, supported by the OSI Directory [X500] in order to support dynamic addressing, global naming and location and other transparencies. File Transfer Access and Manipulation (FTAM) may be also used to support bulk data transfer e.g. for software management in the context of service deployment. Finally, Transaction Processing (TP) is necessary to guarantee the consistency of complex distributed management transactions.

OSI Management projects a fully object-oriented model, with applications in agent roles "exporting" managed objects that encapsulate managed resources while applications in manager roles access these objects in order to implement management policies. Managed objects are formally specified in GDMO [X722], which is a formal object-oriented specification language with emphasis in management. The associated service/protocol (CMIS/P) [X710] has essentially "remote method call" semantics. The separation between manager and agent roles serves the purpose of the model and it is not strong in engineering terms: applications and even objects may be in both roles and this is in fact the norm in a hierarchical layered architecture such as the TMN.

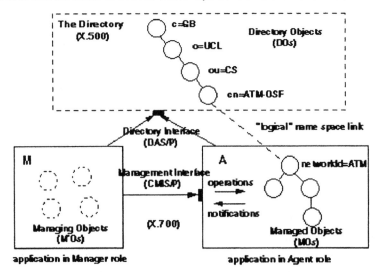

Fig. 1. The combined OSI Management and Directory model

A key difference between OSI management with other ODP-based object-oriented frameworks such as CORBA [CORBA] is that objects always come in "clusters" or "ensembles". Objects in these have a lot of relationships but containment is treated as a primary relationship that yields unique hierarchical names. The cluster is also the unit of global distribution but managed objects and other functional components may also be distributed locally. An object-oriented database capability is offered through facilities known as scoping and filtering. Many objects may be selected through scoping while the selection may be further controlled through filtering, which allows complex assertions on attribute values. This facility may be exercised across the management interface using CMIS but also through object attributes with scope and filter syntax and semantics. In the latter case, their evaluation depends on the managed object behaviour. This capability is used by the intelligent monitoring objects.

Global distribution is based on the OSI Directory which provides a hierarchical distributed object-oriented database. Applications notify the directory of where they are and of their capabilities so that location and other transparencies can be realised. This also allows managed objects to be addressed through global names, which contain information of the logical "cluster" they belong to e.g. a particular TMN Operations System Function (OSF). OSI Management is an object-oriented technology in terms of information specification and access: objects are visible "on the wire" but the internal structure of relevant applications is not dictated and may not even not object-oriented. Platform infrastructures such as OSIMIS [Pav95a] have shown how such an object-oriented specification may be mapped onto a fully object-oriented realisation, providing high-level APIs and various transparencies [Pav94]. The Network Management Forum is currently looking to standardise such object-oriented APIs. The combined OSI Management and Directory model is shown in Figure 1.

3. Monitor Metric and Summarisation Facilities

While the manager-agent model does not in itself restrict the amount of intelligence that may be specified and realised by managed objects, most standard specifications concentrate in providing a rich enough set of attributes and actions which model information and possible interactions with the underlying resource. Notifications are also provided to report significant exceptions but they are usually generic ones e.g. object creation / deletion and attribute value change.

The OSI Structure and Definition of Management Information (SMI / DMI) specifies generic attribute types such as counter, gauge, threshold and tide-mark. Gauges model entities with associated semantics e.g. number of calls, users, quality of service etc. or the rate of change of associated counters e.g. bytes per second etc. Thresholds and tide-marks may be applied to gauges and generate QoS alarms and also attribute value changes, indicating change of the high or low "water mark". Such activities are of a *managing* nature, in fact in the SNMP paradigm they are solely performed by management stations.

Although thresholding functions could be made part of managed objects modelling real resources (in fact, there have been such early object specifications), it does not take long to recognise their genericity. As such, they should be better provided elsewhere so that they become re-usable. The relevant ISO/ITU-T group recognised the importance of generic monitoring facilities and standardised the metric monitor [X738] and summarisation[X739] systems management functions. By making such functions generic, it is possible to implement them once and make them part of the associated platform infrastructure.

The whole idea behind monitor metric objects is to provide thresholding facilities in a *generic* fashion. Monitor metric objects may be instantiated within an application in an agent role and be configured to monitor, at periodic intervals, an attribute of another real resource managed object. The underlying resource may be a physical or logical aspect of a network or service element. It may also be a composite one, realised through an Information Conversion Function (ICF) by accessing subordinate information models. The observed attribute should be a counter or gauge and the metric object either observes it as is or converts the observed values to a rate (derived gauge) over time. Statistical smoothing of the observed values is also possible if desired.

The main importance of this facility is the attachment of gauge thresholds and tidemarks to the resulting derived gauge which may generate quality of service alarms and indicate the high and/or low "water mark", as desired by systems using this function. In fact, the metric objects essentially enhance the "raw" information model of the observed object. The metric monitor functionality can be summarised as:

- *data capture:* through observation or "scanning" of a managed object attribute
- *data conversion*: potential conversion of a counter or gauge to a derived gauge
- *data enhancement:* potential statistical smoothing of the derived result
- *notification generation:* QoS alarm and attribute value change notifications

The metric monitoring model is shown in Figure 2.

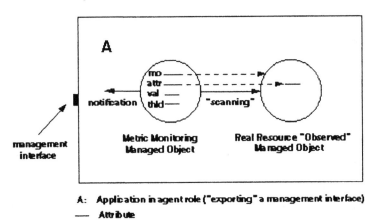

A: Application in agent role ("exporting" a management interface)
— Attribute

Fig. 2. The metric monitoring model

Gauge thresholds are always specified in pairs of values: a triggering and a cancelling threshold. The former will generate a notification when crossed only if the latter has been previously crossed in the opposite direction. This prevents the continuous generation of notifications when the measured value oscillates around the triggering threshold and is known as the *hysteresis* mechanism. Figure 3 shows both high (e.g. "overutilisation") and low thresholds (e.g. "under-utilisation") applied to the observed value. Typically, a normalised value of around 5% is used as the hysteresis interval.

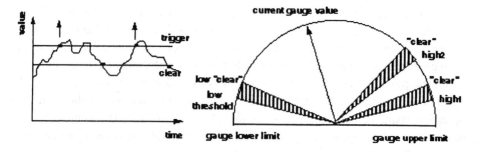

Fig. 3. Thresholding with hysteresis

The metric objects offer the OSI management power through event reporting and logging, even if the "raw" observed management information model does not support such notifications. More importantly, they obviate the use of rates, thresholds and tidemarks in a way tied to specific managed objects but they allow the same flexibility and power dynamically, whenever a managing system needs it. Such a monitoring facility reduces the management traffic between applications and their impact on the managed network by supporting an event-based operation paradigm. It should be emphasised that these facilities are of "managing" nature but offered within a managed object cluster across a management interface.

The summarisation objects extend the idea of monitoring a single attribute to monitoring many attributes across a number of selected managed objects. They offer similar but complementary facilities to metric objects. In this case, there are no comparisons / thresholding but only the potential statistical smoothing and simple algorithmic results of the observed values (min, max, mean and variance). These are reported periodically to the interested managing systems through emitted notifications. The observed managed objects and attributes can be specified either by supplying explicitly their names or through CMIS scoping and filtering. The observed values may be raw ones, modelling an underlying resource, or enhanced values as observed by metric objects. They can be reported either at every observation period or after a number of observation periods (buffered scanning).

The major importance of this facility is that a number of values a managing system needs can be specified once and then reported as mentioned, without the managing system having to send complex CMIS queries periodically or having to perform the statistical smoothing etc. Intelligence in this context has to do with the periodic scanning and reporting of diverse information automatically, after the criteria for its assembly have been specified once. Such criteria may be expressed through CMIS scoping and filtering, providing a lot of flexibility in summarising information of dynamic nature (e.g. traffic across certain ATM virtual path connections etc.) Any other complex computations on the summarised information are left to be performed by the managing system. Such logic is usually of static, pre-compiled nature.

4. Advanced Intelligent Monitoring Facilities

Rationale

While the metric monitor and summarisation facilities can be combined to provide a lot of flexibility, they still leave the task of complex calculations and comparisons to managing systems. In particular, in most cases it is not one attribute value that needs to be monitored and compared to thresholds but the combination of more than one attributes, possibly across different managed objects. The derived value is thus the result of the application of a *mathematical operation* on the observed values. Such an operation could be the $sum(a_1+...+a_n)$, $div(a/b)$, $diff(a-b)$ etc. The combination of more than one operations may be used to construct arbitrary expressions, albeit at the cost of multiple intelligent monitoring objects.

In short, a facility is needed that combines the properties of the metric monitor (thresholding on a derived result) and summarisation objects (observation of arbitrary objects and attributes). This combination is effected through a mathematical operation that is performed on the observed values to yield the derived result. We have recognised the need for such a facility early on and have specified and implemented a number of GDMO classes that provide this functionality, termed collectively *Generic Support Monitoring Objects* [Sanchez]. Since then, the relevant ITU-T standardisation committee has also recognised this need and provided amendments to the metric and summarisation functions offering similar functionality.

In most cases, simple mathematical calculations are enough to express real-world problems. For example, if connection acceptance and rejection counts are kept, the connection rejection ratio is given by the simple formula below. In a similar fashion, if PDU rejected and sent counts are kept, the error rate across that transmission path is given by their ratio. Such arithmetic operations can be either specified statically, through a relevant attribute of the summarisation object, or specified dynamically by combining the observed attribute values with operands. In the latter case, ultimate flexibility is provided in applying arbitrary mathematical operations.

```
Connection Rejection Ratio          Transmission error rate
        Sum(connRej)                        Sum(pdusRej)
-----------------------------         ------------
Sum(connAcc) + Sum(connRej)                Sum(pdusSent)
```

Model

The underlying concepts for the Generic Monitoring Support function are based on the combination of the Metric Monitoring and Summarisation functions. This function provides for the ability to aggregate attribute values, apply operations on them and provide statistical information about the aggregated result of the operation. The function is realised by the Generic Support Monitoring Objects (GSMO) and provides for:

- the ability to monitor information provided by single or multiple attribute types

- the selection of the observed attribute values either explicitly, through the names of the containing objects, or through scoping and filtering

- the identification of an operation (e.g. sum) to be applied on the observed attribute values

- the identification of an algorithm (e.g. uniform weighted moving average) used to smooth the derived result

- the ability to emit notifications when the derived result crosses a threshold or "pushes" a tide-mark or when any of its attributes change

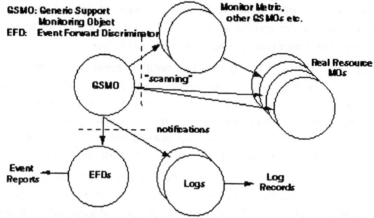

Fig. 4. Intelligent Monitoring and Summarisation model

The model for the operation of the GSMO objects is similar to that of summarisation and is shown in Figure 4. Information is obtained by observing attributes of other objects, including managed objects representing a management view of an underlying network or service resource, metric objects, or other GSMO objects. The attributes of those "observable" objects are accessed at the object boundary, triggering associated behaviour in the same fashion as across a management interface.

A particularly useful type of intelligent monitoring object is the probability estimator. This is a buffered homogeneous object that will scan one observed attribute and will calculate the probability that its value is greater than a preset threshold:

Prob[A>T] = Number of appearances of the event {A>T} / Number of observations

For example, calculating the probability that the connection rejection rate is greater than T involves two intelligent monitoring objects: one calculating the connection rejection ratio as *connRej / (connAcc + ConnRej)* and a probability estimator observing the output of the first one according to the above formula.

While there is a lot of flexibility in specifying attributes and objects to be observed, either explicitly or through scoping and filtering, the full set of available mathematical operations is statically specified and realised when the GSMO information objects become engineering computational objects using relevant platform infrastructure. The introduction of a new mathematical operation implies new specification and subsequent realisation. An additional drawback relates to the fact that any more complex mathematical operation needs to be decomposed to simple generic ones that are supported, necessitating the existence of more than one such objects. This can pose a scalability problem regarding activities at entities with 10's of thousands managed objects (e.g. a mediation function for an access network).

The above observation has led us to research the possibility for specifying dynamically the operation to be applied on the observed attributes. This is possible by merging the relevant operations and operands with the observed attributes in a GSMO attribute with a complex, recursive, tree-like syntax. ASN.1 is very powerful and particularly suitable for such complex representations. The notion of "global attribute" is also introduced as a tuple that consists of a global object name, extending across both the directory and management name spaces, and an attribute type.

Through this type of object, it is possible to specify dynamically any arbitrary mathematical operation as dictated by the management policy, without the need for any additional specification or implementation. The drawback is that the observed "global attributes" need to be explicitly specified as they are referenced in the mathematical operation i.e. the flexibility of using scoping and filtering is lost. This can be compensated by using these in conjunction to the previous types of GSMO objects.

Since managed object names can be global, all the types of GSMO objects can be used for the intelligent summarisation of information which exists logically or physically at different clusters and locations. In this case, the observed objects and attributes need to be explicitly specified. The advantage of monitoring within the same node and reducing management traffic is lost and this defies somehow the purpose of GSMO.

What emerges as a possibility though is the inclusion of further intelligence so that a top-level operation is broken down to sub-parts based on observed values within the same node or domain. Each such sub-part will be delegated to a subordinate GSMO object which will be created in that domain. All the buffered scanning and smoothing will take place locally while resulting values will be reported back to the "master" GSMO object to produce the final value, perform comparisons and trigger notifications. The result of using such a facility is effectively "downloading" intelligence as close to the managed elements as possible, reducing the overhead of management traffic and increasing the observed information timeliness by reducing network latency. The mapping of managed object names to nodes or domains could be through the directory. Further research is necessary to realise such intelligent monitoring objects.

Security considerations

An important aspect of any distributed management environment is security. In the OSI management model, security services comprise authentication, stream / connectionless integrity and access control. The first two rely on information exchanged at association establishment and as such are orthogonal to management information. Access control enables authorised initiators i.e. applications in managing roles or human users driving such applications to have different views of the "exported" management information by the target application. The level of enforced access control could be at the class / instance level or at the individual attribute, action and notification level [X741].

Access control is enforced by special support objects which identify authorised initiators and targets i.e. information to be accessed and the rules of the access control policy. Typically, protected managed objects (targets) are unaware of the access control policy they are subject to. The generic part of the agent application authorises or refuses requests based on the access control objects and the relevant policies. The behavioural logic of managed objects (including the access control ones) is totally unaware of the access control function. This is particularly important, since it allows the access control and management policies to be kept separate.

Unfortunately, this does not hold for intelligent monitoring objects - their sophisticated functionality comes at a price. Intelligent monitoring objects act essentially as managing objects within an application in agent role, accessing other real resource objects at the object boundary. According to the standard access control model, unauthorised initiators may be prevented from creating such objects. If though creation rights are granted, this essentially means that the initiator in question has read access rights to the whole object cluster. This is because intelligent monitoring objects contain attributes whose values point to other objects and attributes and the latter can *not* be checked by the access control function in a generic fashion.

Since it is very common for applications in managing roles to have only partial access to information across a management interface, it is essential to allow for such access control policies in the context of intelligent monitoring. The solution is to make the relevant monitoring objects aware of the access control policy and check that the latter is not violated. This check is performed when they are created or whenever the monitored targets change e.g. through a set operation. The drawback of this approach is that the access control infrastructure becomes non-transparent, though its visibility can be eliminated through object-oriented realisation. This is considered a small price to pay given the rich functionality such intelligent objects provide. In addition, due to their genericity, they are implemented only once and become subsequently part of the relevant platform infrastructure, so the "problem" is eliminated.

Realisation

Implementing OSI management based information specifications can be a daunting proposition without suitable supporting infrastructure. In our case we have implemented standard metric monitoring, summarisation and intelligent monitoring (GSMO) objects using the OSIMIS object-oriented TMN platform [Pav95a]. In OSIMIS, GDMO information objects become C++ engineering objects. A GDMO/ASN.1 compiler was used to produce stub managed object infrastructure, which hides completely all access details, leaving only behavioural aspects to be implemented.

An important platform aspect for realising such objects is the ability to evaluate scoping and filtering locally, to address managed objects by name and to access their attributes at the object boundary. OSIMIS is designed with flexibility in mind and these functions are supported through easy to use APIs. Finally, accessing objects can be through the same API regardless of location, offering useful transparency services. The resultant implementation has become an integral part of the OSIMIS platform. The monitoring objects exist as pre-compiled knowledge at every application in agent role, ready to be instantiated by managing systems to offer their services.

5. Applicability Examples

Let's examine now the applicability of the previously described intelligent monitoring facilities through two representative examples.

FDDI Network Utilisation

As the first example, we will consider a performance monitoring case study for a FDDI Metropolitan Area Network (MAN). The management policy in this case is to continuously monitor the utilisation of the network and generate both early warnings and "red light" quality of service alarms, indicating levels of congestion. In addition, we would like to keep the history of network utilisation over periods of time and, in particular, the highest congestion levels reached over those observation periods.

A FDDI MAN has ring topology and packets are "injected" in the ring at a particular source node, to be subsequently forwarded across the ring by other nodes until they reach the specified destination node. Usually FDDI rings are used as backbones, carrying traffic of other Local Area Networks (LANs) that are attached to them. At every node, the number of octets (bytes) received and sent through the two FDDI interfaces of that node are known. The network utilisation can be thus calculated by averaging the throughput in terms of outgoing traffic through every node on the ring.

Most FDDI rings come with SNMP-based management interfaces at every node. The latter support attributes related to the activity of every physical interface of that node. This can be considered as "raw" information available to managing systems for data conversion, enhancement, thresholding, history etc. In SNMP-based management systems, all these functions take place in centralised management stations which retrieve this information periodically across the network. This mode of operation is in accordance with the SNMP *fundamental axiom* of physically separating managing and managed functionality in order keep network elements simple.

Given the fact that SNMP agents do no support sophisticated intelligent monitoring facilities, there is not much that can be done to "push" intelligence to the network element level. This intelligence has to be deployed instead at the next level of the management hierarchy, which has to be kept as close to the managed elements as possible. In our case study, this is the adaptation / mediation layer which is conceptually part of the element management layer of a TMN hierarchy. We thus operate a generic CMIS/P to SNMP application gateway or Q-Adaptor Function (QAF) [McCar], which provides automatically an OSI view of SNMP managed objects. In engineering terms, one such device will adapt for all the FDDI nodes, being both a TMN Q-Adaptor (QA) and a Mediation Device (MD). This should operate at one of the FDDI nodes, keeping all the monitoring traffic in the FDDI domain.

Averaging the throughput across the FDDI ring is a function that can be easily provided by the intelligent monitoring support objects. An instance of such an object should be created within the CMIS/P-SNMP QA/MD application with the task to monitor locally the FDDI interface objects, calculate the mean, convert it to a throughput gauge, smooth it statistically and compare it to threshold values specifying early warning and red-light (congestion) conditions. In addition, the highest and lowest throughput levels will be recorded through relevant tide-mark attributes. Note that monitoring locally within the QA/MD will translate to monitoring over the FDDI ring across the SNMP interface of the nodes. Despite this, the polling traffic will be limited in the FDDI domain while the result of the intelligent monitoring object will be presented to higher-level functions in an event-driven fashion, through emitted QoS alarm and attribute value change notifications. Event forwarding discriminator and possibly log objects will have to be created to disseminate notifications or log them locally.

The attributes of the monitoring object set by the managing object or application to support this function are:

```
scannerId           "fddi-thput"
granularityPeriod   10 (secs)
algorithmType       EWMA (exponentially weighted moving avg)
estimateOfMeanThld  { Low=0.8 Switch=Off High=0.85 Switch=On
                      Low=0.9 Switch=Off High=0.95 Switch=On
operation           mean      (of observed values)
baseManagedObject   {}        (the top MIT object)
scope               3ndLevel (interface entries are there)
filter              (objClass=interfaceEntry & ifType=fddi)
attributeId         ifOutOctets
```

Note that the interface objects containing the attributes to be monitored are specified using the scoping and filtering facilities. That way, exact knowledge of the FDDI ring topology in terms of its number of nodes is not necessary. An alternative, less flexible way to specify those objects could be through an instance list. Note also that the observed value is statistically smoothed after the calculation in order to reduce problems with rapid fluctuations. Finally, the threshold values are shown normalised: 85% and 95% are the early warning and red-light values respectively, with 5% hysteresis each. In reality, these would have to be specified based on the units of the measured traffic (octets) and the maximum speed of the network (150 Mbits / sec).

The class of monitoring object used above supported a number of operations to be applied on the observed values, from which the *mean* was used. As we already described, the limitation of this approach lies in the fact that the number of supported operations are finite and pre-defined. Our case study was served well by the existing operations, but it would be nicer to be able to supply a coefficient to be applied to the mean so that threshold values can be normalised. Bearing in mind we are observing bytes / sec, the coefficient for FDDI should be $(8 / 150 * 10^6)$. For example, if our FDDI has four nodes, with b the byte counter at each node and c the coefficient, the observed value is given by the expression:

$$
\begin{array}{c}
* \\
/ \quad \backslash \\
\text{mean} \qquad \text{c} \\
(b_1 \ b_2 \ b_3 \ b_4)
\end{array}
$$

e.g. b_1 = { {elemId=fddi-1, interfacesId=null, ifEntryId=2} ifOutOctets }

In this case, the monitoring support object specifies through one attribute both the observed object instance / contained attribute and the operation to be applied on the observed values. Note that the absence of scoping and filtering implies knowledge of the interface object name. This can be obtained through a get operation with scope and filter, before the monitoring object is created.

Summarising, the functionality described for the FDDI case study is a matter of deciding on the management policy and instantiating the type of intelligent monitoring object suitable for the policy in hand. Semantic-free (generic) applications such as browsers, event monitors, loggers etc. may be used to receive the QoS alarms or to retrieve and plot history data and trends. In short, it may be a matter of minutes (or hours at most) to implement sophisticated monitoring policies, as opposed to the weeks (or months) that would be required otherwise for the full development cycle

ATM Network Monitoring

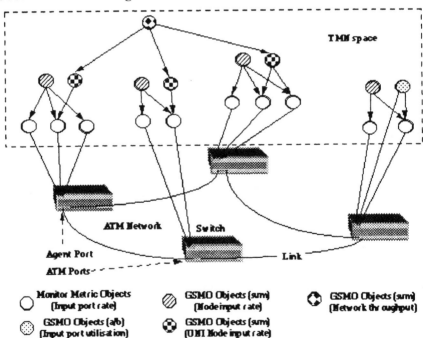

Fig. 5. ATM intelligent monitoring

As a second example, we will consider monitoring the activity of a high-speed network of ATM technology. In this example we will examine how a "raw" attribute (numberOfCellsReceived in a ATM port) that is available at switches with management interfaces can be used to compute the following:

- link / port throughput (rate) and utilisation
- node input rates, including the User-to-Network (UNI) and Network-to-Network (NNI) input rate
- network throughput

As a first step, monitor metric objects should be instantiated at each switch agent (NEF) for the computation of the link or port throughput (numberOfCells over time). As the link/port throughput and capacity are known (the latter is an attribute of the link object), it is possible to calculate the link utilisation through an intelligent monitoring object with a division operation (link rate / link capacity). The overall node input rate can be calculated by another intelligent monitoring object, observing all the link input rates and adding them up (sum operation). Finally, the network throughput can be calculated by a top-level monitoring object, observing all the intelligent monitoring objects that calculate UNI link input rates and adding them up. The relevant hierarchy of monitoring objects on a 3-node network is shown in Figure 5.

Instantiating all these intelligent monitoring objects can be done in a way that does not need pre-defined knowledge of the network topology but "discovers" it from information registered in the directory. Instantiating the intelligent monitoring objects can be done, for example, using an interpreted access API such as the Tcl-MCMIS offered by the OSIMIS platform. A small script e.g. 50-100 lines of interpreted object-oriented logic is enough to start-up the above monitoring function, supported subsequently by generic tools such as browsers, event monitors etc.

6. Discussion

The intelligent monitoring facilities presented try in essence to move intelligence to the agent part of the manager-agent model, reducing network traffic and allowing managing entities to focus in the realisation of management policies, being supported by a rich event-based paradigm. Their generic nature makes it possible to implement them once and make them part of relevant platform infrastructure. Their use can enhance "raw" information models and pays real dividends when used in conjunction with simple information models resulting from SNMP to GDMO translation. Arbitrary expressions combining attribute values are possible, while the use of scoping and filtering provides greater flexibility in specifying information with dynamic properties.

All the above facilities, from the metric monitor objects to the fully dynamic distributable intelligent summarisers, try in essence to break away from the static, compiled knowledge of monitoring intelligence. In fact, they are steps towards fully dynamic intelligence that could be downloaded to the managed resources and operate autonomously as much as possible, reporting back to a master managing object which will have a global view of the domain in which it operates the policy. Such intelligence may have intrusive (control) as well as monitoring aspects in order to react to predefined conditions that express the management policy in hand.

There is a lot of ongoing research towards defining formally arbitrary policy objects. These will contain (part of) the management policy in the form of an attribute containing a script in an interpreted policy language which may access other local or remote objects in a transparent fashion, very much like the intelligent monitors. Temporal constraints should be possible as well as facilities to structure the policy as in programming languages (control loops, procedures etc.) We are actually considering such implementations based on the TCL interpreted language and its object-oriented extensions, while the active policy objects are specified in GDMO [Vassila]. There are a number of problems to be solved though, the main one being concerned with preventing the arbitrary consumption of computing resources that may lead the managed system containing the misbehaved policy object to starvation.

All this work has been based on the OSI management framework. Given the advent of ODP-based approaches, an interesting consideration is to investigate the mapping of those concepts onto OMG CORBA, regarding the latter as the pragmatic counterpart of ODP [X901]. The intelligent monitoring objects are specified in GDMO/ASN.1 from an information perspective and realised as C++ engineering objects, using suitable flexible infrastructure provided by the OSIMIS platform. In theory, it should be possible to map them onto CORBA IDL from a computational perspective and realise them as C++ engineering computational objects on a CORBA platform.

Though this is in theory possible, first considerations reveal a number of hurdles to overcome. First, there is no notion of clusters of objects in CORBA as in OSI agents or access to them with database-like facilities such as scoping and filtering. References to objects are possible but management traffic will be generated, defying partly the purpose. Another more important limitation is that the notion of unique global names does not (yet) exist in CORBA. As such, it is not possible to specify intelligent monitors that span a particular domain administered by one ORB. Finally, it is not easy to map onto CORBA IDL recursive ASN.1 definitions like the one used by intelligent monitors to specify arbitrary operations on observed attributes.

These considerations relate to the current state of CORBA. Trading services may be used in the future to emulate scoping and filtering while the issue of global object references is going to be addressed. The problem of mapping complex recursive ASN.1 syntaxes to equivalent IDL ones remains but there exist workarounds through non-generic mappings. We will report our findings in the future.

7. Conclusions

We explained in this paper the issues behind intelligent remote monitoring and the use of the OSI management access and information models to provide flexible generic support object specifications, the Generic Support Monitoring Objects (GSMO). We also showed the applicability of those objects in the TMN context through two simple but representative case studies. Given the future integration of service execution and management infrastructures in a unifying framework [Pav95b], it is of paramount importance that support for such powerful concepts is maintained.

Acknowledgements

This paper describes work undertaken in the context of the RACE II Integrated Communications Management project (R2059), which is partially funded by the CEU.

References

[M3010]   ITU-T M.3010, Principles for a Telecommunications Management Network

[X701]   ITU-T X.701, Information Technology - Open Systems Interconnection - Systems Management Overview

[X500]   ITU-T X.500, Information Processing - Open Systems Interconnection - The Directory: Overview of Concepts, Models and Service, 1988

[X901]   ITU-T X.900, Information Processing - Open Distributed Processing - Basic Reference Model of ODP - Part 1: Overview and guide to use

[X710]   ITU-T X.710 / X.711, Information Technology - Open Systems Interconnection - Common Management Information Service/Protocol, Version 2

[X722]   ITU-T X.722, Information Technology - Structure of Management Information - Part 4: Guidelines for the Definition of Managed Objects, 1991

[X738]   ITU-T X.738, Information Technology - Open Systems Interconnection - Systems Management - Part 11: Metric Objects and Attributes, 1994

[X739]   ITU-T X.739, Information Technology - Open Systems Interconnection - Systems Management - Part 13: Summarisation Function, 1994

[X741]   ITU-T X.741, Information Technology - Open Systems Interconnection - Systems Management - Part 9: Objects for Access Control, 1995

[CORBA]   Object Management Group, The Common Object Request Broker Architecture and Specification (CORBA), 1991

[Sanchez] Sanchez, J., G. Mykoniatis, J. Reilly, G. Pavlou, K. McCarthy, Intelligent Monitoring Functions in OSIMIS, ICM/WP3/NTUA/0097, 1994

[Pav94]   Pavlou, G., T. Tin, A. Carr, High-level APIs in the OSIMIS TMN Platform: Harnessing and Hiding, in Towards a Pan-European Telecommunication Service Infrastructure - IS&N '94, pp. 219-230, Springer-Verlag '94

[Pav95a]   Pavlou, G., K. McCarthy, S. Bhatti, G. Knight, The OSIMIS Platform: Making OSI Management Simple, in Integrated Network Management IV, pp. 480-493, Chapman & Hall, 1995

[McCar]   McCarthy, K., G. Pavlou, S. Bhatti, N. DeSouza, Exploiting the Power of OSI Management in the Control of SNMP-capable resources, in Integrated Network Management IV, pp. 440-453, Chapman & Hall, 1995

[Vassila] Vassila, A., G. Knight, Introducing Active Managed Objects for Effective and Autonomous Distributed Management, this volume

[Pav95b]   Pavlou, G., D. Griffin, Issues in the Integration of IN and TMN, this vol.

Service Architecture

Session 8 Introduction

Chairman - June Hunt

European Commission DG XIII-B RACE/ACTS
Av. de Beaulieu 9, 4/64, B-1160, Brussels
Tel : +322 295 7936
Fax : +322 296 6272
e-mail jhu@postman.dg13.cec.be

A Service Architecture can be defined as a set of modelling concepts, principles and constraints for the design of the software which realises communication services. An effective Service Architecture needs to impose a structure for modules or components of the architecture and the means for interworking between components. The goal of the architecture must be to identify components which can be re-used, specialised and customised to support the rapid and cost effective construction and deployment of new services.

In the past, many manufacturers have had their own proprietary software architectures for their communication service products, whilst some PNO's have generated architectural procurement specifications of modules/subsystems (e.g. Bellcore's OSCA or the Processor Utility Subsystem of System X). More recently, the thrust has been towards 'open' software architectures for communication systems, starting in the IT world with the OSI and ODP Reference Models and in the Telecom world with the IN and TMN.

In pursuit of an open architecture for telecommunication systems, the work of the Telecommunication Information Network Architecture (TINA) consortium, comprising major PNO's, telecommunication equipment suppliers, computing suppliers and software vendors, has had a major impact. One constituent of this architecture is a Service Architecture. Although the detailed specifications produced by the Core Team are not in the public domain, a considerable amount of information has been made available in published papers and conference proceedings. The first 3 papers in this session describe aspects of the TINA Service Architecture.

However, the modelling principles and constraints implied by any architecture are only effective if they can be utilised by their users (i.e. the specifiers and designers of communication services and software) to produce compliant open systems. User driven experiments are required to verify the approach and feed back the experience to the specifiers of the architecture.

The first paper in this session, 'Session Control Model for TINA Multimedia Services' describes part of the TINA Service Architecture and gives an overview of a

TINA Auxiliary Project which is applying this model to the implementation a Joint Document Editing multi-media conference service.

Another application of aspects of the TINA Architecture is described in the paper 'TINA based Advanced UPT Service Prototype: Early Introduction of TINA through the IN Domain'. In this case, the target is a supplementary service, implemented on top of an existing experimental service, which provides personal mobility services. The paper also deals with interworking between TINA-compliant systems and non-TINA systems.

The third paper 'Use of Atomic Action Principles to Co-ordinate the Interaction between TINA Service Managers' postulates a mechanism for co-ordinating complex interactions between invocations. The use of this mechanism within the TINA architecture will be validated as part of an implementation of an experimental tele-education service in the ACTS programme.

One of the inputs to the TINA architecture was the work done by the RACE I project ROSA. In parallel, the RACE II project CASSIOPEIA has also been defining a Service Architecture, using ROSA as a source. The last paper in this session 'A Comparison of Architectures for Future Telecommunication Services' compares the principles of the 2 architectures. As will be seen in the conclusions of the paper, there is much similarity at the conceptual level and one of the themes of the ACTS programme will be to harmonise the two approaches.

To gain universal acceptance, an open Service Architecture must be 'fit for purpose', consistent and usable. Over the next few years, we all need to work together to ensure these objectives are met.

Session Control Model for TINA Multimedia Services

Fernando Ruano, Telefónica I+D (Spain) ex-representative to TINA-C
Cristina Aurrecoechea, Columbia University (NY - USA)
e-mail: cris@ctr.columbia.edu

Abstract: This paper describes the Session Control model in the TINA Service Architecture. The Session Control is a service independent abstraction that supports the generic control aspects of the service session and provides a base on which services can be built. Information and computational models are proposed, which extend the capabilities of the session beyond the traditional call concept. An example of a JDE-Conference service is given to show the use and flexibility of this model.

1. Introduction

TINA (Telecommunications Information Networking Architecture) is an architecture [1] for services provided on public and private networks. Services in TINA are deployed as objects running on a distributed processing environment enabling the transmission of the control information and the multimedia streams between them. In this paper we introduce the concept of Session Control Segment, as part of the TINA Service Architecture. In TINA, the functions needed to build a service can be grouped into three different Segments:

- **Service Control Segment**: comprising the *specific aspects* (information, logic, access) of a Service.
- **Session Control Segment**: providing generic procedures of controlling *parties* and *resources* and their interactions. These interactions are termed generic, because they can take place in a broad range of services. The objects included in this segment are considered service independent. The entities comprised in this segment are represented by a *Session Graph*.
- **Resource Control Segment**: providing for the configuration and manipulation of the resources required to provide the service. The *communication resources* are represented by a *Logical Connection Graph* [2].

The Session Control Segment offers service interfaces to the Service Control Segment, and makes use of the Resource Control Segment. In this paper we discuss the interface offered by the Session Control Segment to the Service Control Segment and define the computational objects in it. We present an object oriented model for a generic Service Session according to the TINA modeling methodology [3]. In it, parties and resources relate via operational and stream interfaces.

The Session Control Segment will support the execution of new services by defining relationships which go beyond the pure communication aspects. In that sense, this Session Control model covers and extends the traditional *Call Control* and *Call Model* concepts, specifying service-independent mechanisms to describe the state and the behavior of a generic service session. The capabilities of the Session Control Segment defined here cover, among others, the following aspects: party attachment to a service session, resource management, stream binding, role assignment to participants and service composition.

Section 2 will elaborate on the information modeling concepts using OMT notation [4]. The computational viewpoint will be briefly introduced in Section 3. Finally, Section 4 presents an example based on a JDE-Conference Multimedia Service to show the use and flexibility of the model.

2. Information Modeling concepts

In this section we introduce the following concepts:

> *Service Session and Party,*   *Party - Service relationship*
> *Session Member,*   *Session Relationship,*   *Session Graph*

2.1 Service Session Model

The concept of *Service Session* has been defined in TINA Service Architecture to represent an instance of a Service. Here we narrow down the scope of the term to include only the non-specific service aspects of the Session. A user or user application will instantiate at some point, after the necessary controls (subscription, security, etc.), a new Service Session. By doing so, it becomes the *owner* of the session.

To create a Service Session, firstly a relationship object called *Primary Relationship* (PSR) is instantiated. When a user wants to participate in a Service Session, it creates a *Party* who will join this Primary Relationship. This relationship relates the parties with the Service, and is called *Primary*, because once it is created, the Party can create new ones in the context of the session. A Service Session is a container of the relationships established among the entities taking part in the service: the *Session Members*. The Session Member (SM) class is specialized into a Party class and a Resource class:

- A *Party* is an entity with *negotiating capabilities* taking part in the Service Session. It may be any stakeholder (i.e customers, end users, service providers, network providers .
- A *Resource* is a source of support for the execution of the Service. It can be a Video-server, a Mail-box, a Directory Server, a White-board, etc.

The relationships created in a Service Session are called Session Relationships (SR).

2.2 Session Relationship

Session Relationships (SR) are generic associations established among Session Members (SM) -Parties and Resources- and (recursively) Session Relationships during the lifetime of a Service Session. By generic associations we mean they are applicable to, or can take place in, a large number of telecom applications. SRs are binary, semantic associations, which are information objects characterized by the attributes, the related entities types and the constraints they impose on the behavior of the session. SRs integrate association and negotiation aspects (which are orthogonal) to model the session. A SR will be *negotiated*, only when it relates (or it can potentially relate), to several parties.

An SR instance is composed of the following two classes of information objects:
- *Terminus*: containing the state of a Session Member (party or resource) taking part in the SR.
- *Link:* containing the SR attributes shared by the members in this SR. It maintains the references of the termini taking part in the SR.

The generic constraints supported by a basic SR -kept in the Link- are:

- *Cardinality:* refering to the range of values specifying the minimum and maximum number of termini allowed in one instance of the SR, on each side. SRs are bidirectional: they can be navigated starting from any of the members.
- *Referential Integrity:* any SR instance guarantees that the members taking part on it, exist.
- *Ownership:* every SR has exactly one owner: the Party that creates it. The exact meaning of the "ownership" attribute depends on the SR definition. The owner of a SR is not necessarily a member of it.

As an example, the Primary Relationship *Party-Service* is the first SR established and joined by a Party. A terminus in this relationship has an attribute about the general state of the party session: active, suspended, etc. The cardinality of the Primary Relationship is N:1 because N Parties can take part simultaneously in the Service Session.

A SR will also specify:
- *Member Type*. The types of the members restrict the domain of the entities to which it can be applied. They can be subclasses of Party, Resource or Session Relationship (see below SR base classes).
- *Member Role*. If the member is a party (not a resource) then it plays a *role* in the relationship. If a SR is *plain* there will be a single role applicable per member type, otherwise there will be several. The semantics of the role determine the type of interactions that are possible in that association. Members in a SR can change roles dynamically during the SR lifetime.

A Terminus will contain information of the member type and role attributes of a SM in a SR.

Stream binding SR (SbSR)
Stream Interface Binding, or Stream for short, is a clear example of a generic SR established among parties or resources, with cardinality N:M. The Stream concept is modelled at the Session Control level as a Session Relationship, being the SbSr a specialization of a base SR. The result of this specialization is:

- The member's types are enriched with new attributes expressing:
  - the directionality of the stream, two values are possible: sender/receiver. The SbSr establishes the constraint of having exactly one sender and one or more receivers.

- characteristics of the stream end point, like data format, protocols, buffering, etc.

• Several termini of this SR can correspond to the same member, because a stream can be established between two interfaces belonging e.g. to the same party.
•
• The SbSR Link is enriched with new attributes describing the stream, like traffic and quality of service.

2.3 Session Graph

The SRs created in a Service Session are connected through the Session Members taking part in them, and constitute the Session Graph. The Session Graph represents the state of a session at a certain point in time. It shows the Parties and Resources taking part in the Service Session and the SRs between them.

The Session Graph concept will be used to control and manage the session. In the next section the necessary computational objects are introduced, these objects handle the Session Graph information object through operational interfaces. Examples of a Session Graph are shown in section 4 (scenario). The mapping to a Logical Connection Graph is out of the scope of this paper.

3. Computational Modeling concepts

Figure 1. shows an example in which the Session Control segment establishes -after internal negotiation- a SbSR by acting as a client of a CSM (Communication Session Manager) and a SRM (Special Resource Manager). These computational objects have been defined in theTINA Architecture [2]. In the figure, two computational objects are shown: the USP (User Service specific Part) and the SM (Session Member), service specific and service independent objects respectively. There is a mapping 1:1 between Session Members in the Information and Computational Viewpoints.

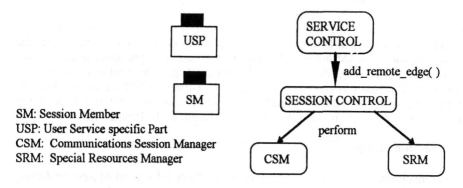

SM: Session Member
USP: User Service specific Part
CSM: Communications Session Manager
SRM: Special Resources Manager

Fig 1: Example of the computational viewpoint of a Session Control Segment

Two alternative models of decomposition for the computational model can be chosen: *distributed* or *centralized*.. In the distributed model, autonomous/peer entities take part in the session, each object holding a local view of the session. In the centralized, there is only one central controller of the session, and around it, subordinate objects which take care of local aspects. In the Session Control Computational Model presented here, the Distributed Control Model is adopted. We believe this model allows a bigger degree of flexibility in the design of the service, bigger reliability, less scalability problems and no additional federation mechanisms. On the other hand, it requires concurrency control support to guarantee the information consistency.

To deal with it, a transaction support is offered in this computational model. This service allows to group several operations into an atomic unit or transaction, such as if the entire chain of operations is not completed -due to: a lack of resources in the network or in affected servers, a lack of agreement among the parties, a fail condition or because the transaction is cancelled-, then a rollback occurs (i.e. the all-or-nothing characteristic of transactions).

The interactions held between parties (now as computational objects) for controlling a service session can be *negotiated* or *not-negotiated*. If the interaction affects only the local context of a party, it will be *not-negotiated*, but if the interaction affects the shared context of the session, then it will be negotiated or not, depending on the state of the session and the role of the party in the service. For instance, the conductor of a conference can be allowed to add or drop a new party without negotiation with the other parties.

In a negotiated interaction, the operation requested is considered an *offer* to other parties involved. The session, comprising the set of SMs, will take this request and will ask the implied parties. The session will then collect the answers and will decide what to do according to them and to some previously specified voting system (unanimity, majority, etc.). If the operation is approved, then it will be executed. In both negotiated and not-negotiated interactions, depending on the request and the conditions established in the session, the result of the interaction will be notified to the other parties.

3.1 Session Members

The Session Control Segment is decomposed into a set of *Session Member* (SM) objects. Each of these SMs represent a member (Party or Resource) taking part in the Service Session. An SM holds and maintains the state of the member in the Service Session (its local view). SMs are specialized according to the kind of members they represent. The first obvious specialization is SM-Party or SM-Resource. An SM-Party may have negotiation capabilities (i.e. it may support the negotiation interfaces). Its lifecycle is equivalent to its attachment to the Primary SR.

3.2 Session Edges

The computational structure of the information object SR is a set of *Session Edges*, each representing a Session Member (party or resource) taking part in the relationship. A Session Member (SM) will contain Session Edges (SEs) representing its participation in different Session Relationships (SRs).

Each SE bears a type and supports an interface (derived from) SrEdge. The type of the SE determines the SRs it can be attached to. The interface supported by any SE allows its represented member to control the session relationship the member is involved in. The SEs in a SR are not hierarchically organized (i.e. with a central edge and subordinated ones), but in a peer-to-peer way, each SE being an autonomous entity.

The information contained in an SR, classified previously as Terminus and Link, is distributed into the component SEs in the following way:

- A SE contains its corresponding Terminus, i.e. the information which is local to its represented SM. An SE will have a role, if it is contained in a SM-Party.
- The Link information object is repeated on each SE in the SR. The consistency of all the copies is maintained by means of notification and negotiation mechanisms.

4. Scenario based on the JDE-Conference Service

This is an example of the use of the Session Control segment for a multiparty, multimedia conference service with WYSIWIS and Common Pointer facilities. The Joint Document Editing (JDE)-Conference service will be composed of a set of application specific objects which interact during the session with the Session Control segment by using the interfaces previously described.

It is not the objective of this paper to focus on the design of the application, but to show how the Session Control features are used, and how their use makes easier the implementation of complex facilities at the application level.

4.1 Application interaction with the Session Control

The interaction is described by showing how the application facilities map with Session Control.

Action 1: Schedule a Conference:

This action will invite possible participants to take part in a future conference. The date of the conference can be discussed (possibly with a range of values) as well as the agenda.

Interaction with SC: The initiator of this action will create a Primary SR (PSR) with the following attributes:
- Application name
- List of starting participants (each participant will or not confirm its attendance)
- Date (negotiable)
- Agenda (negotiable)

The PSR will be created in *scheduled* state and will keep the agreement reached among initial participants starting the session at the specified date. When a user confirms its attendance, it becomes a party in the (non-running) session. Date and Agenda will be discussed sequentially.

The parties will have assigned a role for the session (which is modifiable). This role refers to general aspects of the session, like the capabilities to invite or exclude other members and the general *rights* held on the session. For example, the roles of *actor* or *spectator* refer to the speaking rights the member has. A computational representation of the Session Graph is shown in Figure 2.

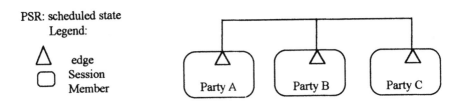

Fig 2: SG after action 1

Action 2: Conference Start-up.

The session will start (not-running state) automatically at the scheduled time. At that moment, a party will set the initial configuration and will change the state from not-running to running.

Interaction with SC: The initiator of the conference will build a (minimum) starting session graph representing the initial conference configuration. This session graph will be offered to the parties which had previously accepted to take part in the conference (in action 1). The parties will negotiate the session graph as a whole (in a single interaction). Once agreed, the session will start running. The state of the Session Graph before the negotiation process is shown in Figure 3. We assume the parties agree on it.

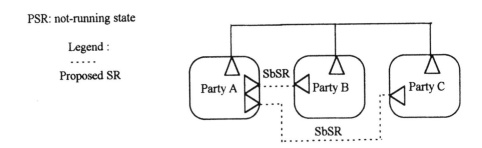

Fig.3: SG being negotiated in action 2

Action 3: Audio-visual interparty stream.

At any time when the session is running, any party can setup a stream to send or receive other parties' voice and/or image.

Interaction with SC: The party A will create two new SbSRs with C with the following attributes:
- Audio, G.711, bandwidth, QoS
- Video, H.261, bandwidth, QoS

These two SbSRs will be synchronized by means of a lip_synchronization_SbSrs (SR group) established between them. Once created, this compound SR will be negotiated among the implied parties. Each party will accept or refuse the connection.

In the SbSR negotiation, if the party with the role of "sender" refuses the SR, then the SR will be rejected; if a party with the role "receiver" refuses the SR, then it will not take part in it. If a party wishes to receive the live audio/video from another party, which is already taking part as "sender" of an SbSr with the same attributes (bandwidth, encoding, etc.), then the party will join the existing SbSR instead of creating a new one.

Action 4: Open document.

At any time, any party may open a document for editing. The edition of the document can be shared with other parties.

Interaction with SC: The party will create a new SR of the type AccessRightsSr with the file. If the file is already opened, the party will join (if allowed) the existing SR. The attributes of this SR will be:
- access permits: read/write
- sharable: the party has to be consulted for any action concerning the resource
- semaphore: a semaphore used to lock the access to the resource will be included

Action 5: Invitation of new parties:

Once the session is running, new members can be invited to join.

Interaction with SC: A party will add a new member to the Primary SR, in the general case, after negotiation with the other parties. The Session Control will interact with the corresponding User Agent [1], and, upon that party's acceptance, a new session member will be instantiated. This session member will have assigned a role for the session.

Action 6: Exclusion of parties:

A party may be obliged to leave the session.

Interaction with SC: A party with the proper role may destroy the attachment of a Session Member to the Primary SR, excluding the party from the session.

Action 7: Request for joining:

Once the session is running, a user may ask to join the conference.

Interaction with SC: The user will try to add a new member to the Primary SR by creating its Session Member object for the session.

This proposal will be discussed between the members of the session and, upon acceptance, the new SM object with the specified role will be instantiated. Once created, the object will interact with the corresponding User Agent to get configured.

Action 8: Leaving the conference:

Any party may leave the session at any time.

Interaction with SC: The corresponding SM object will be destroyed, and consequently, all the edges in all the SRs the user was taking part in. This action may imply closing the entire session if there are no more members on it.

Action 9: Switch from not-conducted to conducted mode:

The rights in a conference can be managed in two different ways:
- Conducted: One of the parties acts as a *Chairman* of the conference. The chairman grants/suppress the rights (of speaking, of moving the common pointer, etc.) to the other parties.
- Not-conducted: There is no floor control, all the parties (with the role of actor) have permanently all the rights.

Interaction with SC:
The rights of the parties for different aspects of the service can be managed separately. For instance, let us consider Common Pointer and Speaking rights. If both facilities are considered common to all members of the session and are managed by the same chairman, then they can be controlled jointly by the PSR. Otherwise, they will be managed by separate SRs, as in the case where several parties have the right of speaking, but only one has the right to move the pointer, or in the case where both types of rights are managed by different chairmen (the group of parties using the common pointer facility may not be the same as the group of parties using the video conference facility).
In our example, A is using the common pointer facility with D (joined in action 6) and with B. One of the attributes of this AccessRightsSr SR will indicate the mode: Conducted/Not conducted. If conducted, one of the parties will take part in the SR with the role of chairman. D is the chairman of this facility.

To switch from conducted to not-conducted mode, one of the parties in the SR will make this proposal, including the name of a candidate for chairman. This proposal will be negotiated. If accepted, the roles of the parties will be modified according to the new situation, with the attribute "holder" containing the name of the party which holds the rights (the token) at each moment.

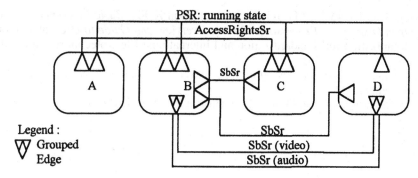

Fig.4: SG after action 9

Action 10 Closing the conference:

The conference can be (implicitly) closed when all the participants have already left it, or it can be (explicitly) closed by the chairman (in conducted mode). The second case is described by this action.

The chairman will destroy the Primary SR, each party will be informed that the session is over.

5. Conclusions

A new object oriented session control model for distributed services has been presented. This model is being specified by the Core-Team of the TINA Consortium as part of the TINA Architecture development effort. Its goal is to help the provision of new distributed applications, by supporting complex relationships between the different participants and servers in a multi-stakeholder environment. Its main characteristics are the use of generic relationships to characterize the state of the session and the non-hierarchic interactions between the objects. The major benefits, like its flexibility and wide coverage of service needs, have been presented.
This model is currently being implemented by Telefónica I+D in the framework of the SECRETO TINA Auxiliary Project, whose objective is to create a service creation environment based on the results of TINA.

6. References

1. *An Overview of the Telecommunications Information networking Architecture,* TINA95, Melbourne, Australia.
2. *TINA-C Connection Management Components,* TINA95, Melbourne, Australia.
3. *TINA-C DPE Architecture and Tools,* TINA95, Melbourne, Australia
4. Rumbaugh et al., *Object-Oriented Modeling and Design,* Prentice Hall, Englewood Cliffs, N.J., 1991.

5. ISO/IEC 10746-2.2 / ITU-T Recommendation X.901, *Basic Reference Model of Open Distributed Processing - Part 1: Overview and Guide to Use,* International Organization for Standardization and International Electrotechnical Committee, June 1993.
6. *Association Service Specification,* OMG TC Document Number 93.11.10, November, 1993
7. *Relationship Service*, OMG TC Document 93.11.9, November, 1993

TINA based Advanced UPT Service Prototype: Early Introduction of TINA through the IN Domain

Javier Huélamo, Eugenio Carrera and Han Zuidweg
Alcatel Corporate Research Center

Abstract: This paper presents a TINA based Service Architecture for an advanced UPT service called PSCS (Personal Service Communication Space); a type of service which is typically addressed in the IN domain. The TINA principles have been incorporated in the implementation of the computational and information objects needed for the PSCS service operation and management (in IN parlance, SCF, SDF and SMF).

1. Introduction

The increasing worldwide demand for IN services, currently applicable to enhanced telephony services, manifests the great business opportunities for this service class. The advent of the B-ISDN providing a vast range of multimedia multi-party services enlarges the application field of the IN services allowing a great deal of new IN features. The IN architecture was conceived to interact with existing network technology, introducing the required network adaptations to support supplementary services. Therefore, one of the proven advantages of the IN technology is that it can easily be incorporated into the present network infrastructure enhancing its added value. The centralized control proposed by IN-CS1 is not the most adequate for global IN services. Although the IN-CS2 will adopt a more distributed control, the current IN recommendations do not support the property of distribution transparency for the service components.

The emerging Telecommunication Information Networking Architecture (TINA) is being developed by the TINA Consortium (TINA-C) taking advantage of the most recent advances in telecommunications, distributed computing, and object orientation techniques. The main goal of TINA-C is to provide a consistent reference architecture for open telecommunication architectures encompassing operational and management services, integrating IN and TMN domains. TINA has incorporated a lot of management concepts from TMN. Although TINA also supports IN services, the influence from IN has been less relevant because of its functional approach. A smooth migration path from the current networks toward fully TINA compliant networks can be by means of the introduction of TINA concepts in the IN domain keeping the part of the IN architecture (CCF and SSF) that facilitates the interaction with the existing networks and signalling protocols, and changing the IN aspects incompatible with the TINA's object orientation, such as the separation of SCF and SDF. The figure 1 shows the IN's functional entities (SMF, SCF and SDF) where the TINA principles can firstly be adopted. This will allow the early exploitation of TINA results in the current networks providing more competitive IN services based on advanced technologies, and make easier the IN evolution toward long term architectures as TINA.

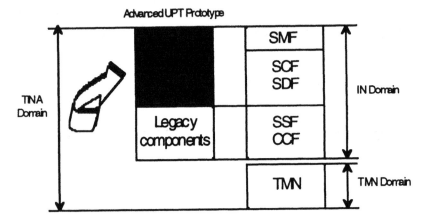

Figure 1. Introducing TINA through the IN domain.

This paper presents a TINA based Service Architecture for an advanced UPT service called PSCS (Personal Service Communication Space) based on an experimental B-ISDN ATM network. PSCS is a type of service which is typically addressed in the IN domain, that allows users to organize their communication according to personal criteria such as time and location. Besides the personal mobility, this service also supports session mobility. The proposed architectural concepts for this PSCS service can be applied easily to other IN services. The TINA principles have been incorporated in the implementation of the computational and information objects needed for the PSCS service operation and management (in IN parlance, SCF, SDF and SMF). According to this approach, the functional entities SSF and CCF can be developed using IN-CS1 technology and current signalling protocols. An object called a Network Agent is used to enable the interaction between the SSF and CCF, and the rest of the TINA based objects.

A prototype is being developed within the RACE (R2104) PERCOM project to validate the proposed architecture and will be available at the end of 1995. The main service features supported are: authentication, scheduled registration/deregistration for incoming/outgoing communications, service profile handling, session mobility, service provider management, and PSCS - non PSCS users/terminals intercommunication. The personalization options may be applied to an unlimited set of Personalizable Basic Services (PBS). In the PERCOM demonstrator, the following PBS will be supported: file transfer service, telephony service and fax service. The PSCS service components are located in PSCS service nodes incorporating a Distributed Processing Environment (DPE) to achieve the distribution transparency. The service provider access directly the service nodes to operate and manage the service. The experimental ATM network used (provided by TRIBUNE RACE project) has a general call control based on standardized signalling protocols. Some extensions have been required in this call control to identify personal identifiers required for the personal mobility and to manage session identifiers required for the session mobility. Several PSCS terminals without a DPE are also being developed allowing users and subscribers to manage their own service profiles, with different priorities and right to use.

In the paper, the most relevant computational objects, including some of their handled information objects, and DPE servers are described to present the proposed PSCS Service Architecture. Finally, several session mobility service scenarios are presented to illustrate the interaction between the diverse computational objects.

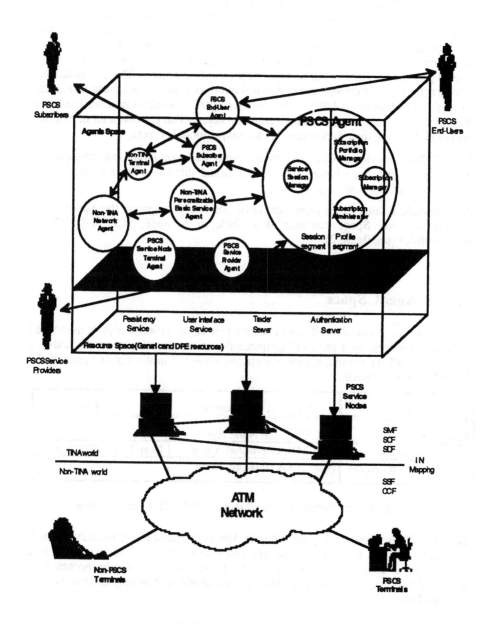

Figure 2. TINA based PSCS Service Architecture

2. TINA based PSCS Service Architecture

The PSCS Service Architecture prototyped is mainly based on TINA concepts with some refinements on the TINA Universal Service Component Model (USCM) and extensions to cope with problems raised whenever a TINA application is to be accommodated in a non-TINA world [1]. During the design, the ODP (Open Distributed Processing) paradigm has been adopted by defining models for the ODP's five viewpoints. This paper focuses on the computational model that defines the architecture by means of entities or objects that interact with each other according to predefined rules. These objects, encapsulating data and processing, are able to provide services or set of capabilities to other objects. The computational objects offer services through computational interfaces and can be distributed across the system (wherever a DPE is the underlying communication infrastructure). Moreover, they may manage local information objects that provide a set of external operations to make their data accessible to computational objects.

In PERCOM service nodes, the information and computational objects are implemented in C++, whereas the computational interfaces are specified using Interface Declaration Language (IDL).

The proposed architecture is composed of two basic object spaces: the Agent Space and the Resource Space that map as a whole to the SMF, SCF, and SDF IN functionalities (see figure 2).

3. The Agent Space

The Agent Space contains all the existing agent computational objects associated with stakeholders, with physical components, or representing services. Table 1 summarizes all the computational objects implemented in the Agent Space of the PERCOM Service Node.

Stakeholders	Physical Components	Services
End User Agent	PSCS Service Node Agent	PSCS Agent
Subscriber Agent	Non-TINA Terminal Agent	Non-TINA PBS Agent
Service Provider Agent	Non-TINA Network Agent	

Table 1. Agent Computational Objects in PERCOM Service Node

End User, Subscriber and Service Provider Agents represent inside the service nodes the actual players involved in the use and management of the service. These computational objects embody stakeholder information that can be accessed by means of IDL interfaces, and implement the set of actions that may carry out by the stakeholders.

Some computational objects, those related to non-TINA world, play a very relevant role in the PERCOM service nodes implementation since they represent the non-TINA components of the service. These are the computational objects running on a

DPE which are capable of understanding those components outside the DPE domain. These agents allow the TINA and non-TINA infrastructures to interwork. More specifically, in the PERCOM service nodes, the non-TINA Network Agent is in charge of allowing the interaction between the TINA world and the non-TINA network, in particular with the IN's SSF and CCF. The non-TINA Network Agent is continuously listening to receive frames from the network, dispatching the received frames to the other computational agents, and sending answer frames to the network. The non-TINA terminal agent interacts with physical terminals by implementing the proprietary protocol between service nodes and terminals, whereas the non-TINA Personalizable Basic Service (PBS) communicates with the control of the basic service in the ATM network i.e. TRIBUNE Control Server.

The PSCS service agent represents the service itself. It consists of a set of computational objects which can be divided into two segments:

- The *Profile Segment* is composed of the computational objects (subscription manager, general subscription manager, and subscription administrator) associated with the profile handling, containing the information objects, dealing with subscriptions and registrations, needed for the service management.

  There is a Subscription Manager per subscription. A subscription is characterized by a PBS identifier, a subscriber identifier, and a period of time (start time and end time). This object controls the end user profiles composed of a collection of both incoming and outgoing registrations. The end user profile is modelled as an information object containing the so-called registration scheduler information object, which manages the registration information objects according to time criteria. Figure 3 depicts the structure of the Subscription Manager including the information objects handled.

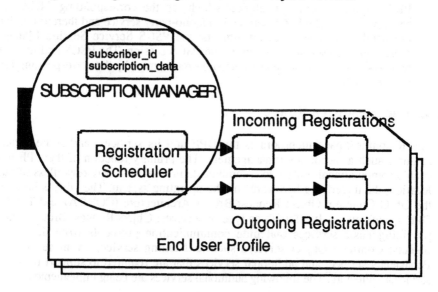

Figure 3. Subscription Manager

The General Subscription Manager (one per end user) handles the PSCS session registration information objects, as shown in figure 4.

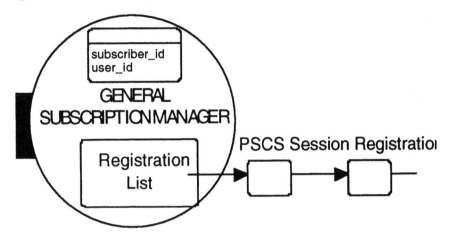

Figure 4. General Subscription Manager

Each PSCS service, supplied by a PSCS service provider, has its own Subscription Administrator, responsible for managing all its PSCS subscriptions.

- The *Session Segment* includes the computational objects associated with sessions termed PSCS Service Session Manager which handles the associated PBS session information objects which are the corresponding TINA user session objects. The PSCS service is a mono-party service and therefore it does not need several user session objects; the PSCS Service Session Manager represents the user participation in the PSCS session. The PSCS end-user can handle several personalized PBS sessions represented by the corresponding PBS information objects.

4. The Resource Space

The Resource Space incorporates the DPE and its services, and some generic resource such as accounting management. The primary function of the DPE is to enable computational objects to communicate in a uniform way regardless of their location, the underlying hardware and the operating system. The DPE is based on the OMG Common Object Request Broker Architecture (CORBA) and TINA-C DPE specifications. It offers three type of services: Object Request Broker (ORB) providing basic services that hide communication protocols from distributed objects; Common Object services such as Naming service, Event service and Persistency service; and additional services that are required to support the PSCS service architecture. The following additional services are being implemented:

- An Authentication service allows keys, characterized by an agent type, identifier and password, to be authenticated against an authentication database. It includes a separate interface for authentication database management.

- A Trader imports and exports service offers which are defined by an interface identifier and a set of attribute values. It makes use of both the Dynamic Invocation Interface and the Interface Repository of the ORB. The main advantages of this trader with respect to the CORBA naming service are that, in this case, only the IDL identifier needs to be known instead of a name for every instance, and the conditional selection of sets of references based on attribute value criteria is also possible.

- A User Interface service supports dialogues between users and computational objects through a window based interface. This service creates windows that are dynamically adapted to a client request, it retrieves and sends back user input to the client which is a menu driver. Three basis types of windows have been defined: a question menu window for the user to select an item, a labelled menu window for the user to fill in one or more inputs, and a text window that prompts a message.

5. Service Scenarios

This section presents several scenarios [2] on session mobility are presented to illustrate how computational objects interact each other within the PSCS service nodes. These service scenarios show how to suspend and resume a PBS session,

5.1 PBS Session Suspension

Although there are two possible scenarios depending on whether the PSCS end-user requesting session mobility is the calling or called party, only the first situation is included in this paper. Before the PBS session suspension request, at least the following phases have already taken place: authentication, PSCS service session opening, PBS selection, and PBS session opening. Figure 5 shows all the distributed invocations between computational objects calls caused after the PBS session suspension request arrive in the Service Node - Network interface.

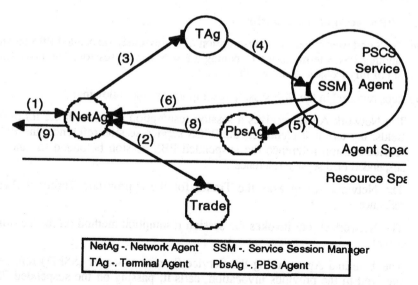

Figure 5. PBS session suspension (requested by the calling PSCS end user)

The following is a short description of each invocation:

1. The Network Agent receives a PBS session suspension request from a PSCS terminal. The parameters included in this invocation are the originator terminal identifier, and the PSCS and PBS references involved in the session.

2. The Network Agent ask the Trader for the reference to a Terminal Agent whose identifier is the one received in the call.

3. Using the reference returned by the Trader, the Network Agent invokes the session suspension method on the Terminal Agent, passing on the PSCS Service Session Manager's reference and the PBS reference.

4. As indicated by the Service Session Manager reference, the Terminal Agent forwards the session suspend order to the Service Session Manager, with the PBS session reference as parameter.

5. The Service Session Manager asks the PBS Agent for the PBS Service Node address (the TRIBUNE Control Server).

6. The Service Session Manager asks the Network Agent to set up a connection with the PBS Service Node. The Service Session Manager receives the reference to that connection.

7. The Service Session Manager invokes the PBS Agent's session suspension method, with parameters PSCS service session, PBS session, and connection reference, and terminal identifier.

8. The PBS Agent transfers the order to the Network Agent.

9. The Network Agent sends the suspend session frame to the TRIBUNE Control Server.

5.2 PBS session resumption

The PSCS end user requests the resumption of a previously suspended PBS session from another PSCS terminal after opening a PSCS service session. The invocation diagram of this scenario is shown in figure 6.

The steps required in the PBS session resumption are the following:

1. The Network Agent receives a session resumption frame with the following fields: the terminal identifier, the PSCS service session reference, the PSCS service session reference the suspended PBS session belonged to, and the suspended PBS session reference.

2. The Network Agent asks the Trader for the appropriate Terminal Agent reference.

3. The Network Agent invokes the session resumption method on the Terminal Agent.

4. The Terminal Agent, using the Service Session Manager (SSM1) reference received in the previous invocation, calls it; passing on the suspended PBS session reference and its former PSCS reference.

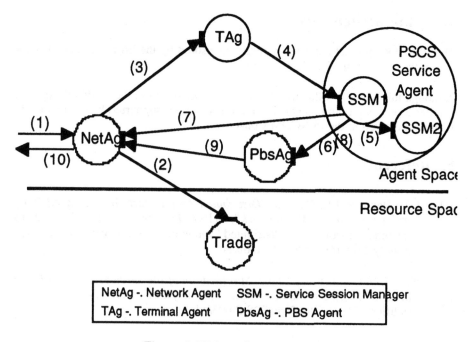

Figure 6. PBS session resumption

5. SSM1 notifies to the Service Session Manager under which the PBS session was suspended (SSM2) that it takes over control of the suspended session, so that SSM2 returned the PBS session data to SSM1.

Similarly to the suspend session scenario, SSM1 interacts with the Network Agent and the PBS Agent in order to set up a connection to the TRIBUNE Control Server and send to it a frame ordering the resumption of the session.

6. Conclusions

A Service Architecture has been developed adopting, extending and refining TINA principles for the selected PSCS service features. The paper shows how TINA concepts can be used to implement the IN functional entities SCF, SDF and SMF where the distribution transparency is more necessary. The proposed architectural concepts can be applied to different IN services, implementing the IN service operation and management (associated with SCF, SDF and SMF) using TINA principles. The IN-CS1 technology and current signalling protocols can be retained in the first phase of the introduction of TINA. The paper has described how TINA based IN service can be provided within a non-TINA network. The interaction between the TINA components and the non-TINA components is carried out through non-TINA network agents, non-TINA terminal agents, and non-TINA PBS agents. Several DPE servers (Trader, Authentication Server, User Interface Server) has been developed to complete the selected CORBA compliant DPE.

7. Acknowledgments

The PERCOM project has been partially sponsored by the European Commission within the scope of the RACE program.

Although this paper is influenced by the work carried out in the PERCOM project, it reflects our own opinions which are not necessarily supported by the rest of the partners involved in the PERCOM Consortium.

8. References

[1]   J. Huélamo and E. Carrera, *"Distributed Architecture for Advanced PSCS Services in an ATM Network"*, TINA'95 Conference: "Integrating Telecommunications and Distributed Computing - from Concept to Reality", February 13th-16th, 1995, Melbourne.

[2]   PERCOM Consortium, 1994, *"Specification of the Services provided by the PERCOM Service Node at its External Interfaces"*, Deliverable D05, CEC Identifier R2104/SESA/WP3/DS/P/005/ed1.

Use of Atomic Action Principles to Co-ordinate the Interaction between TINA Service Managers.

Guy Reyniers, Patrick Hellemans
Alcatel CRC Antwerp
e-mail: grey@rc.bel.alcatel.be

Abstract: TINA provides an architecture to support the Telecommunication Services of the future [2,3]. This architecture consists of a number of Service Managers (e.g. Service Session Manager, User Service Manager) that co-operate to provide Telecommunication Services to the users. In certain cases, complex interactions are required and this paper suggests the use of Atomic Action (AA) transaction principles to provide the required co-operation mechanisms.

1 Introduction

The aim of the TINA architecture is to provide an architecture for efficient creation, deployment, operation and management of Telecommunication Services on a world-wide basis. It consists of a number of Service Managers that co-operate to provide Telecommunication Services to users.

An issue still under consideration within TINA is an effective mechanism for co-ordinating the interactions between Service Managers. In the specifications released during 1994 , these interactions are modelled as unrelated synchronous method invocations. The problems with this approach are elaborated in section 2.

Within TINA, new session control mechanisms are being worked out to introduce grouping of different operation invocations. In this paper, we describe a method based on Atomic Action (AA) principles, which is based on work performed in the RACE MAGIC project [1,4] associated with enhancements to B-ISDN signalling for Multimedia Multiparty Services. Whilst developing this approach, it was found necessary to extend AA principles to include the concept of transferring mastership over an AA. This is further described in section 3 of this paper.

It is our intention to validate the approach presented in this paper in the ACTS VITAL project, where we will be developing a MultiMedia MultiParty (MMMP) service within the framework of a TINA based distributed architecture.

2 Problems with the Current Approach

The major computational objects in the TINA Service Architecture are the Service Session Manager (SSM), User Service Session Manager (USM) and Communication Session Manager (CSM). An instance of service execution (a session) is realised by one SSM and some USMs (one per involved user). The SSM supports service execution, joining of users, and negotiation among users. It manages information and resources shared by users in the service execution, while the USM takes care of only those for its user. The CSM controls the connectivity and the allocation of resources related to the communication. More information can be found in references [2,3].

In the current TINA approach, the co-operation between an SSM, USMs and CSM is defined in terms of simple operations on the Service Managers. For example, in order to add another "Party" and a "Medium" (e.g. audio, video) to the Service Session, two unrelated operations are needed. If the "creation" of the medium fails for some reason (e.g. the user has no permission to add media to the Service Session), the user himself is responsible for requesting the removal of the requested additional Party.

A drawback of this approach is that it does not handle partial failures of related operations automatically. In general, following a partial failure, it is usually essential that the state of the system is brought back to the state before the requests were issued. When something goes wrong with any of the individual operations, the 'client' has to cover each individual failure and rollback scenario.

The example of a bandwidth manager for a trunk shows that it is not always possible to undo the effects of related requests. Suppose that the user's request implies a bandwidth increase (or allocate) for one connection (VC) and a bandwidth decrease (or deallocate) for another one. Take also into account that it must be possible to undo these operations because of the failure of an operation on some Service Manager. Suppose that the bandwidth increase is not possible without the decrease because of bandwidth limits on the access link. The following scenario causes a problem for the current TINA approach: deallocate bandwidth for connection1, another request (unrelated, for another user) allocates bandwidth for connectionX, allocating bandwidth for connection2 fails because there is insufficient bandwidth left. Reallocating bandwidth for connection1 can not be done so the 'undo' is not possible. If the decrease and the increase were performed in one AA the other Service Manager could not have seized the bandwidth.

Another example is the fact that the service creation (by a Service Factory) and service requests are unrelated. Suppose that the service session creation succeeds, but a user does not accept participation in the service session (a service request by the session owner). In this case the owner must request to delete the service session. Again this problem can be avoided by combining the service creation and the initial service request. This illustrates the need for a more elaborated co-operation mechanism between not only the service related managers (SSM, USM) but also between those and other computational objects (e.g. the factories).

An additional drawback is that operations are carried out one by one, not as a group. The additional delay introduced by each remote method invocation obviously has a negative impact on the expected performance of the system.

3 The Atomic Action Mechanism

Atomic Actions are groups of operations performed by multiple agents either as a whole or not at all. Performing an AA consists of 2 phases. In the first phase, the AA tree is constructed and agents indicate that they are willing to perform the requested operations. In the second phase, agents are instructed to commit (the requested operations must be completed) or to rollback (the system must be brought back in the state before the AA started).

The proposed Atomic Action mechanism is based on the Commitment, Concurrency and Recovery (CCR) Protocol. Some other transaction mechanisms use a different terminology, but the principles are the same. An Atomic Action method should be considered as an envelope that encapsulates primitive operations.

In the first phase, BEGIN operations are invoked on all agents that have to confirm the AA. Each BEGIN contains the operations specific for the agent (AA Segment). Every agent that needs the assistance of another agent before giving its approval shall, in its turn, invoke a BEGIN method on that process. Every involved agent can be depicted as a node within an AA tree. An invoking agent is the superior to the invoked agent (subordinate). The agent that controls the AA is referred to as the Master of the AA (top agent in the tree).

The first phase terminates successfully when all subordinates of the Master have accepted the AA (are in the ready state). An agent can also reject its AA segment by invoking a REFUSE method on its Superior. Refusal of the AA by a single agent causes an unsuccessful termination of the first phase.

If the first phase was successful, a COMMIT method is invoked by the Master on all its Subordinates. They propagate the COMMIT if they are themselves superior to other agents. As a result the different agents have to perform actions needed to perform the requested operations.

An unsuccessful first phase results in a ROLLBACK method invoked in a similar recursive way.

In general, an AA is started and controlled by the same agent. However, in some cases it is desirable that the agent that starts the AA does not keep control over it (e.g. the user process starts the AA, but a network process is responsible for allocating network resources, so it is desirable that the network agent is the master of the AA). We call this addition to the AA mechanism "transfer of mastership". Since the semantic of this procedure is similar to invoking READY, (i.e. the subordinate accepts the AA if the operations in the READY invocation are performed), we name it TREADY. This procedure also optimises the number of AA invocations (2 instead of 3) when there are only 2 agents involved in it.

Note that there is a big difference between an AA and the traditional state of the whole system. AAs represent transient phases during the lifetime of the service invocation. The AA concept is generic; it is independent of the nature of the operations it encompasses. This makes it possible to apply the AA mechanism between completely different agents; only the operations contained in the AA Segments differ. In the RACE 2044 MAGIC project the operations were "create", "modify" and "delete" operations on "Call and Resource Objects". In the VITAL project we will develop an Object Model for the MMMP Service and use the same primitive operations to compose the Service Session configuration (the actual call).

4 A Possible Way to Apply the AA Mechanism to TINA

4.1 Principles

As part of the VITAL MultiMedia MultiParty (MMMP) Service, we will develop an object model that can be used to negotiate the components of the desired Telecommunication Service and use Atomic Action principles to group primitive operations on these objects and to co-ordinate the co-operation between the involved Service Managers (Factories, User Agent (UA), USM, SSM, CSM ...). In this paper we focus on the AA mechanism.

The key principle is to make the involved Service Managers (or computational objects) AA aware. Therefore the interface to these Service Managers has to be defined in terms of AA methods (Begin, Ready, Commit ....).

An example is the request from the SSM to the CSM to setup a unidirectional connection. With the existing TINA approach, this request requires a number of operations:

- *create_logicalvertex(LV1)*,
- *create_logicalport(LP1)*,
- *create_logicalvertex(LV2)*,
- *create_logicalport(LP2)*,
- *create_logicalline(LL1,LP1,LP2)*.

While the AA version appears as a single operation:

Begin(AAID,
*   create_logicalvertex(LV1)*,
*   create_logicalport(LP1)*,
*   create_logicalvertex(LV2)*,
*   create_logicalport(LP2)*,
*   create_logicalline(ll1, LP1, LP2))*.

The SSM keeps track of the global state of the service invocation. It communicates with the different USMs to handle the negotiations between the different users and with the CSM to implement the connectivity requirements of the service invocation. Therefore, the SSM is the best candidate for mastership over the AA. Mastership is transferred to it starting from the agent that initiated the AA.

An outline of a proposed method is as follows:

- A Service Session is created by the End User Application (EUA) by invoking a Tready method on the Generic Session End Point (GSEP). Parameters are create operations for objects that determine the initial service configuration.

- The GSEP transfers the request to the User Agent (UA) in the network with a Tready. The UA performs some checks (e.g. subscription) and when these are successful it requests the USM & SSM Factories to create a USM and a SSM, otherwise it rollbacks the AA.

- The UA then transfers the control to the USM that builds its local data structure based on the create operations in the request. It then transfers control to the SSM.

- The SSM, in its turn, builds its own data structure and analyses which users need to confirm the AA. It requests this confirmation by first invoking a Begin on the UA's of the involved users.

- Again a UA does some checks and when these are successful, it creates a USM. It invokes a Ready on the SSM to indicate this.

- The SSM can then request co-operation of the USM by invoking a Begin on it.

- The USM has no relation with an EUA yet, so it invokes a Begin on the GSEP that creates the EUA and routes the request to it.

- The EUA requests the (human) user to confirm the AA containing the initial configuration.

- The EUA indicates acceptance by the user by invoking a Ready on the GSEP.

- This Ready travels up the AA tree to the SSM. When all subordinates of the SSM have accepted, it can request the required network resources from the CSM.

- When this completes, the SSM can start the second phase of the AA processing. It invokes commit methods on all its subordinates. This commit travels down the AA tree to the EUA's. The communication can then start.

The same mechanism can be applied for service session modification. The difference is that GSEP, UA and Factories are no longer involved since after the service session is created, the EUA has a reference to the USM and it can invoke a Tready on the USM directly.

4.2   Example: Setup of a 3-party Association

An example will clarify the approach. In order not to have a too complex example, the intention is to establish a 'Call Association' only.

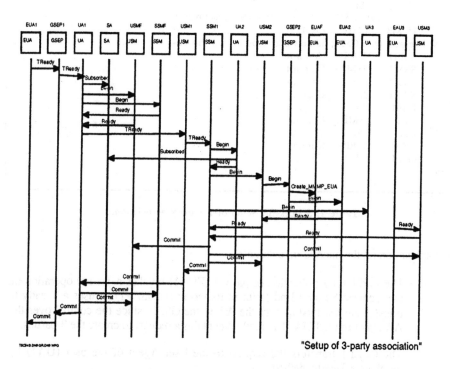

Fig.1: Setup of a 3-party association.

The method invocation flow is depicted in figure 1. More details on the invoked methods are shown in table 1.

```
                    // Start of a MMMP Session; only parties, no connections

Index    From    To      Method                Primitive Operations
--------------------------------------------------------------------------------
1        EUA1    GSEP1   Tready                +MMMP(+Party(1),+Party(2),+Party(3))
2        GSEP1   UA1     Tready                +MMMP(+Party(1),+Party(2),+Party(3))
3        UA1     SA      Subscribed(MMMP) // returns T/F
4        UA1     USMF    Begin                 +USM(1)
5        UA1     SSMF    Begin                 +SSM(1)
6        SSMF    UA1     Ready
7        USMF    UA1     Ready
8        UA1     USM1    Tready                +Party(1),+Party(2),+Party(3)
9        USM1    SSM1    Tready                +Party(1),+Party(2),+Party(3)
10       SSM1    UA2     Begin                 +Party(1),+Party(2),+Party(3)
11       UA2     SA      Subscribed(MMMP)
12       UA2     SSM1    Ready
13       SSM1    USM2    Begin                 +Party(1),+Party(2),+Party(3)
14       USM2    GSEP2   Begin                 +MMMP(+Party(1),+Party(2),+Party(3))
15       GSEP2   EUAF    Create_MMMP_EUA() // returns EUA(2)
16       GSEP2   EUA2    Begin                 +Party(1),+Party(2),+Party(3)
17       SSM1    UA3     Begin  // Similar to Party(2)
18       EUA2    USM2    Ready
18'      EUA3    USM3    Ready
19       USM2    SSM1    Ready
// USM interacts with CSM first ("reserve Access resources")
19'      USM3    SSM1    Ready
// SSM interacts with CSM; no network resources for this case
20       SSM1    USMF    Commit
21       SSM1    USM3    Commit
22       SSM1    USM2    Commit
23       SSM1    USM1    Commit
24       USM1    UA1     Commit
25       UA1     SSMF    Commit
26       UA1     USMF    Commit
27       UA1     GSEP1   Commit
28       GSEP1   EUA1    Commit
// end
```

Table 1: Successful Setup of a 3-party association.

The explanation on the method invocations:

- The End User Application of party 1 (EUA1) invokes a Tready operation on the Generic Session End Point in its terminal (GSEP1). It uses a Tready to transfer the mastership over the AA to the GSEP, since the control over the AA is not in the EUA. The method contains a request to create the 3 parties.

- The GSEP1 transfers the request to the User Agent of the user (UA1) by invoking a Tready method.

- The UA1 checks the subscription by invoking the Subscribed method on the Subscription Agent (SA). This illustrates that it is still possible to use synchronous method invocations for simple question/answer interactions within the context of the AA.

- When this check is successful, the UA1 request the User Session Manager Factory (USMF) and the Service Session Manager Factory(SSMF) to create the USM for party 1 and the SSM by invoking Begin methods on them.

- When a Factory has no problems creating a manager, it invokes a Ready method.

- The UA 1 then transfers control over the AA to the USM1. Again the request contains primitive create operations for the 3 parties.

- The USM1 in its turn also transfers the control to the SSM 1; the request contains the party create operations.

- The SSM1 is the computational object that is in control of the service, so this object is the master of the AA and hence the control over the AA will not be transferred any further. The SSM1 invokes a Begin method on the UA for Party 2 (also on the UA for party 3 (17)).

- The UA2 checks the subscription and when the check is successful, it creates the USM2 (using the factory, not shown).

- It invokes a Ready operation on the SSM1 to indicate successful completion.

- The SSM1 then invokes a Begin method on the USM2, containing the create operations for the Parties (the initial configuration), in order to get confirmation of participation in the Service.

- The USM2 has no reference yet to the UAP2, so it invokes a Begin method on the GSEP2.

- The GSEP then invokes a create operation in order to create the EUA2 (we assume the interface to the EUA Factory in the terminal is not AA based).

- The Begin method invocation on the EUA2 is the request for confirmation of the (human) user. Note that both the confirmations from user 2 and user 3 is needed for the successful setup of the service. The method invocation flow for the request to user 3 is similar to the flow for user 2 ((17),(18')), hence it is not shown completely. These flows can be executed in parallel for efficiency reasons.

- When the user 2 accepts the participation in the Service, the EUA2 invokes a Ready method on the GSEP2, that in its turn invokes the Ready on the USM2. Note that this Ready method can contain values for attributes of the configuration objects as parameters.

- The USM2, in its turn, invokes a Ready method on the SSM1.

When the SSM1 has received all confirmations of the participating parties (users 2 &3, USM 2 & 3 have invoked the Ready method), it starts the second phase of the AA processing. Since everything went well, this is done by invoking Commit methods. This Commit invocation flow follows the AA tree that has been established during the first phase of the AA processing ((20-29)). Note that, in general, the SSM will invoke a method on the CSM to establish the needed network connections before starting the second phase. However, in this example, no resources are required.

5 Conclusions

In this paper we have identified a problem with the operational interfaces currently specified within the TINA Architecture. We have presented the basic principles of Atomic Actions and have shown how they can be used as the interaction mechanism between Service Managers. We will develop a demonstration that uses a similar approach and advocate the use of AAs to the TINA Core Team.

6 References

[1]   "Stage 2 specification", R2044/BTL/DP/DS/P/06/b1, September 1993.

[2]   "Service and Management Architecture in TINA-C", TINA '95 Melbourne, February 1995.

[3]   "TINA-C Service Components", TINA '95 Melbourne, February 1995.

[4]   G. Reyniers, P. Hellemans, "An Object-Oriented Approach to Controlling Complex Communication Configurations", ISS '95 Berlin, April 1995.

A Comparison of Architectures for Future Telecommunication Services

Hendrik Berndt, TINA-C core team, Deutsche Telekom AG,
hberndt@tinac.com
Martin Chapman, BT Laboratories, Martlesham Heath, IP5 7RE UK,
martin@drake.bt.co.uk
Peter Schoo, GMD — FOKUS, Hardenbergplatz 2, D - 10623 Berlin,
schoo@fokus.gmd.de
Ingmar Tönnby, Ericsson Telecom AB, S - 12625 Stockholm,
tonnby@ericsson.se

Abstract: TINA-C and the RACE CASSIOPEIA Project have defined architectures for supporting designers of telecommunication services in a broadband environment. The scope, approaches, and models defined by both projects are compared. Similarities provide a common interpretation of how to build services in the future. Differences suggest items for further study or refinement. At a time of increasing importance of the application of distributed systems technology in telecommunications, this paper addresses issues of service design that may be important for researchers, designers, and managers of information and telecommunication services.

1 Introduction

In this paper we compare and contrast two architectures that are being defined to suit the needs of the future telecommunication services marketplace. Within RACE, the project CASSIOPEIA worked on an Open Services Architectural Framework (OSA) [1,2,3,4] since 1992. The Telecommunication Information Networking Architecture Consortium (TINA-C) is a collaboration between global partners, and has been working on a future telecom architecture [5,6] since 1993. The work within TINA-C and CASSIOPEIA on service architecture have both been influenced by the previous RACE project ROSA (RACE Open Services Architecture) [7,8].

ROSA aimed to define an architecture for the specification, design, construction, and implementation of a broad range of service types, including conventional telephony and advanced information services i.e. incorporating voice, data, images and video. Although ROSA did not deliver a prescriptive architecture, it did show the feasibility of conceptual modelling [9], used to formalise key separations and relationships; it showed the applicability of object-orientation in design and specification phases; and that both are suitable for the development of telecommunication services.

TINA-C and CASSIOPEIA have been working in parallel and have had some common partners. Since both projects have a similar ancestry and similar objectives it seems a useful exercise to see where the two projects have converged and diverged and for what reasons. Similarities can be built upon, and used to encourage the industry as a whole. Differences can be used to identify further areas of work.

Both projects contain a wide area of work. For the purpose of this paper we will focus our attention on service architecture, communication architecture, and component models. The distinctions between both architectures and their models are elaborated in Sec. 4 along basic principles and separations of concerns (introduced in Sec. 3) commonly characterising the modelling of services. The paper is not comparing the actual architectural elements, as this would go beyond the size of this paper. The interested reader is referred to further references in Sec. 6. Above all, the next section introduces and contrasts the scope and technical approaches being taken in TINA-C and CASSIOPEIA.

2   Scopes and Technical Approaches

Both TINA-C and CASSIOPEIA, distinguish for the future telecommunications environment a network technology based part, comprising the infrastructure resources which enables data transport, a part comprising computing capabilities to access and control the infrastructure and execute software modules which reside in the third part. The third part comprises information and telecommunication services being constructed as software modules and finally deployed in a computing environment supporting services.

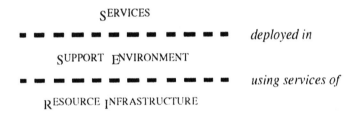

Fig. 1: Structure of Environment for Distributed Telecom Applications

This logical division is illustrated in Fig. 1: an environment computationally supporting distributed telecommunication applications which offer services, superposing the telecommunicaton network resource infrastructure. This division allows technical solutions in each of the problem areas to evolve independently, while making appropriate abstraction used to relate the parts to each other and reason about their properties.

2.1   TINA-C Approach

TINA aims to define an overall software architecture (Fig. 2) to design and introduce new telecommunication services, as well as the ability to manage these services and the networks that support them in an integrated way. The technical approach of TINA therefore is fourfold. First it aims to define a distributed computing architecture for telecommunications, including an object model. This will be based on the developments of the Object Management Groups (OMG), enhanced to include telecommunications requirements. Secondly, it aims to define a service architecture, which will enable the construction of services in a distributed environment and is applicable to a wide range of services. Thirdly, it aims to define a network architecture that provides a technology independent representation of transport network for use by services. Fourthly, it aims to define a management architecture, which will enable the management of distributed computing environments and software modules, services and networks to be done in an integral manner. In each of these four areas, ODP principles have been applied. Thus TINA is aiming to provide an information, computational and engineering specification of distributed computing environments, services, networks and management systems.

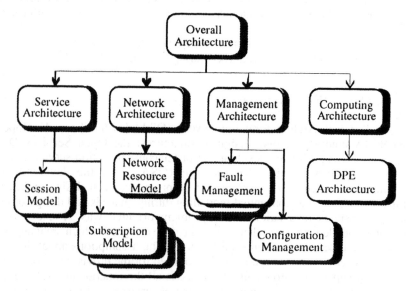

Fig.2: TINA-C Architecture Overview

The ultimate goal is to specify a software architecture. This will consist of object definitions, interface specifications, and rules for composition and usage. This should result in a set of specifications that can be used by developers for programming infrastructures or services.

For the purpose of this paper the focus will be on the service architecture and the abstractions presented to services by the network architecture. Seven components can be identified for the discussion in this paper, Terminal Application, User Agent, Terminal Agent, Subscription Agent, Service Session, User Service Session, and Communication Session.

- Terminal Applications: software that executes on end-user terminals providing user interfaces and facilities for controlling and interacting with services.
- Terminal Agent: an entity representing a terminal. Its role is to represent the capabilities of the terminal, facilitate terminal/network connections, and provide local resource management during service execution.
- User Agent: an entity representing a user. Its role is to handle incoming and outgoing call setup on a user behalf, and to allow personal profile and subscription information to be browsed and modified.
- Subscription Agent: a component that manages subscription records for users.

- Service Session: an instance of an executing service. A service session exists through the duration of a call. It maintains state about the progress of the call, e.g. users involved, and embodies service logic that users can invoke.
- User Service Session: a part of Service Session that maintains state about a users involvement in a specific service, e.g. current charge. One is created for each user that joins a service. A user (or more precisely a terminal application) interacts within a service session via a user service session.
- Communication Session: a component that represents the network connections and resources used during a session.

2.2 CASSIOPEIA Approach

The Open Services Architectural Framework (OSA) [1] has been developed in CASSIOPEIA comprising two design architectures: the Open Services Design Architecture (OSA_{app}) for the development of telecommunication services; and the Open Services Provisioning Architecture (OSA_{sys}) for the development of environments in which these services will be provided. Each design architecture encompass means for the conceptual modelling during analysis and design of either services or systems in a suitable and problem oriented way. These two peer architectures complement each other: services can be designed, which are independent of specific resource infrastructures, although making assumptions about them; and service support environments can be designed, used for the provisioning of services. The notion of support environment encompasses all the functionality related to the deployment, execution, management and usage of services; and it ensures access to and use of telecommunication resource infrastructures.

In the CASSIOPEIA approach services are specified at different levels of abstractions in terms of OSA-components. In a realization process these are mapped onto deployable components, defining the necessary computational object to provide the service in a selected system environment. The decomposition and refinement of the OSA-components are supported guided by OSA_{app}, aiming to achieve reuse and reusability of the designed services. CASSIOPEIA is not normative in what functionality should be provided in any particular system environment, but allows the environment to be populated with any suitable set of deployable components which can be used and combined to realize other services, and enables this set to grow with the deployment of new services.

An OSA-component represents the providing of a service. It is conceptually structured into mission, representing the service core, and ancillary facets, describing the preparation of a service for being used, existing in a supporting environment and using other resources (services in terms of OSA-components). Its semantics is expressed according to an object model, i.e. several model expressions in refinement relation to each other. The realization process maps OSA-components, i.e. objects are refined, until eventually object configurations are elaborated, which are sustained in the decided service support environment, such that the component is deployable.

The service design architecture is structured to reflect in each part the different problem domain concerns (Fig. 3). Each part contains specific concepts, guidelines, recipes and components to support the designer, allowing appropriate abstractions by restricting the focus on essentials, thus leading to an increase of reusability. The architecture's main part is the Telecom Service Domain, which contains support for the development of information transportation services, that are controlled and managed as required by a using application. It is surrounded by service domains for Enterprise Support and Resource Control. The former reflects constrains by requirements emerging from the enterprise and its policies a service is used in; the latter comprises service abstractions of the resource infrastructure capabilities.

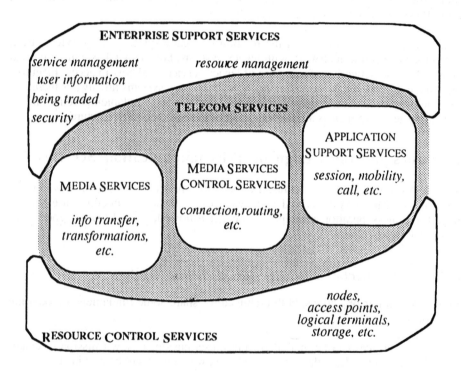

Fig.3: Structure of OSAapp

OSA_{sys} prescribes properties of an OSA-system and system oriented aspects of services to be deployed. An OSA-system models a service support environment which can be spread over several ownership domains, and is based on distributed processing techniques [10]. In realising services, a designer considers OSA-systems from the computational viewpoint, regarding functionality enabling deployment and provision of services. Those mechanisms ensuring distributed processing are considered by the system designer from the engineering viewpoint.

2.3 Comparisons of Scope and Technical Approach

The scope and technical approach of TINA-C and CASSIOPEIA largely overlap. Common to both is: a service driven approach, as opposed to being driven by network and resource technology; the separation of service design and, on the other hand, system or DPE design; integrating management and management systems with services and resource infrastructures; and applying object orientation. However some important differences, which lead to different approaches and different ways to express the architectures, can be noted.

TINA-C has the explicit goal of leading to a software architecture and to address software engineering issues. CASSIOPEIA goal has been more focused on requirements engineering of services. In consequence, the TINA approach has much more focus on computational issues, while OSA is more oriented to problem decomposition on a more conceptual level.

Differences can also be seen in the relation to legacy systems and services. Both recognise the need to interwork with legacy systems, but TINA takes a more unified approach to the network, deferring use of legacy services and how to access them to a later stage. CASSIOPEIA has taken the view that legacy systems are always there and that the need to take the legacy services into account is a fundamental requirement. Legacy systems are seen as an important part of the resource infrastructure.

3 Basics for the Design of Telecommunication Services

The TINA and CASSIOPEIA architectures embody a priori knowledge on how to build telecommunication systems and services. Key to these architectures are sets of modelling and separation principles which, if followed, should led to "good" and flexible designs.

3.1 Fundamental principles for modelling service

Driven by similar objectives, both architectures share some fundamental modelling principles.

- Services can be modelled from a user/customer view (mission oriented) and be independent of any system in which they will be provided.

In an advanced telecommunication environment, the terms 'user', 'service' and 'resources' become relative. Specifications of services are made with the general view that its users may be other services, not necessarily humans, and its resources may be other services as well.

- Services can be defined and provided using nested services, i.e. one service can be based on a set of other services.

Services are be expressed and built in software as discrete components which can eventually be mapped onto an environment supporting the deployment and provisioning of the designed and implemented service.

- Management and control can be designed coherently and with the same design framework.

It is a fact that services need to be managed. The current approach to considering service management separately from the services themselves is not effective. Consequently, control and management can not be taken as separate task in the service design, but have to be integrated to ensure the design of manageable services.

- Every user shall be able to have a set of private services, which can be different from other users services.

Apart from problem oriented service solutions, users and customers must be able to set their preferences for the use of 'their' services. The service design must be sensitive to this need. Also, users and customers can take advantage of their premises' equipment. This impacts also the thinking about the access to customized or personalized services.

3.2 Fundamental separations of concern

A set of separation principles have been identified by both projects which ensure the identification of correct problem domain structures and increase the possibility of re-use.

- separation of media services from their control and management

Services for information presentation, generation and transport (media services) are distinguished from those controlling, coordinating and managing media services. While media services focus on effectivity, media services control services emphasize on flexibility.

- separation of service access from the service core

Service should allow different arrangements to access a service; and many services may use the same access arrangements. This is to support mobility, and, also, to allow new access models for existing services, or setting new services on existing access modes.

- separation between application/session oriented and resource/communication oriented problems

The resource oriented problems address allocation, monitoring and management of communication resources while application oriented problems addresses the support of users and applications in their use of communication services.

- separation of local and global aspects of services

A communication service involves by definition two or more user entities, having a local view of the communication service in which it is involved. The local view of a service is part of the service contract between the customer enterprise to which the user belongs and the provider enterprise. To coordinate and mediate between user entities, the service provider needs to take a global view of the service. This must be reflected in the designed service.

- separation of local resource handling and global network resource coordination

The execution of a service will require the allocation and use of resources. Central control of resources is unrealistic in an environment where different organisations will be involved in the delivery of a service. For example, customers should be responsible for resource management of their own equipment.

4 Support for the Design of Telecommunication Services

This section discusses the basic modelling principles and separations of concerns presented in the previous section. It explains how TINA and CASSIOPEIA have considered these principles and separations in their approaches.

4.1 Mission oriented and system independent modelling of services

In TINA this principle is interpreted in two ways. Firstly, TINA defines ways to control the execution of services but does not restrict the functionality of service logic. Provided the service logic interacts with other components in a TINA conformant way, it can be programmed to do anything within the limitations of available resources (in contrast with SIBs in IN). Secondly, abstractions of the resource infrastructures and computing machineries are defined, which provides technology independent interfaces on which services can be built.

In CASSIOPEIA this principle is reflected in various ways. The OSA-component is structured in two major parts, the mission and the ancillary facets, with the intention that a modelled mission can be combined with various ancillary facets. To achieve system independent modelling, OSA decouples service and system concerns by the division into OSA_{app} and OSAsys, and contains in OSA_{app} system independent abstractions of resources.

4.2 Nested services for defining and providing new services

The execution of a service in TINA is accomplished by the creation and control of a service session via user agents. In principle, it is possible for two service sessions to interact with one another, where one service session acts like a user agent. This model can be applied recursively. Although from a composition point of view it is relatively easy to chain services together, conflicts may arise in the combined use of resources and in other types of policies (such as billing). This is very similar to the feature interaction problem in current IN systems. TINA has not yet addressed such issues.

This principle has been a very important guide in the CASSIOPEIA work, and nesting has been divided into two separated (although linked) sets of problems: that of defining services based on definition of other services, and providing services through reuse and combination of other services. It is recognized that services will be reusable for these purposes only if they are carefully selected and designed to perform useful services, independent of which modelling technique is used to define them. The definition of services of an OSA-component can be expressed in terms of services offered by other OSA-components, without restrictions on that services shall be provided in such a structure. Concepts and rules in OSA_{app} intend to promote reusability of designed services. To support nesting of services during provisioning the service machine, populated with services offered by deployed components and services offered by the resource infrastructure, contains services which can be used by other services. The service machine evolves with deployment of new services.

4.3 Coherent design of management and control

The consistent treatment of management and control in TINA is achieved by two mechanisms. Firstly, no syntactic distinction is made between management and control operations. The interfaces offered by objects in a TINA system are defined in the same way, irrespective of whether the interface will be used for management or control. Designers of components should include operations to allow the component to be managed. Secondly, management applications should be offered by encapsulating them into service sessions. A manager will interact with a management service session through their user agent, in the same way end-users interact with information services. A manager's user agent will probably need special privileges and access rights over an end-user's agent, but the basic control structures are identical.

In CASSIOPEIA no fundamental difference is seen between management services and control services, the difference is rather seen as pertinent to who and for what reason the services are used. By consequence no difference is made in how these services are modelled. CASSIOPEIA contains no specific support for designing management applications, but supports, e.g. through the design ancillary facets, that services are designed such that they are manageable.

4.4 Customization and personalisation of services

In TINA several opportunities are given for customization and personalisation. There is no predefined set of services in a TINA environment. A service may be sold to many customers, or conversely may be developed for single customers. A consistent subscription and deployment architecture that can handle either case is being defined. User agents can be customised to deliver call set-up notifications to certain terminals based on combinations of time-of-day, service type, terminal capabilities, and terminals currently in use. It is possible for a user to customise the presentation of a service, for example the font and colour background of a conference menu. This is achieved by downloading preferences from the user agent to user service sessions.

As for TINA, CASSIOPEIA has no predefined set of services in an OSA-system. It is assumed that customization is the rule, not an exception. OSA contains the concept of service configurations with components, which through use of architectural guidelines, are designed to be easily configurable and combinable for use by individual users or particular groups of users. The strict recognition of local and global views of services is of vital importance to this end.

4.5 Separation of media services from their control and management

Multi-media information is exchanged between entities in TINA systems through the use of streams. During a session the streams that need to be established between parties are negotiated. Once agreed (or subsequent modifications agreed), communication management is requested to build the streams.

This separation is directly reflected in the problem domain structure of OSAapp, where media control services are seen as a separate domain from the media services themselves, recognizing the need for maximum flexibility in the control while aiming for large stability in the media services. CASSIOPEIA does not aim to give support for engineering of the media services, i.e. services involved in the actual propagation and transformation of streams, but sees these as capabilities of the resource infrastructure. The media service components in OSA_{app} are system and technology independent abstractions of such services.

4.6 Separation of service access from the service core

Access to services in TINA is achieved through terminal and user agents. User agents can launch, or join in, service sessions, which embody the service logic. This provides a flexible model whereby service designers need only provide service sessions, and access providers do not have to restrict the types of services that can be offered.

Accessing a service is seen in CASSIOPEIA as an issue for providing special access services. The objective of access services is to minimize limitations of the actual communication between the using party and the access service. Thus service access problems are modelled separately from the services themselves, provided e.g. by user agents.

On the other side, service may be designed to support particular access services, e.g. reflecting enterprise policies. Guidelines to support such design goals belong to the enterprise support domain of OSA_{app}.

4.8 Separation between application/session oriented and resource/communication oriented problems

In TINA this separation is embodied in the distinction between service sessions and communication sessions. Communications sessions describe the connectivity between the parties of the session, whereas service sessions are used to control connectivity requirements and other service specific activities. An example is a video conference. The service session is used to negotiate the membership of the conference, and the roles of the participants (chairman, etc.). The requirements of connections between participants, which may not include all parties, is represented in communication sessions. One advantage of this approach is to reflect the scenario of parallel sessions in a real-life conference.

In OSA_{app} application support services (supporting modelling various session, mobility aspects) are separated from the primary connectivity and communication problems supported by the domains for media services and their control and management.

4.9 Separation of local and global aspects of services

There are two areas where TINA recognises global versus local views. The first is in the area of subscriptions, the second in the area of executing services.

For subscriptions it is possible to define nestings of groups of people where capabilities are defined by the containing group, and subgroups can extend or refine the capabilities. An example would be a company that subscribes to a phone service, where all employees are prevented from phoning premium rate numbers and furthermore, some subset of employees is prevented from making international calls. An executing service will comprise two types of session. The service session represents the combined activity and state of a service. Each user in the session has a corresponding user service session, which represents an individual view of the state of the session, such as the users current incurred charge.

In CASSIOPEIA this principle is also reflected in the use of global and local sessions, such that a user (session party) will only need to have a local view of the used services. Regarding service customization, user specific data and service functions can consequently be located in separate entities of a realised service, giving higher flexibility to physical distribution.

4.10 Separation of local resource handling and global network resource coordination

In TINA, streams between terminal resident objects are established by communication sessions, via service sessions. The communication session manager (CSM) accepts requests to build a stream between two objects. To process such a request, the CSM identifies the relevant terminals the objects are on, and divides the request into three, one to the connection management software to establish a network connection between the two terminals (which may subsequently be broken down into requests to different network providers), and one to each terminal to instruct it to establish an internal binding between the network access port and the object. These interactions have to be ordered carefully to ensure the relevant information is available (e.g. port numbers). Terminals are therefore responsible for their own internal resource management of ports (or sockets) and bindings, and connection management is responsible for the management of network resources.

In CASSIOPEIA the basic view is that the network resources belong to the resource infrastructure, and the services of these resources are reached within an OSA-system via resource components. The local resource handling is regarded as being handled within the resource infrastructure itself, while global coordination is handled by the media service control and management services, such as connection management services.

5   Conclusion

The original plan for this paper was to take up a number of particular architectural features of TINA-C and CASSIOPEIA and compare them. However we found that a comparison of the architectures as they are organized and expressed, and the techniques and terminology used would be difficult to do in a conference paper. Therefore it was decided to examine scope, goals and basic approaches. The principles described in Sec. 3 were originally intended as a set of CASSIOPEIA principles to be opposed to a corresponding set of TINA-C principles. During the work we found a surprising commonality. Therefore the course of the paper was shifted to follow these principles and provide views on how they are reflected in the architectures.

We have found more similarities between the approaches than originally expected. The main differences between the architectures are due to the differences in scope, where CASSIOPEIA is emphasizing requirement engineering while TINA-C has more focus on computational issues and software engineering. We have found no major inconsistencies between architectures at the level of detail we examined. In fact services modelled using the OSA framework can probably be implemented using the TINA-C architecture, and similarly a TINA-C service would not violate the main rules of OSA.

The common inheritance from ROSA is of course one of the reasons for the similarities. However it can be seen that the TINA-C work appears more related to ROSA than the CASSIOPEIA results. The reason for this is probably that TINA-C has refined the components identified by ROSA towards a computational world, while CASSIOPEIA has amalgamated the ROSA work into a new conceptual framework where they are less immediately recognizable.

6   References

[1]   RACE Project R2049 (Integrated Service Engineering — CASSIOPEIA) Deliverable R2049/FUB/SAR/DS/P/023/b1, CASSIOPEIA, Open Services Architectural Framework for Integrated Service Engineering, March 1995 (also available at URL http://www.fokus.gmd.de/step/cassiopeia/)

[2]   G. Bruno, J. Insulander, U. Larsson, F. Lucidi, A Service-driven Vision of Integrated Broadband Communications: the OSA Approach, in [11], pp 529 - 538

[3]   J. Insulander, P. Schoo, I. Tönnby, S. Trigila, An Architectural Approach to Integrated Service Engineering for an Open Telecommunication Service Market, Proceedings of the International RACE IS&N Conference on Intelligence in Broadband Services and Networks, November 23-25, 1993, Paris, France

[4]   D. Prevendourou, I. Tönnby, G.D. Stamoulis, T. An, Providing Services in a World of IBC Resources: An Architectural Approach, in [11], pp 343 - 353

[5]   TINA-C, Overall Concepts and Principles of TINA, Document Label TB_MDC.018_1.0_94, February 1995

[6]   G. Nilsson, F. Dupuy, M. Chapman, An Overview of the Telecommunication Information Networking Architecture, Proceeding of the TINA'95 Conference, February 1995, Melbourne, Australia

[7]   RACE Project R1093 (ROSA) Deliverable 93/BTL/DNR/DS/A/005/b1, RACE, The ROSA Architecture, Release Two, Version 2, May 1992

[8]   A. Oshinsanwo, et.al., The RACE Open Services Architecture project, IBM Systems Journal, Vol 31, No. 4, pp 691-710, 1992

[9]   M. Chapman, I. Tönnby, P. Schoo, Suggestions for Object-Oriented Modelling from ROSA, Third Telecommunications Informations Networking Architecture Workshop (TINA'92), Narita, Japan, January 21-23, 1992

[10]   ITU-T | ISO/IEC Recommendations X.901 -- X903 | International Standard 10746-1 -- 10746-3 : ODP Reference Model Part 1 -- 3

[11]   H.-J. Kugler, A. Mullery, N. Niebert (Eds.), Second International RACE IS&N Conference on Intelligence in Broadband Services and Networks — Towards a Pan-European Telecommunication Service Infrastructure, September 7-9, 1994, Aachen, Germany, LNCS No. 851, Springer-Verlag, 1994

Poster Sessions

The following abstracts represent those papers which were presented as poster sessions during the conference.

The posters vary from descriptions of IS&N applications to descriptions of tools for user interface design.

External Access to TMN: An ODP Specification

Marco D'Aurelio, Mauro Peirgigli (Sirti, Italy)

PRISM (Pan-European Reference configuration for IBC Service Management) is a RACE II research project dealing with service management for broadband networks in a Pan-European multi carrier environment. The objective of the project is to provide a reference architecture for the service management that can be used by system designers in order either to model their requirement and to track the quality of the system modeling. In order to be, as much as possible, near to real world the service architecture is validated via two case study (VPN and UPT) which have two major goals: to check the effctiviness of the rules and the procedure expressed by the service architecture and to provide feedback for the architecture refinement and stabilization.

Working in a context where the standards play a very important role, and also market deregulation, the PRISM project aims to be compliant with the TMN standards and try to consider the TMN itself as one of the main actors involved in the telecommunications scenario. The TMN is not seen just as set of rules but as a real application, running in a multi domain environment, following its own policies and offering management services to its user. Moreover the TMN management applications are very critical for the organization because they allow to control the telecommunications resources. An unauthorized modification of the information or the corruption of the software applications can result in heavy losses of money for the organization. On the other hand the enterprise owning the TMN wants to provide its services, and hence also the management services, in a customer oriented way. The user needs very simple and easy procedures to be followed in order to ask and utilise the service.

In PRISM the functionalities necessary for assuring a high level of integrity, confidentiality and availability to the TMN resources are put together in the External Access Administration (EAA). It is a software application in charge of controlling if persons or applications trying to access the TMN have the necessary permissions to use the TMN management services they ask for according to the organization's internal policy or to the customer contracts. When the access is forbidden, the EAA takes the security measures required by the active internal security policy; on the contrary, if a successful access attempt takes place, the EAA also performs a dispatching action allowing the requestor to use the requested management service.

This paper aims at providing the architectural and functional specification of the EAA concept using the Open Distributed Processing (ODP) methodology. ODP is a joint ISO/CCITT international standard that provides a framework for the specification, design and implementation of distributed systems. Nowadays the importance and complexity of distributed systems is continuously growing. ODP provides a method for better dealing with the complexities of such systems by considering them from different points of view.

The analysis carried out in this paper only uses the first three viewpoints (out of five) defined in ODP since it aims at providing general results that are expected to be applicable to a wide range of situations. We start modelling from the Enterprise Viewpoint where the generic company which owns a TMN and offers telecommunication services is modelled as a community i.e. a number of objects sharing a common target or policy. The TMN itself is seen a community whose goal is to the provide Management Services supporting the telecommunications services. The management services are modelled as activities, the management service components as subactivities and the management function as chain of actions. A set of agents and artefacts are in charge to perform all the actions necessary to support the services. The TMN community also follows its own policy; this policy is formed by the rules and the structure/functional constrains defined in the M. series of the ITU-T recommendation.

Within the TMN community a special agent, the External Access Administration agent, has the task to control, following the parameters and the rights written in the contract that the user and the company previously agreed, the access to the TMN assuring the proper level of security and dispatching features in order to allow the user to access the desired Management Service. In the Enterprise Viewpoint the focus is on the relationship among the actors, the roles, the contracts and between the latter and the policies/domains defined by the company or by some international regulator like ONP.

In the Information Viewpoint the entities identified in the Enterprise Viewpoint are better detailed and refined while in the Computational Viewpoint is given a picture of how these entities interact allowing the predisposition to distribution transparency, migration, fault tolerance and so on.

The Implementation of PSCS Features in a RACE ATM Demonstrator

Hajo Bakker, Jens Heile Heilesen (DELTA, Denmark)
Gottfried Schapeler (Alcatel SEL, Germany)
Miltiades Anagnostou, Christos Georgopoulos (NTUA Greece)

1 Background

The specification and standardisation process of UPT/PSCS services and features is continuously ongoing within ETSI and RACE but up to now only a few practical implementations of this new concept have been carried out. The RACE PERCOM project will prototype selected PSCS features in the existing RACE TRIBUNE ATM testbed enabling Personal Mobility and personalisation of services to end-users. Practical experience will be gained in the field of PSCS service elements implementation by the execution of experiments, demonstrations and measurements.

Major achievements of the PERCOM prototype are expected in two areas:

- the combination of several (broadband based) services to support Personal Mobility,

- the implementation of the PSCS Service Node following a TINA-C like approach.

The paper is grouped into five paragraphs from each a brief summary can be found in the following.

2 System Overview

This section of the paper describes the PERCOM demonstrator which consists of the TRIBUNE network elements and the additional PSCS related items (painted in grey):

- The PSCS Service Node (PSN) is responsible for the control of all PSCS services and the administration of all PSCS users. The node ismainly based on TINA concepts adopting the ODP paradigm.

- The extension of the TRIBUNE Control Server (TCS) separates PSCS and non PSCS calls. As soon as the TCS identifies a called number to be a personal number within the ATM signalling protocol, this request will be transferred to the PSN. The PSN responds the physical address of the terminal which the PSCS user is registered to.

- The PC-based PSCS terminals offer a GUI which puts the user in a position to handle and control all available P SCS services. Via these PCs only data communication can be carried out. So TRIBUNE terminals which offer a wider range of services will be placed next to the PSCS terminal, whereby both terminals are acting as one logical PSCS terminal unit and enable the use of additional services (video, fax...)

Also part of this section is a description of the overall architecture comprising the PSCS Service Node, the extended TCS, and the PSCS terminals. In addition to the detailed system description, the identified restrictions and limitations caused by the practical implementati on of the TRIBUNE testbed are indicated as well. In principle those restrictions might also occur in other networks since PSCS services will be introduced there. TRIBUNE Control Server: PSCS Specific Extension of Call and Bearer Control Functionality

In order to support the PSCS services implemented in the PSCS Service Node the TCS has to be modified and extended, especially in the field of Call and Bearer control. The system specification written in SDL complies with the requirements for enhan ced Call and Bearer Control functions in PERCOM. The PSCS requirements are further broken down in MSC (Message Sequence Charts) diagrams. The activities carried out to validate the PSCS Call and Bearer control extension serves the following purposes:

- Validation of the SDL specification aiming at the detection of errors and in consistencies done by automatic test of all possible SDL specification execution parts.

- Verification of SDL system consistency with the MSC

- Interactive simulation of the SDL system in order to investigate its dynamic behaviour.

Upgrading of TRIBUNE Terminals to PSCS Terminals

Another area of work within PERCOM deals with the developmentof a terminal unit, the PERCOM-PSCS-terminal, in order to offer demonstration and usage of the specified personalised services. The PERCOM-PSCS-terminal is build up of two main components:

- a TRIBUNE PC equipped with an ATM interface board (PCS)

- an original TRIBUNE terminal

The TRIBUNE PC provides local PSCS access control, a GUI with respect to the PSCS related functions, and allows interaction between the PSCS user and the PSN in a user friendly windows environment. The TRIBUNE PC also provides file transfer and broadband applications (video on demand, video conference, co-operative work). The TRIBUNE terminal (e.g. Multi-Service Te rminal, MST) offers video phone and ISDN telephone services. Each PERCOM-PSCS-terminal can be identified by a single Logical Terminal Address (LTA). As counterpart, the PSN is able to associate the LTA physical addresses to application capabilities of the respective terminals. The control information flows through Ethernet links while the data streams follow ATM paths.

3   PSCS Service Node

The PSCS Service Node (PSN) implementation follows the TINA-C approach. The realisation is done according to the Open Distributed Processing (ODP) approach, which in principle enables a distribution of the service and mobility provisioning. Within the PERCOM project the PSN is implemented on a workstation and incorporates an OMG-CORBA compliant platform as Distributed Processing Environment (DPE).

Implemented PSCS Service Features, Scenario Description

In its last paragraph the paper describes the implemented PSCS service features included in the PERCOM demonstrator, these are:

- Identification, authentification of PSCS users

- Registration, deregistration for incoming and outgoing calls

- Flexible Service Profile handling (r outing depending of the time of the day, routing depending on different tele-services, filtering of incoming calls)

- Session Mobility (the participation of a user in a session can be suspended and resumed i.e. the network allows connections to be changed without dropping a session)

- PSCS - non PSCS user interworking (sessions between an PSCS and non PSCS users will be supported)

- The simultaneous use of combinations of services by a PSCS user.

A scenario description shows how these PSCS services are used in the PERCOM demonstrator (usage of ATM signalling, interworking between the PERCOM-PSCS terminals and the PSN as well as the interworking between the PSN and the TCS).

Simulation Model for Intelligent Network Based Personal Communication Services

Martin Ostrowski, Reginald Coutts, University of Adelaid, S. Australia

This paper presents a simulation model for the evaluation of the signalling traffic in a pedestrian orientated PCS system with a user base and behaviour representative of the mixed mobility offered by PCS. The aim of the model is to predict the time and spatial distribution and state of pedestrians within the simulation space. Such data is valuable in the dimensioning of networks where empirical pedestrian flow data is not available, or where planning is to take into account future projections.

In addition to the time averaged spatial distribution, an analysis of instantaneous local variations in pedestrian traffic is presented. The analysis is founded on well established results on pedestrian flow, density and speed relationship. 'Hot-spots' are found to occur at the intersections of streets and in the vicinity of transport facilities. The distribution of traffic at these hot-spots is given.

The model introduces the basic concept of a pedestrian trip as a series of transitions between states and uses observed flow signatures for different types of space to predict user flow in the simulation space. The signalling traffic generated as a result of physical and state space movementis predicted by calculating the occupancy and transition rates between states, and the flow rates between physical locations.

1 State Exchange Model

The pedestrian traffic originating from a cell in the simulation space is a product of the flow rate for a space type s at time t, and the relative density of that space type in that cell. A space type is classified by function, and we make the approximation that the flow signatures for all occurrences of a given space type in the simulation space are identical. Superimposing traffic for each type of space in that cell gives the total traffic from that cell.

The direction of this traffic will be distributed proportionally to the polar density distribution of destination space, weighted with distance.

The effect of this is that, for example, most lunchtime shopping trips from an office will be towards a major shopping centre if one is nearby, but the proportion of this traffic will decrease and shift towards smaller facilities nearby if the distance increases. Similarly, commuters are likely to choose the railway station nearest their office for the forward and return trips (the maximum likely walking distance to the office from the station is called its sphere of influence). While walking distance probability distribution data is available for different types of trips, we use the exponential distribution as a reasonable approximation.

What remains is to determine the proportion of total traffic emanating from a cell that is destined towards each type of destination space. For example, the majority of traffic from office space during lunchtime will be distributed between food, shopping and recreational facilities, and only a negligible percentage will be towards transport. This attraction factor is a consequence of movement through state space, and is related to the varying proportions of the state transitions at different times of the day.

The pedestrian flow through a cell in the simulation space is then a superposition of the flow through that cell from every other cell in the simulation space. Some traffic from other cells will stop at destinations in the direction of the cell before that cell is reached. Therefore, the proportion of traffic passing through that cell is 1 - (ratio of traffic in direction of cell terminating before the cell and the total traffic in that direction), written as p/Ptot.

2   Signalling Traffic from State Space Movements

In addition to signalling directly related to physical movement, represented by handovers and location updates, traffic can be generated by both the existence of users in a certain state, and the transitions of users between states. This is particularly applicable in the IN supplementary service case, where the invocation of services is related to the state of the user rather that their physical location.

Using the simulation model, it is possible to calculate the total rates of transition between all pairs of states as well as the number of users in a given state at any time.

Estimations can be made about the traffic generated by transitions. For example, call screening is likely to be invoked when users leave the office to go to lunch, and removed on their return. This kind of behaviour has important implications for the signalling load in the network, as will be shown in the next section.

The signalling traffic observed at the SCP/SSP level is a superposition of the traffic generated by users existing in a certain state at time t, and the traffic generated by users moving between states at time t:

What is especially useful is that if a new service can be characterised in terms of usage probability in different types of states and transitions, as is the case for many of the services described in the current IN recommendations, then the impact of the new service on the signalling traffic at the network level can be evaluated, for an arbitrary level of service penetration.

3   Manhattan: A Case Study

To verify the model against empirical data, results from a hypothetical PCS system in the Manhattan CBD area are presented, and are found to be consistent with other predictions and observed data. The parameters for this systems are derived from penetration predictions of terminal mobility and various services. The results include the offered traffic distribution, flow rates (from which handover and location update signalling traffic is derived) and the distribution of SCP and database queries resulting from the usage of IN based services.

Integration Path of IN and TMN Architectures - Using UPT as a Case Study

Pedro SantAna, CPRM Marconi, Portugal
Oscar Bravo, Telia, Sweden
Amador Martin, Telefonica, Spain

This paper tries to cover the evolutionary steps towards an integrated IN/TMN architecture for UPT, analysing the advantages and disadvantages of possible new areas for integration. Starting from a short-term UPT architecture, where IN and TMN are only joint (not integrated), it will be proposed and analysed a sort of medium-term solution based on the integrated architecture being analysed by the standardisation bodies but extended in some key areas in order to ease the transition to a completely integrated architecture in a target scenario. The objective is to have a more open, flexible and distributed architecture (a sort of ODP-based integrated architecture) for UPT and UPT management. Notice that although mainly concerned with the case of UPT, this paper also tries to give contribution and feedback to the general problem of integrating IN and TMN. The suggested areas for work are:

1 Mapping Between IN FEs and TMN FBs

As it is agreed to adopt TMN as the generic management architecture to manage all telecommunication networks, the approach of the standardisation bodies to integrate IN and TMN architectures is based on the mapping of IN FEs into TMN FBs. However, some functional entities, such as service processing FEs, only includes some minor management aspects. So it has to be analysed whether an integrated architecture should be entirely based on encapsulating FEs by FBs, or if in some cases it should be the other way around (meaning a sort of mixed architecture which includes both a mapping of IN FEs into TMN FBs and a mapping of FBs into FEs).

2 IN Service Creation Aspects In An Integrated Architecture

The SCEF FE allows a service to be defined, developed, and tested in a IN structured network, and is the only IN FE which has no direct relation to a TMN FB. This is due to the gap arising from the fact that service creation aspects are not clearly covered in TMN. In the proposed integrated architectures of the standardisation bodies (1), no reference is made to the mapping of this FE in the TMN architecture. This area should also be covered in an integrated architecture and a first proposal is to consider the implementation of the SCEF as an OSF at the Service Management Level (SML). This means to introduce a new q reference point between the SCEF and the OSF (defining the functionality of a SMF in IN). In practical terms, this means that the implementation of the SCEP may be allocated to a specific OS connected to the OS at the SML by an interface similar to the Q3.

3 Evolution of FEs Performing Control to FBs Performing Management

Some of the FEs of IN performing control of UPT service execution contain functions that may be regarded as management functions. In particular the case of the SCF may be investigated, as some of the functions described for this FE can be either seen as part of a NEF or/and as part of an OSF. This FE is the key element of the management versus control subject in an integrated architecture, and as such the advantages and disadvantages of the possible mapping approaches in a functional architecture and in the correspondent physical architecture should be investigated. It is important to state the requirements of the possible implementations of a SCP, in particular the implementation as an Element Manager OS co-ordinating the co-operation of SSPs, SDPs and IPs both for the UPT service execution and the management tasks assigned by the TMN to the Element Management Level.

4 Integration of Protocols: INAP and CMIP

The evolution of SCPs to perform both UPT service execution and management tasks has the consequences of evaluating an integration path of the protocols of IN, in particular INAP and management protocols (e.g. CMIP). As protocols for management and IN are different, the possibility of a unique protocol for an integrated architecture should be analysed. However performance requirements are a critical issue. For the time being, the INAP is a simple protocol with high real time performance requirements. As the functionality of the SSP, SCP, SDPs (Network Data Bases) and IPs tends to increase incorporating more complex operations, the require ments on INAP will increase, without relaxing the real time protocol capabilities. On the other hand, a similar requirement is put on the management protocols, in terms that more intelligence is required to the MOs and more complex operations are required to be solved by the protocols.

5 Data Integration

Data integration may be achieved using one of the following strate gies, both possible candidates to be applied in an integrated architecture:

Migration of MIB to Network Data Bases

In this strategy the data contained in the MIB are allocated as far as possible to the data contained by the implementation of the SDF. It can be regarded as an strategy in which the actual definition of data required for management purposes together with the functionality required by the TMN are described in a collection of Managed Objects while the data is actually implemented and stored in the Network Data Bases. The NDBs are an implementation of the SDF and they store in addition data which are required by the UPT service execution but not relevant to management. Because the SDF does not contain real objects, the functionality required by management operations and notifications are defined and implemented in the MOs defined within the TMN.

Migration of Network Data Bases to MIB

In this strategy the data contained in the NDBs (implementing the SDF) are totally or partially allocated to the MIB. This means that the MOs are defined to contain enough data to cope with both the requirements of the UPT service execution and the management requirements. If a total integration approach is attempted, the performance requirements for access to and response time from MOs should be achieved. In a partial integration approach the migration may begin with those data items requiring less real time performance on the implementation. In the first case the NDBs does not exist anymore, the SDF is totally integrated in the MIB of the TMN.

6   Applying ODP Concepts to an Integrated Architecture

Today's architectures are targeted to be more open, flexible and distributed. Much of the work being done in the field of architectures for Telecommunication services and systems (e.g., PRISM and TINA-C) is nowadays based on ODP. ODP may be considered as a general object-oriented framework for distributed systems. So it may be studied how some concepts from ODP, specially in the information and engineering viewpoints, may be applied to the integrated IN/TMN architecture, so that some correspondences with concepts regarding deployment and communications (both protocol and information) in ODP may be suggested.

Dedicated Server Multicast Routing for Large Scale ATM Networks

H.R.Sharobim, Saga University, Japan

Video distribution services such as telemedicine services, teleconferencing and Near Video on Demand (NVoD) will constitute a large portion of the future customer's requirements. A common factor among these services is the need to connect a single conference user or video provider with a multiplicity of other users or consumers.

These point-to-multipoint services are known also as multicasting services.

To provide multicasting it is necessary to prepare "copies" of the message to be distributed. The copy function could be incorporated in the ATM switches or could be provided by specialized or dedicated centers known as copy servers. In addition to point-to-multipoint requirements, there are also point-to-point connections such as normal telephone ser vices. Thus the traffic in the network is expected to rise sharply and may lead to congestion, queues and buffer overflows. Hence new routing algorithms, multicast connection finding algorithms are necessary to prevent congestion, as well as to maintain the contracted grade of service (GoS) set up at call request time.

In this paper we propose a network architecture and routing algorithm tosupport multicast services using dedicated servers for large scale ATM networks.

The architecture consists of a backbone network of ATM switches connected together, with dedicated servers distributed around the network. The role of the dedicated servers is to support the copy function needed in multicast services. The need for such an architecture arises due to the high cost and complexity of incorporating the copy function within the switches.

In this kind of networks the ATM switches do not have the copy function capability. However there shall be a number of dedicated servers which provide a copying function to assist in multicasting. The servers are not necessarily connected to all ATM switches.

The dedicated server network architecture takes into consideration:

- Dedicated servers provide for the copy function and distribute the information streams.

- ATM switches to provide for switching and connection.

- Variable traffic demand by the destination nodes.

- Capacity constraints on the links.

- Cost of the links based on a per unit traffic.

Application of IS&N for Efficient Aircraft Pre-Design

F.Arnold, Daimler-Benz Aerospace Airbus GmbH, Germany

The efficiency of the aircraft pre-design can be substantially increased by the intelligent usage of the current European broadband networks. In the EU-projects PAGEIN (RACE No. R2031) and ADONNIS (ESPRIT No. 9033) a software platform has been developed, which enables distributed supercomputing via ATM high-speed networks combined with audio/video conferencing support. The interactive usage of distributed soft and hardware resources in a broadband network implies substantial advantage in design time compared to conventional working scenarii.

Three-dimensional numerical methods for the computation of aerodynamic flow over aircraft have an essential impact on the design cycle. The numerical simulation supports the solution of difficult problems with respect to the design, e.g. of laminar wings and high by-pass engine integration, and thus reduces time/costs and increases quality of aircraft development. A consortium of Daimler-Benz Aerospace Airbus (DA), DASA-Dornier, DLR, GMD, IBM etc. develops a new 3D Navier-Stokes Method to simulate the viscous complex flow over complete aircraft. The consideration of distributed processing by the massively parallised algorithm of this method brings numerical flow simulation to a new order of computation. About six millions of unknown flow variables can be solved by the recent supercomputers (e.g. Cray-YMP) within a couple of hours for much simplified aircraft configurations. The new distributed parallel processing will increase the possible number of unknown variables by approximately ten without increasing the computing time for complete aircraft configurations. The solution of such a flow simulation gives a data volume of about 100 Mbyte. Using the intelligent software platform COVISE developed by the Rechenzentrum Universitaet Stuttgart (RUS) this data set will be filtered in advance. Afterwards only the reduced data volume for the requested visualization will be transmitted by multi-casting via broadband network to the involved engineers at different sites. Based on the master/slave hierarchy each of the involved engineers can define and control pre-processing, processing, and post-processing tools by the combination and distribution of software modules to different hardware resources. Everybody can see the same visualization of the computed results and can control them. The session is supported by audio, video and whiteboard display.

COVISE has been applied on the 34Mb/s Datex-M network of the Deutsche Telekom AG and on the international ATM trial to communicate with french aerospace partners. The background and application of COVISE, the DA relevant network infrastructure, and the impact of IS&N to modern aircraft design will be illustrated in the posted presentation.

- The traffic demand of the specific node.

- Total traffic demand by all nodes.

- Capacity constraint on the links.

The server allocation problem is also discussed. Several problems are highlighted such as the number of servers, number of links per server, distribution of the servers around the network. A simple mathematical treatment of the server allocation problem shows that the shortest path algorithm also serves as a tool to provide potential nodes to where the servers may be connected. By connecting the copy servers to these nodes lower network cost and balanced traffic distribution are expected.

It is shown that potential server locations depend on several factors:

- Starting node link cost to be of a low value. The starting node is the node to which the server is to be connected.

- About equivalent number of nodes on both sides of the starting node. (The equivalent number of nodes is the number of nodes modified by the traffic demand of the node)

- Nodes with high degree, or nodes with many branches may serve as potential locations of servers.

As the heuristic is based on network divisions, it might not lead to an optimal feasible solution. Nevertheless it has the following merits:

- Applicable to large networks due to network divisions.

- Aligned to actual networks, where management of servers is done on sub network basis.

- The algorithm may also be modified to a network where the copy function is installed inside all or some of the switches.

While several problems need to be solved, the present work indicates:

- Incorporating the copy function in dedicated servers may serve as an interim solution to the introduction of full scale ATM services.

- The routing algorithm , based upon network division and the shortest path algorithm may be applied to large scale networks.

- The algorithm takes into consideration the capacity constraints.

- The algorithm may be applied to other networks which employ dedicated servers for copying of original messages.

For the proposed network architecture, a multicast routing algorithm is designed with the objective of minimizing the total multicast cost, subject to the traffic constraints. The network architecture is modeled by using graph theory where, switches are modeled as nodes and links as edges. With each edge two parameters, cost and traffic demand are associated. Traffic is defined on a connection basis and cost is defined on a per unit traffic basis.

The designed algorithm is based upon:

- A new network division technique that enables multicast routing for large scale ATM networks.

- Utilization of the shortest path algorithm for node-server connection.

- Interconnection of the servers is similarly done by the shortest path algorithm.

- Testing for the capacity constraints by modifying the shortest path with the next shortest path.

The network division technique is based on that large networks are usually difficult to manage and often require complex routing schemes. Hence there is a need for smaller network divisi ons where simpler routing algorithms can be applied. Basically the network is divided into several smaller divisions or groups, where a copy server would be responsible to prepare enough copies for its group only. The same routing algorithm applied for all network divisions groups the destination nodes among suitable servers, in such way that the multi cast cost is minimized. Within each group the copy server would be responsible for copying the message to its destination group. Since the copied traffic follows a path from the server to each destination, then the shortest path ensures a low cost network. Traffic flow along the links is checked against the capacity constraints. If the constraints are not violated then the algorithm finds a feasible solution. However if the constraints are violated then the next shortest path is attempted.

The algorithm also solves the unconstrained problem. The unconstrained problem is similar to the constrained one however with the capacity constraint reduced to zero.

The unconstrained solution is necessary since it provides for comparison with the constrained one. Also in the early phases of introducing ATM technology, the traffic volume is expected to have low value.

Associated with the unconstrained problem is the problem of choosing additional servers or server selection cr iteria. A technique for deciding upon additional servers is introduced by comparing the multicast cost from any node to two servers.

In general it is found that server choice depends upon:

- The relative path cost of a node from both servers. Usually there is a tendency to connect the node to the nearer server.

Developing Alternative Metaphors for Special B-ISDN Services

C.Stephanidis, C.Karagiannidis, A.Koumpis, ICS, Crete, Greecs

The emerging B-ISDN environment offers new opportunities for the socio-economic integration and independent living of people with special needs (distant learning, tele-working, tele-shopping, sophisticated alarm systems, etc). Facilitating access to B-ISDN services to people with disabilities poses, however, many additional problems. To address these problems, interventions may be required at one (or more) of three levels, namely modifications at the terminal level, provision of special service components, and development of special services (i.e. services that are accessible by people with special needs).

Developing new special services for each particular user group, however, may be prohibitively expensive. In this respect, special services might be better developed as the composition of the 'conventional' B-ISDN services-kernels, originally developed for able bodied users, together with additional 'special shells' that are built on top of these kernels, and provide the semantic enhancements that are necessary for the various disabled user groups.

The advantage of this approach is that special services are not developed in isolation from the conventional ones. This also implies that advances and improvements in conventional services propagate directly to the special services as well.

Central to the enhancements that are required in this direction, is the integration of new metaphors, that utilise the available user I/O channels, and provide a suitable interaction environment per user category. Methodologies that assist the development of new interface metaphors, therefore, are considered of great importance for the provision of special services.

In this paper, a model is proposed, which assists the specification of interaction metaphors independently of the specifics of the target implementation environment. The proposed model is based on a formalism that captures application and interface semantics. It consists of two submodels, namely the metaphor specification and metaphor implementation submodels. The application of the model does not require expert programming skills, and can be utilised by multi-disciplinary groups of professionals.

The model has been applied for the development and evaluation of two new metaphors, namely the narrator and the watchdog metaphor, that aim to address the requirements of blind users in the context of interactive multimedia retrieval applications.

The narrator metaphor has been applied for the case of an electronic library application, resulting in a 'talking book' special service that provides oral interaction functionality to blind users. The metaphor shares the characteristics of a (human) narrator, i.e. the user may ask for modifications of the voice characteristics and the speech of narration, or for a summary of a particular chapter or the entire book, etc.

The watchdog metaphor has been applied on top of a conventional document browsing service. The functionality of the resulting special application is based on the 'help facilities' provided by a watchdog to blind people. Thus, while a watchdog provides, for example, blind users with feedback regarding the surrounding environment (e.g. the condition in the street), the 'computerised watchdog' provides blind users with feedback regarding the document to be browsed (e.g. whether the end of the document has been reached). The watchdog may provide passive help (i.e. upon request of the user), or active help (e.g. suggesting when a document should be saved, whether to save the current document before quiting the application).

Current work focuses on the development of formal schemas and methodologies that facilitate the automated transition from the specification to the implementation of interaction scenarios. Also, several alternative metaphors that address the requirements of people with special needs are currently under examination.

Tool Based User Interface Construction Facilitating Access to Users With Disabilities

C.Stephanidis, Y.Mitsopoulos, A.Stergiou, A.Koumpis, ICS, Crete, Greece

Problems faced by people with special needs in accessing B-ISDN services have been addressed1 through an in-depth analysis of their interaction requirements based on human factors issues and ergonomics criteria; several barriers have been identified which prevent people with special needs from having access to information available through the B-ISDN. The identified barriers are related to the accessibility of the B-ISDN terminal, the accessibility of the anticipated conventional B-ISDN services as well as the perception of the service information.

In order to cope with these difficulties, different types of solutions are proposed, which address the specific user abilities and needs, at three different levels:

- Adaptations within the user-to-terminal and the user-to-service interface, through the integration of additional input/output devices and the provision of appropriate interaction techniques, taking into account the abilities and requirements of the specific user group.

- Service adaptations through the augmentation of the services with additional components capable of providing redundant/transduced information.

- Introduction of special services, only in those cases where the application of the two previous types of adaptation are not possible or effective.

In the context of the RACE R2009 IPSNI-II project, a framework for the design and development of User Interfaces for future B-ISDN services and applications has been designed aiming to facilitate accessibility of B-ISDN services by "all users", including people with disabilities.

Based on the developed framework, a user interface construction tool called INTERACT has been implemented which builds on the notion of separating a service / application in two functional components, namely the application functional core and the user interface component. This allows the provision of multiple user interfaces and the support of "high-level" design of the interaction dialogue by means of abstract interaction objects, i.e. independently from the presentation details and operational constraints of a particular technological platform.

From a software engineering perspective, INTERACT primarily aims at accelerating the development time and reducing the required resources for the development of both early computer mock ups as well as the final user interface of a service, providing at the same time reuse of the abstract interaction objects for different end user categories.

INTERACT supports most of the characteristics of state of the art User Interface Builders providing also means of achieving accessibility of user interfaces by people with disabilities. In this perspective, and in addition to the standard I/O communication channels supported by existing toolkits for developing graphics based applications, INTERACT provides enhanced user interface customisation possibilities to enable the selection of different interaction styles through the utilisation of alternative human and computer interface channels. More specifically, INTERACT introduces additional features regarding:

- the media and modalities used during interaction,

- the employed interaction techniques,

- the utilised I/O devices,

- navigation in the user interface,

- feedback provided to the user,

- etc.

For instance, in order to support the development of user interfaces accessible by blind users, INTERACT enhances the interaction objects with additional attributes (e.g. presentation in auditory or tactile form) and provides facilities for switching between the applications, as well as facilities for the exploration of the graphical objects of the various applications.

At present, INTERACT supports devices and interaction techniques appropriate for blind, low vision and motor impaired users. To this effect INTERACT provides the designer with a "solutions space", which provides a series of alternative interaction styles, including:

- tactile based user interfaces,

- speech based user interfaces,

- combined speech and tactile based user interfaces,

- large size widgets,

- large size widgets with auditory cues,

- large size widgets and speech.

Special effort has been given in the design of INTERACT to provide "easy-to-use" interactive facilities in such a way that non-computer experts, human factors specialists, rehabilitation experts, etc. could easily and effectively design the user interface of services accessible also by users with disabilities.

Preliminary tests with different user groups have confirmed the practical value of INTERACT. Work currently under way seeks to augment the supported user interaction styles through further exploitation of the haptic and audio channels. In addition, design assistance is foreseen by means of a module that provides the user of INTERACT with suggestions for lexical aspects of the User Interface such as the presentation of interface objects with respect to the user group under consideration.

Author Index

Springer-Verlag
and the Environment

We at Springer-Verlag firmly believe that an international science publisher has a special obligation to the environment, and our corporate policies consistently reflect this conviction.

We also expect our business partners – paper mills, printers, packaging manufacturers, etc. – to commit themselves to using environmentally friendly materials and production processes.

The paper in this book is made from low- or no-chlorine pulp and is acid free, in conformance with international standards for paper permanency.

Lecture Notes in Computer Science

For information about Vols. 1–920

please contact your bookseller or Springer-Verlag